Integrated STEAM Edition

25 Classroom Projects

The ARDUINO® Classroom

Inside the Book
- **25 Classroom Tested Projects**
- **Standards Alignments**
- **110+ Project Extensions**
- **Materials Lists**
- **Tested Code**
- **Technical Diagrams and Photos**
- **Assessments**

Integrates
- **STEAM**
- **C / C++ Programming Basics**
- **Digital Prototyping**
- **Engineering Design**
- **Project Management**

Free with Website Registration
- **Project Updates**
- **Community Ideas**
- **Project Submissions**

Premiun Website Membership
- **Videos of Projects**
- **Classroom Presentations**
- **Images, Diagrams, and Code**

C / C++ Programming

Digital Design and Prototyping

María Isabel Mendiola Ramírez
and
Peter G. Haydock

www.thearduinoclassroom.com

STEAM EDITION
VOLUME ONE

Integrating programming, design, and prototyping into Science, Technology, Engineering, Arts, and Math (STEAM) classrooms with Arduino® based projects.

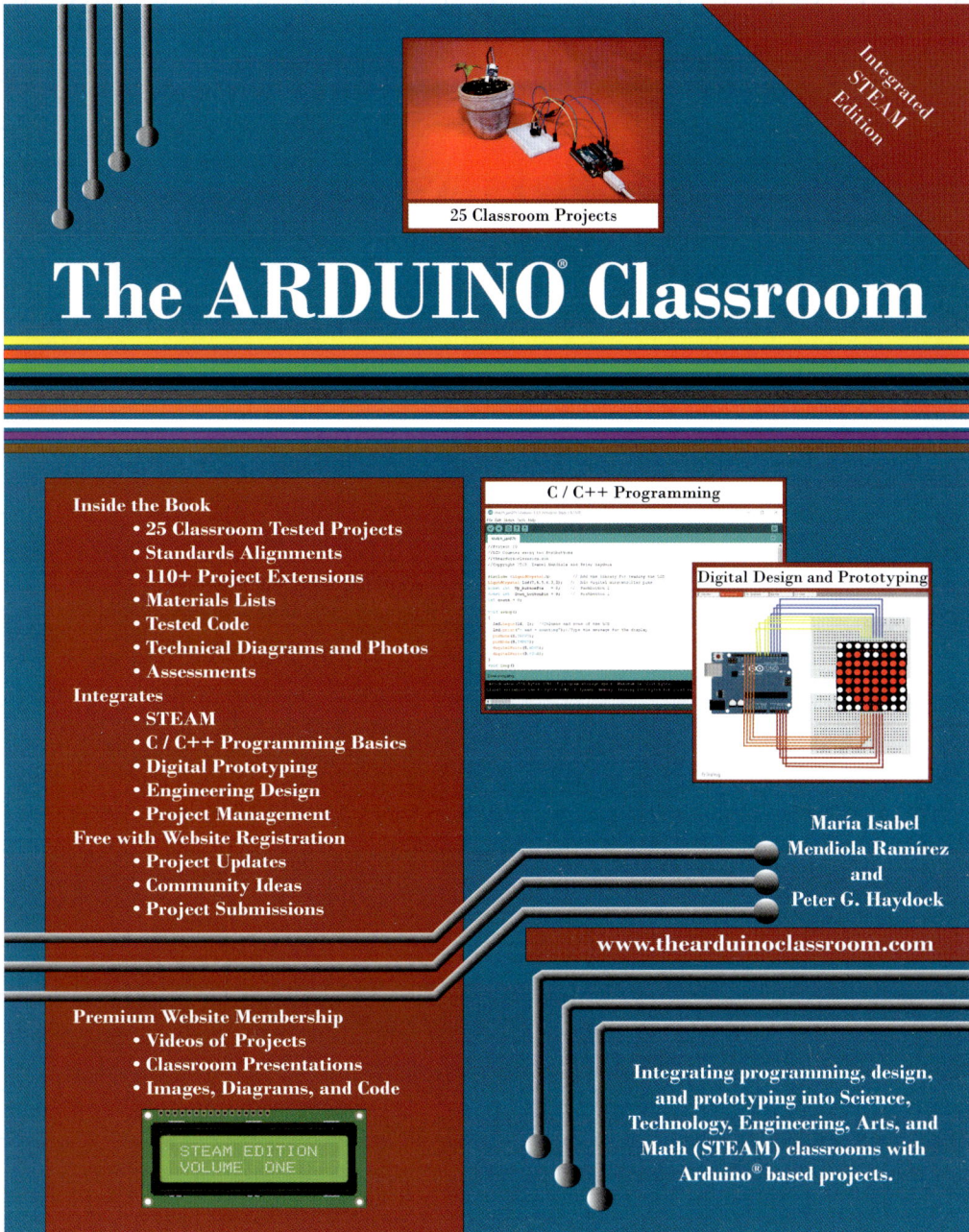

Published by
Gearbox Labs Inc.

Written by
María Isabel Mendiola Ramírez, M.Ed.
and
Peter G. Haydock, MLIS

The Arduino® Classroom: STEAM Edition, Volume 1

The ARDUINO® Classroom: STEAM Edition, Volume 1 presents 25 standards-aligned Arduino® UNO projects for grades 5 and up. The classroom tested projects supplement a STEM/STEAM curriculum or each can be implemented as standalone or extension activities.

Isabel Mendiola and Peter Haydock, with a combined 50 years of experience in education, have both taught at a variety of educational levels and in multiple content areas. They have facilitated winning projects in numerous student competitions and won international awards for developing award-winning classroom resources.

They have written this book to aid educators, homeschool families, informal learning spaces, and hobbyists in the integration of projects based on the Arduino® UNO microcontroller platform into core school content. The book presents computational thinking, engineering design, problem solving, and project skills for the 21st century.

Published by Gearbox Labs Inc., Hingham WI, USA

© 2019 María Isabel Mendiola Ramírez and Peter G. Haydock

Permissions may be addressed in writing to support@thearduinoclassroom.com.

Book and Covers design by Peter Haydock

Images
Front Cover The ARDUINO® Classroom by Peter Haydock, Project 20 photo by Peter Haydock, Project 10 Sketch screen shot from Arduino® IDE program by Peter Haydock, Fritzing Screen shot made by Peter Haydock

Back Cover Professional Development photo by Peter Haydock

All images, diagrams, or technical art found in this book are taken or created by Peter Haydock or Isabel Mendiola, are in the public domain, or used under license and credited on the page as required. For a full list of image credits see the book web site.

Printed on paper which is made of 10% post-consumer waste.

Printed in the USA by Times Printing, Random Lake, Wisconsin

Arduino® is a registered trademark of Arduino AG
Tinkercad® is a registered trademark of AutoDesk®
Fritzing is open-source software.

ISBN-13: 978-0-578-55418-1

Library of Congress Control Number:2019911111

10 9 8 7 6 5 4 3 2 1

Table of Contents

Introduction

There are SKILL, PROJECT, and SUPPORTING CONTENT pages in the book.

SKILLS

SKILLS pages start the first three sections and precede the projects. These are one to three page tutorials that build general knowledge before implementing projects. These are shown with blue headers and start with the word "SKILLS." Skills are listed in the table of contents and the section dividers. The Skills are:

Basic Circuits	Electricity	Arduino® UNO Microcontroller Board
Arduino® IDE Software	Sketches and Coding	Tinkercad®
Assessment	Data: Serial Monitor and Plotter	
Project Management	Mathematical Operations and Conditional Statements	
Libraries	GitHub	Fritzing
Troubleshooting Projects	Open Source Projects: The Next Step	

PROJECTS

PROJECTS pages are indicated by red headers with the project number and project title. The 25 projects are organized in four sections based on technologies presented, skills, and difficulty level. Projects are listed on the next page, in the table of contents, and the section dividers. Look at page 6 for more information about project features.

The first page for each presents pre-project activities and knowledge. This page includes:

Project Photo	Lesson Integration	Groupings (Sizes)
Level	Time to Complete	Objective(s)
Prerequisite Skills	Purpose and Skills	STEAM Connections
Key Vocabulary	Project Introduction	Anticipatory Sets

The next page starts the "Step-by-Step" instructions on how to implement the project. This page includes:

Standards (ISTE*, NGSS**, Common Core***)		Materials List
Digital Prototype	Project Notes	Pinout Diagrams
Prototype Code	Project Build Instructions	Project Photos
Project Code	Sketch Screen Shot	Connections

The last two pages of the project are the "ABC" pages. The reflections and summary activities, discussion questions, project extensions and real-world problem encourage creativity, connections to the real world, and if time allows, challenges the learners to develop their own investigations. Carefully review the options for implementation as additional materials or resources may be required. There are three levels of questions and extensions:

"A"	"B"	"C"
Essential	Recommended	Optional

The second page of the "ABC" section and the last page of the project details possible answers (indicated in blue italics) to the questions for the warm-up questions at the beginning of the project and the extensions and activities found on the previous "ABC" page.

SUPPORTING CONTENT

Supporting content (ex. Project Assessment Rubric, Glossary, and Materials List) is shown with a purple header.

Project materials are sold separately. See the book website for materials source recommendations.

*ISTE Standards Students, ISTE Standards for Students, ©2016, ISTE® (International Society for Technology in Education), https://www.iste.org/standards/for-students. All rights reserved
**NGSS is a registered trademark of Achieve. Neither Achieve nor the lead states and partners that developed the Next Generation Science Standards were involved in the production of this product, and do not endorse it.
***© Copyright 2010. National Governors Association Center for Best Practices and Council of Chief State School Officers. All rights reserved.

Register for the website and gain access to resource links. Purchasing a premium member on the website will give you access to project videos, presentations, photos, and diagrams.

Projects Overview

Legend: ✓ = Full match to standards · ✓(E) = Supports standards with extensions · ✓(P) = Supports standards with problem

Section	Project	Page	Time to Complete (hr:min) Project / Extensions	Difficulty	Earth (Science)	Life (Science)	Chemistry (Science)	Physics (Science)	Technology	Engineering	Allied Arts	Math
Section 1 Building Blocks	UNO Heartbeat	18	0:15 / 0:30	Starter		✓		✓	✓	✓	✓(E)	✓
	Lighting Effects	24	0:45 / 0:45	Starter		✓	✓(E)	✓	✓	✓	✓	✓
	Buzz Me	32	0:30 / 0:45	Starter		✓		✓	✓	✓	✓	✓
	Dimmer Switch	40	0:30 / 0:45	Starter		✓		✓	✓	✓	✓	✓
	Rainbow Light	48	0:45 / 0:45	Intermediate	✓	✓(E)	✓(E)	✓	✓	✓	✓	✓
Section 2 Displays, Inputs, and Controls	Spin It Up	68	0:30 / 0:45	Starter				✓	✓	✓	✓	✓
	Count on Me	76	0:45 / 0:45	Starter				✓	✓	✓	✓	✓
	LCD Billboard	84	0:45 / 0:45	Intermediate				✓	✓	✓	✓	✓
	Lights On/Lights Off	92	0:30 / 0:45	Intermediate				✓	✓	✓	✓	✓
	Pushbutton Counting	100	0:45 / 0:45	Intermediate	✓(E)	✓	✓(E)	✓	✓	✓	✓	✓
	Music Maker	108	0:45 / 0:45	Intermediate	✓(E)			✓	✓	✓	✓	✓
	Calculator	116	1:00 / 0:45	Advanced				✓	✓	✓	✓	✓
Section 3 Sensors	Night Light	128	0:30 / 0:45	Starter	✓	✓		✓	✓	✓	✓	✓
	Sensing Jolts, Bumps, and Rattles	136	0:45 / 0:45	Starter	✓	✓		✓	✓	✓	✓	✓
	Polychrome	144	0:45 / 0:45	Intermediate	✓	✓		✓	✓	✓	✓	✓
	Digital Ruler	152	0:45 / 0:45	Intermediate	✓(P)	✓		✓	✓	✓	✓	✓
	Turbidity Sensor	160	0:45 / 0:45	Intermediate	✓			✓	✓	✓	✓	✓
	My Heartbeat	168	0:30 / 0:45	Intermediate		✓		✓	✓	✓	✓	✓
	pH Properties	176	0:45 / 0:45	Intermediate	✓			✓	✓	✓	✓	✓
	Soil Moisture	184	0:45 / 0:45	Intermediate	✓			✓	✓	✓	✓	✓
	Gas Sniffer	192	1:15 / 0:45	Advanced	✓			✓	✓	✓	✓	✓
Section 4 Combining Inputs and Outputs	Cool It!	202	1:15 / 0:45	Intermediate	✓	✓(E)	✓(E)	✓	✓	✓	✓	✓
	Emoji Me	210	1:15 / 0:45	Advanced				✓	✓	✓	✓	✓
	Did You Hear Me?	218	1:15 / 0:45	Intermediate	✓(E)			✓	✓	✓	✓	✓
	Robot Race	226	2:00 / 2:00	Advanced	✓(E)	✓(E)	✓(E)	✓	✓	✓	✓	✓

✓ = Full match to standards   ✓ = Supports standards with extensions   ✓ = Supports standards with problem

Section 1
Building Blocks

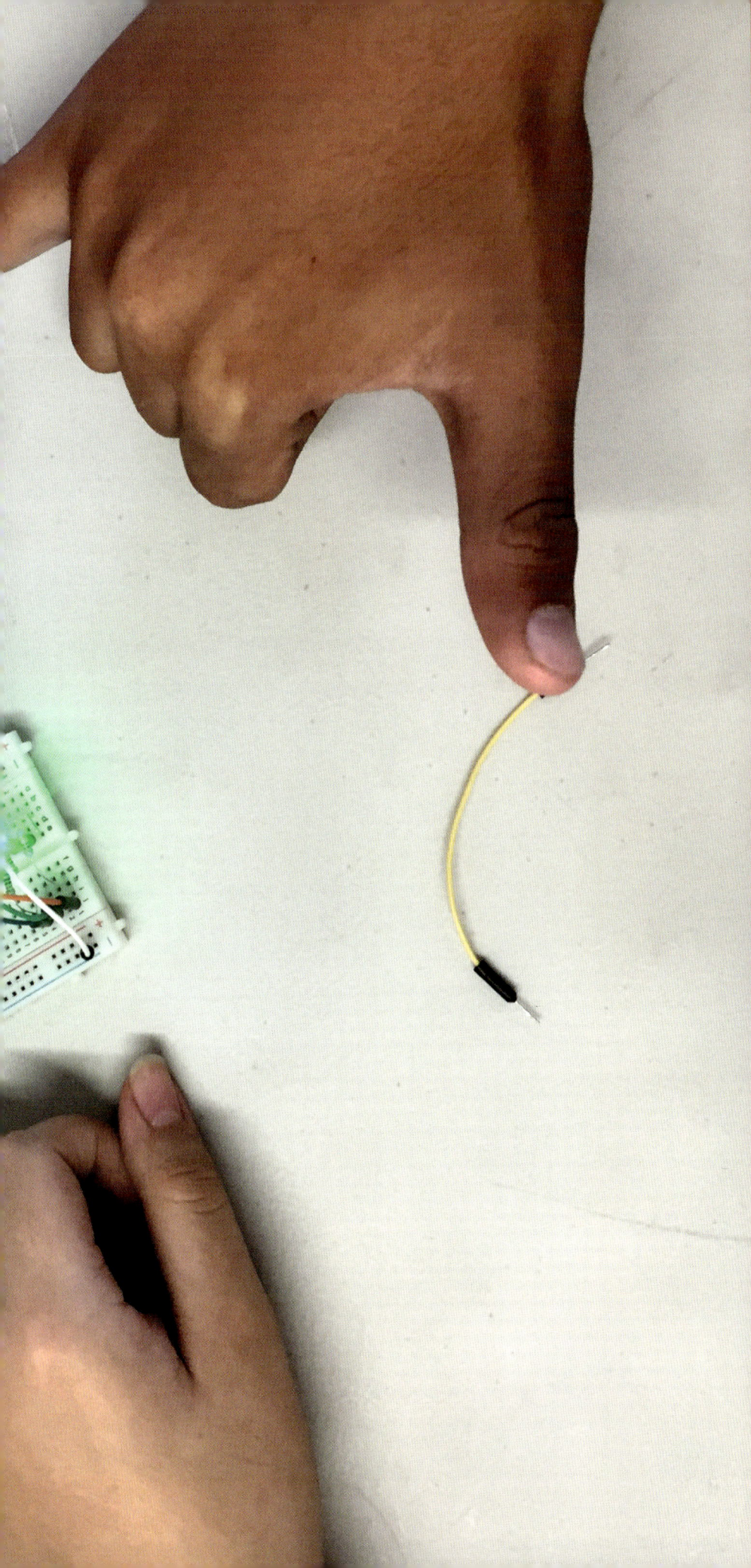

PROJECTS

This page will help you successfully navigate each element of the book, identify each section, and complete each project.

Each project starts with a page that defines teaching and learning elements including:

- Lesson Integration
- Groupings
- Level
- Time to Complete
- Objective(s)
- Prerequisite Skills (with page references)
- Purpose and Skills
- STEAM Connections
- Key Vocabulary
- Project Introduction
- Anticipatory Sets

The second page of each project includes:

- Standards Alignments (Purple Box Light Purple Fill)
- Materials List (Green Box, Light Green Fill)

Step-by-Step 1-15

The Step-by-Step instructions (with number of steps) start with how to build a digital prototype with Tinkercad® (pages 15-16) or Fritzing (pages 64-66).

Steps are numbered. **1**

Code is in a red box with white fill.

Important notes are in red boxes with red fill and white text or indicated in-line with red text.

Useful notes are in blue or purple boxes.

Green check-marks indicate stopping points for groups to check in before proceeding to the next steps.

Screen captures of the code are show from Tinkercad® and/or the Arduino® IDE.

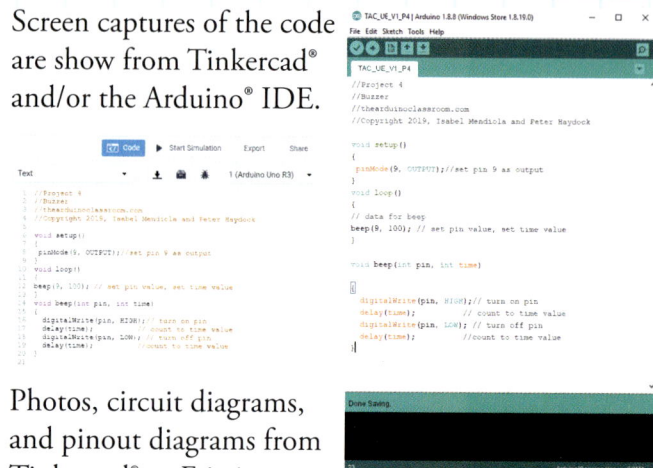

Photos, circuit diagrams, and pinout diagrams from Tinkercad® or Fritzing are provided alongside text descriptions of what to do.

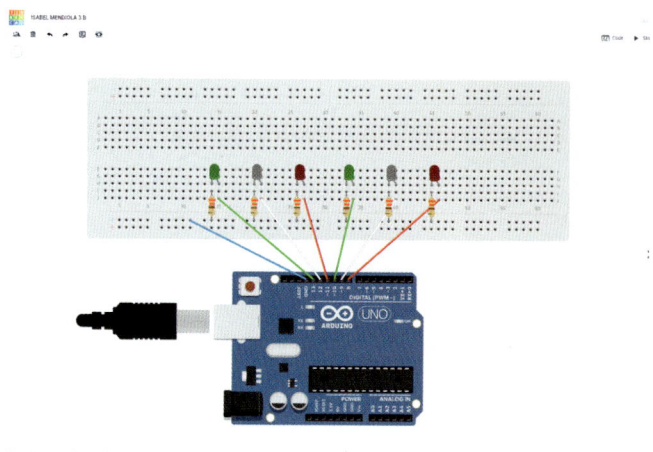

More complex projects will have red tables to guide jumper wire placements.

Look for hexagonal boxes with the word "Connection" in it for interesting facts connected to the project.

Connection

The second to last page of each project provides reflection activities, discussion questions, project extensions, and a guide for a problem-based learning activity.

The last page of each project provides answer guidance for questions and extensions posed in the project.

At the heart of every project is an electric circuit. No sensor would work, no light would turn on, and sounds or actions would fail. A complete circuit being in place allows everything to happen.

This page explains how electricity flows in a project. Understanding a circuit diagram will also help with troubleshooting if the project does not work at first.

For a circuit, three things are needed:
1. source of electricity
2. something that resists (uses) electricity
3. closure of the circuit through a ground

In this direct current example, the source of electricity is a battery (but could be the USB port from the computer). The electricity flows clockwise (from the anode) though the circuit passing through the light and resistor and ending at the negative side (cathode) of the battery.

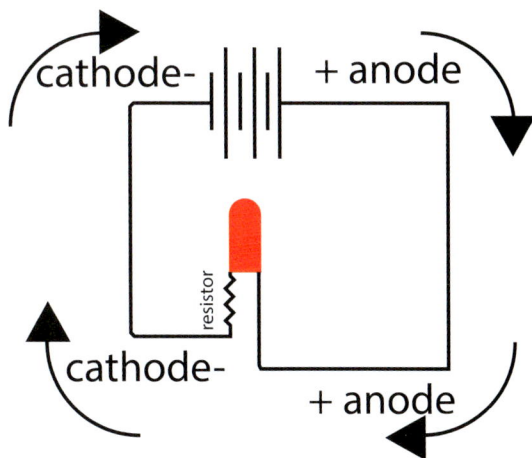

Searching the Internet for UNO projects will reveal many projects with only a circuit diagram for guidance. This may be because an engineer submitted the project.

To more easily read a circuit diagram, break the diagram into smaller less complex circuit segments. The basics of the circuit do not change, so reducing the circuit to smaller less complex sections works. Printing the diagram out and tracing each circuit with a highlighter can also be helpful.

There is an online tools that will help make project prototyping easier, called Tinkercad® (pages 15-16). Tinkercad® creates pinout diagrams and verifies and simulates the code after it is entered. It verifies that the circuits are complete. Pinout diagrams are much easier to read and help you connect wires and parts correctly.

> Connection - The word "circuit" has the same root as circle, circus, circumnavigate and circumference. All of these words share the original Greek word, *kirkos,* which means "ring or circular line." In Roman times, races and entertainment happened in a circular performance area called a circus. In a direct current (D.C.) electrical circuit , electrons move in a "circle" from anode to cathode.

A few notes as you work with your UNO board:
- The final connection for your circuit on the UNO board is marked as GND (ground). This is the cathode(-) end of the circuit on the board. There are three ground pins, one on the digital side of the UNO board and two on the analog side. You can generally use any of these as your cathode(-) end.
- Use one of the breadboard columns marked in blue (-) to ground your circuit. Then connect the column to a ground on the UNO board.
- Pay attention to the technical notes on your inputs and outputs. Some have leads (the wires entering or exiting the device) that are meant to be connected to the anode(+) or cathode(-) side of the circuit. For example, LEDs have a longer leg that needs to be in the anode direction of the circuit. If you turn the LED around, the LED will not light.
- You can power multiple devices in your project by using the red column (marked with a "+") on your board.
- It can be confusing that the "+" is the source of electricity and the "-" is the outlet. Just think of electricity of flowing from where there are more electrons to where there is less and you will be OK. This also goes for batteries.

Always use caution with electricity. Electricity, even in low voltages can shock or injure. Replace worn or broken materials immediately. Never use near liquids unless instructed to do so.

As presented in the previous section, a circuit is the path electricity takes as it moves through a device. In a direct current circuit, the anode(+) end connects with the cathode(-) end through at least one part like a light or resistor. The wiring through the different parts in the device and how the parts and wires are interconnected determines the path of the electricity.

Interconnections come in two forms, series connections and parallel connections. In a series connection the electricity flows from one part of the circuit to the next part in order. In a parallel connection, the electricity divides and each path receives portions of the electricity at the same time. Series and parallel connections can be combined to form more complex circuits.

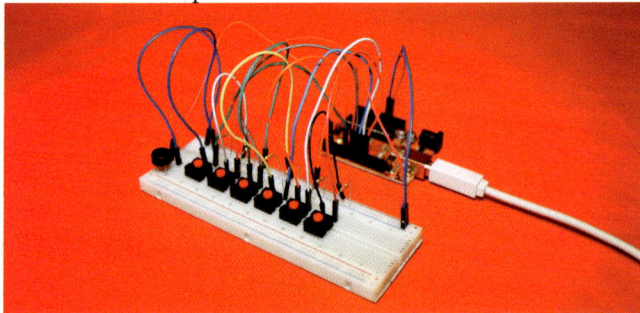

For the projects in this book, we are working with direct current (DC) which flows from anode(+) to cathode(-). On the UNO board the electricity can start in any of the pins except for those labeled as ground (GND).

No matter how complex the circuit is (the number of series and parallel segments), it can be reduced to a simple circuit of an input voltage (anode +), one equivalent resistor, and the ground (cathode -).

However it is not enough for an engineer or coder to simply have a diagram of the flow, they must also have a grasp of the physics and underlying mathematics of what is happening in the circuit.

The math of the circuit is governed by several formulas which help the engineer or technologist pick the right parts to implement in their project. The most important formula is Ohm's Law.

$$V = IR$$

V is voltage in Volts
I is current in Amps
R is resistance in Ohms

Another equation that is important in a circuit can be summarized as, the sum of the voltages across each part is equal to the total voltage of the circuit.

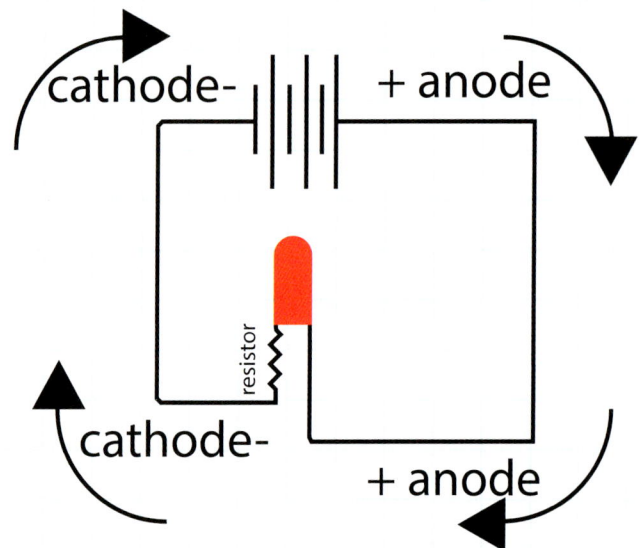

Looking at the first diagram we see that the three components of a circuit are in place; a source of electricity, a resistance (LEDs in this context have no measurable resistance), and a ground as a simple circuit. Ex.

$$V_C = V_{LED} + V_R$$
$$V_C = 0 + V_R$$
$$V_C = V_R$$

V_C is the voltage of the whole circuit or the output voltage and is the sum of all voltages in the circuit.
V_{LED} is the voltage across the LED (which is assumed to be 0. 0 is substituted in for V_{LED}).
V_R is the voltage across the resistor.

Alternating current (AC), the type of electricity delivered by the outlets in your home or school, is a bit different, but the key concepts of series and parallel circuits presented here are the same.

The formula would be rewritten to:

$$V_C = IR_C = IR_R$$

Once the known values are placed in the formula the equation can be resolved or solved.

SERIES CIRCUITS

In a series circuit the total voltage of the circuit is the sum of the sum voltages across each resisting part.

$$V_T = V_1 + V_2$$

V_T is the voltage of the whole circuit / output voltage.
V_1 is the voltage across the first resisting part.
V_2 is the voltage across the second resisting part.

There are two understood parameters in a series circuit. First, the current (I) is the same across all of the circuit. Second, the circuit resistance is the sum of all of the resistances. The resultant formula for resistance would be:

$$R_T = R_1 + R_2$$

R_T is the resistance of the whole circuit in ohms.
R_1 is the resistance of the first part in ohms.
R_2 is the resistance of the second part in ohms.

Recall that resistors are color banded and you can determine their value by comparing the resistor to the band chart. See the inside front cover for the chart.

You also know V_T because the output of the UNO board is either five volts (5v) or 3.3 volts (3.3v). You also have value of resistor (by reading the color bands) and you should be able to determine the resistance of any other parts from its technical specifications. With this information you can put the values into the Ohm's Law equation and solve for the current (I) in amps.

PARALLEL CIRCUITS

The second type of circuit is the parallel circuit. This circuit connects in parallel the parts of the segment of the circuit and distributes the current across all the parts at the same time. Look at the circuit diagram on

this page and see how each anode(+) leg is connected separately and in series with its own resistor. The cathode(-) legs then are joined back together. If each light were turned on at the same time the current would be divided among the six (6) LEDs.

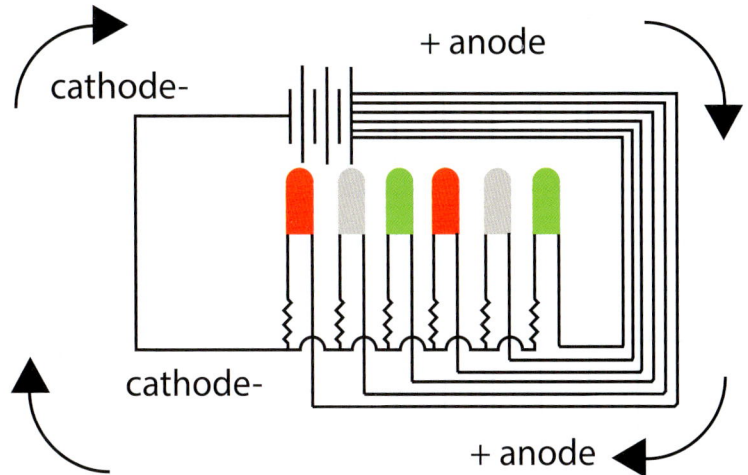

The equations that govern a parallel circuit are a bit different than the series circuit because the current (I) is not necessarily the same across different parts of the circuit.

The overarching equations still apply.

$$Vc = IR \text{ and } V_T = V_1 + V_2$$

Like the series example, we know the resistances of the parts involved in the circuit and still want to solve for the current value. With a bit of algebra, we resolve two formulas.

$$I_T = I_1 + I_2 \text{ and } 1/R_T = 1/R_1 + 1/R_2$$

This shows that the total current is the sum of the individual currents from each parallel part of the circuit and that the resistances can be determined using the sum of the reciprocals.

MIXED CIRCUITS

While beyond the scope of this text, mixed circuits of parallel and series elements can be solved. This is done by solving each series section that is distinct and each parallel section and combining until the whole circuit has been simplified into one simple circuit.

SKILLS
Arduino® UNO Microcontroller Board

Take this tour of the UNO microcontroller board. Every sketch runs on the board and interacts with the hardware devices connected.

Digital Pins 0, 1, 2, 4, 7, 8, 12, 13
Strictly digital pins that can only communicate a one (1) or zero (0). In a sketch or code the "HIGH" logic value is equal to a digital one (1) and "LOW" logic value is equal to a digital zero (0).
GND pin is a ground pin (cathode -)
AREF pin allows an alternate voltage (<5v) input to be used.

Digital Pins 3, 5, 6, 9, 10, 11 (PWM~)
These digital input and output pins can both send and receive regular digital signals and those with Pulse With Modulation (PWM). PWM allows for signals with varying length of pulses. This then is interpreted by the UNO board as a value other than a digital one or (1) or zero (0).

Reset Button
Restarts the uploaded code on the board.
USB Connection
The connection that allows a computer and board to communicate. It also powers the UNO board when connected.
Barrel Jack An alternate electrical input that accepts electricity from 9 volt power sources (ex. 9v battery with a battery adapter).

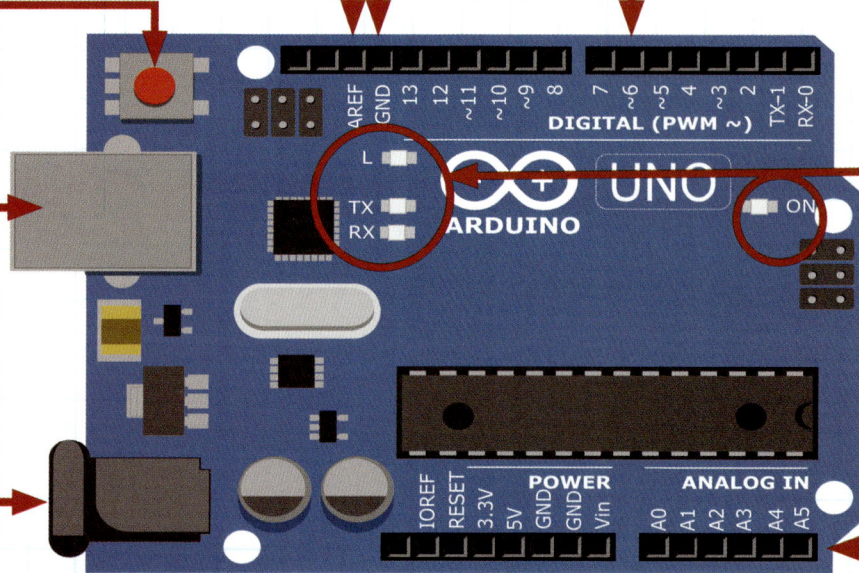

L is the LED controlled by digital pin 13.
TX indicates sending data.
RX indicates receipt of data.
ON indicates that the board is connected to a power supply.

LEDs locations may differ from the diagram. The USB connection may be a "B" or "micro" connection.

Breadboards label columns with letters (in this example a- j) or charge, anode(+) or cathode(-).
Rows are numbered (in this example 1-30). Pins in the same row on the same half (a-e or f-j) are connected. Pins in the anode(+) or cathode(-) columns are connected.

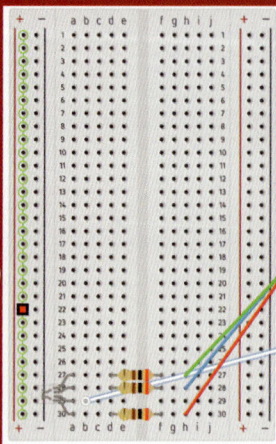

Power Pins
Vin A pin for connecting an external voltage (Input voltage). Usually nine volts (9v).
GND are ground pins (cathode -).
5V is the five volt (5v) power supply pin. Anode(+)
3.3V is the three point three volt (3.3v) power supply pin. Anode(+)
RESET Allows for an external reset button connection. Operates like the red button on the board.
IOREF sends a signal to an external connection about the voltage used on the board.

Analog Pins
The six analog input pins A0 through A5 can send and receive values other than one (1) and zero (0). These pins signals connect to analog devices like many humidity and temperature sensors or inputs like the 10k potentiometer.

Drivers for the UNO board may need to be installed on the computer. See the book website for links to board drivers.

SKILLS
Arduino® IDE software

Install the Arduino® IDE software for your computer from https://www.arduino.cc/en/main/software

File (p. 12)
The menu for starting a new project (sketch), saving a sketch, printing the sketch's code or saving the sketch with a new name.

Edit (p. 12)
This menu gives access to word processing like commands to change the sketch.

Sketch (p. 13)
With the code entered, it can be verified, compiled, or uploaded to the UNO board.

Tools (p. 13)
Launch the data monitors (very useful), connect the board to the correct COM port, and manage the code libraries.

Help
Link to help guides, troubleshooting documents and the main Arduino® website.

Write and edit a sketch, check and save it, and upload it to the board via the Arduino® IDE software.

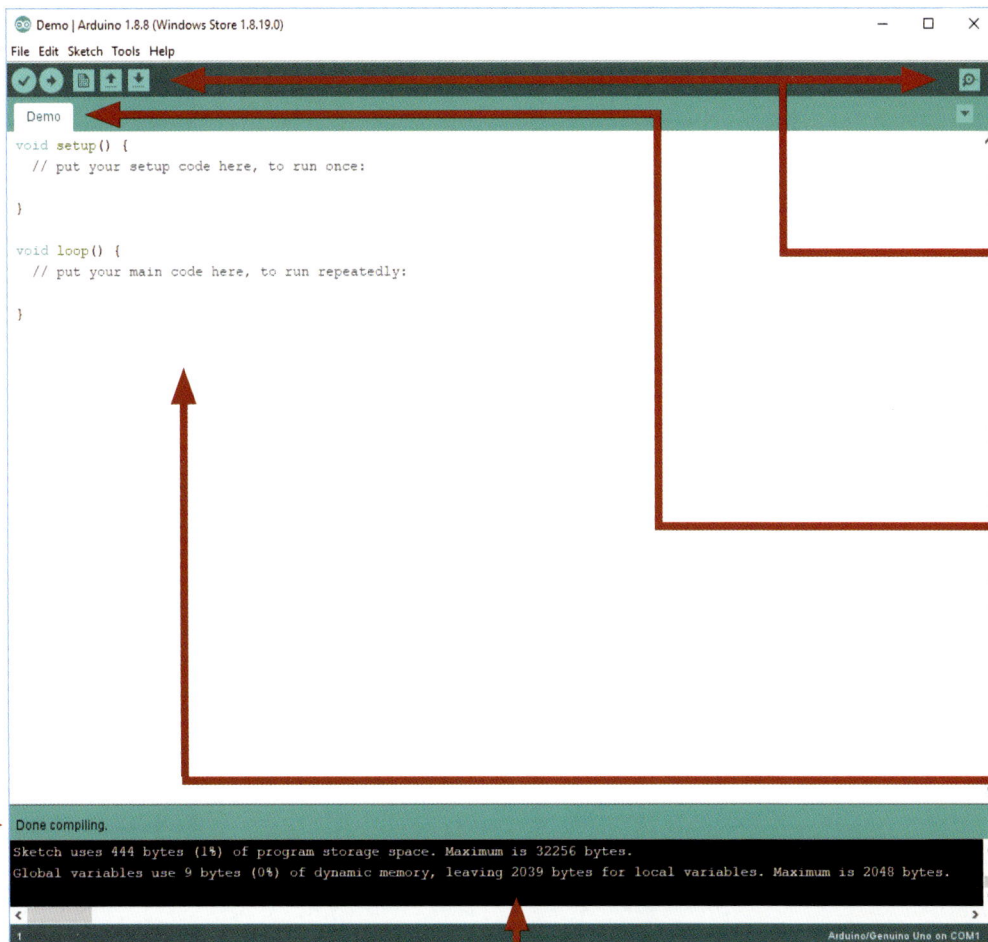

Demo | Arduino 1.8.8 (Windows Store 1.8.19.0)

File Edit Sketch Tools Help

Demo

```
void setup() {
  // put your setup code here, to run once:

}

void loop() {
  // put your main code here, to run repeatedly:

}
```

Done compiling.

```
Sketch uses 444 bytes (1%) of program storage space. Maximum is 32256 bytes.
Global variables use 9 bytes (0%) of dynamic memory, leaving 2039 bytes for local variables. Maximum is 2048 bytes.
```

1   Arduino/Genuino Uno on COM1

Shortcuts
Verify Check the sketch for errors before uploading.
Upload Send the sketch to the board.
New Start a new sketch.
Open Open a sketch.
Save Save the sketch.
Serial Monitor Open the Serial Monitor.

Sketch Tab(s)
Each open sketch is shown with its name (Demo in this case) in a tab.

Sketch (Code)
The location to enter code.
Each new sketch starts with the code seen here.

Current Task
IDE software displays what it is currently doing.

Task Details
Here, the IDE software displays a history of its verifying or uploading a sketch. Errors are detailed here. Board memory used is the last detail shown in a successful upload.

Board and Connection Information
The IDE software shows the type of board connected and the COM port settings for the sketch.

Take a tour of the IDE software online

The Arduino Classroom § 11

File Menu (Keyboard Shortcuts shown to the right)

New - Start a new sketch (program or code).
Open - Open a sketch.
Open Recent - The IDE software remembers the most recent sketches it opened.
Sketchbook - A larger list of the most recent sketches opened in alphabetical order.
Examples - Dozens of sketches to look at, draw inspiration from, implement, or edit.
Close - Close the IDE software and if needed, save any changes to your sketches.
Save - Saves the current sketch. The default name is the date with a letter added at the end.
Save As - Save the sketch with a new name.
Page Setup - Defines the size and orientation of the printer paper.
Print - Print the sketch to a printer.
Preferences - Controls a large number of default settings for the IDE software including, size of font, to the preferred save location and even some network options. For the most part you can leave these alone.
Quit - Close the IDE software without saving any work done on your sketches.

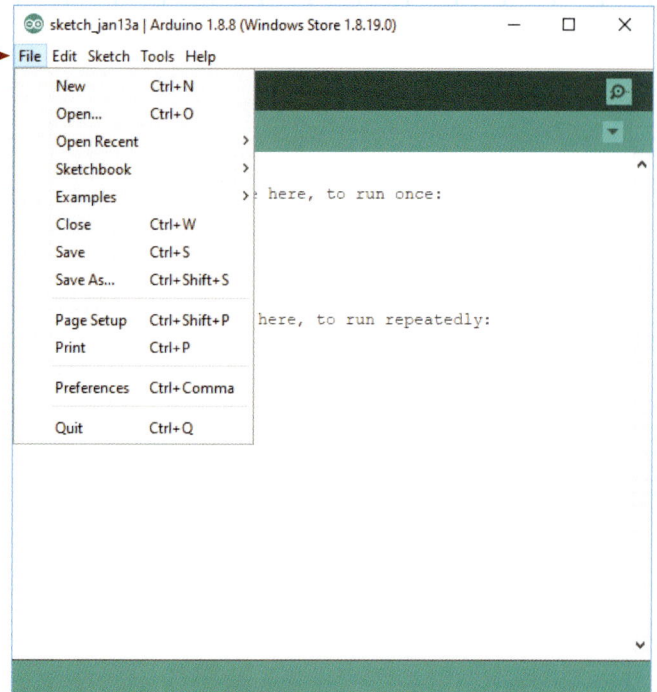

Edit Menu (Keyboard Shortcuts shown to the right)

Undo - Undo the last edit made to the sketch.
Redo - Restore the last undone edit in a sketch
Cut After selecting text from sketch, use this to cut or remove the text. Puts the text into the computer's clipboard.
Copy - After selecting text from the sketch, this will copy the text to the computer's clipboard. Use with the paste menu option.
Copy for Forum - After selecting text from your sketch, you can copy it for use in a website forum.
Copy as HTML - After selecting text from the sketch, this will copy it for use on a web page.
Paste - Place the cut or copied text into the sketch.
Select All - Selects all of the text in a sketch to be cut or copied.
Go to line - If the code is large, use this feature to jump to a specific line number.
Comment - Allows the insertion of comments into the sketch for documenting a line or section of code.
Increase / Decrease Font Size - Change the font size of the sketch text.
Find / Find Next / Find Previous - Allows the search of the sketch for a particular word, phrase, or number. Can be used to search for any sequence of letters and numbers.

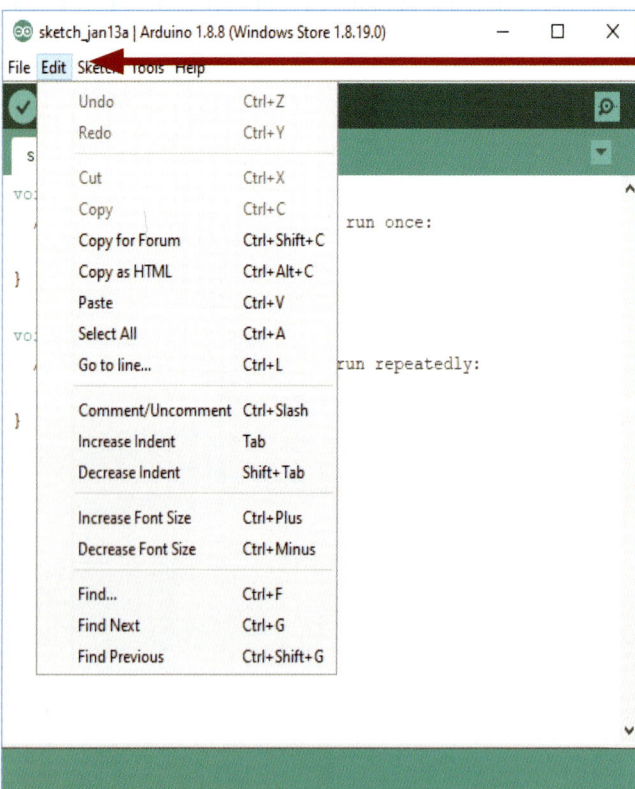

```
sketch_jan13a | Arduino 1.8.8 (Windows Store 1.8.19.0)          —   □   ×
File  Edit  Sketch  Te    
        Verify/Compile            Ctrl+R
        Upload                    Ctrl+U
  sketch  Upload Using Programmer  Ctrl+Shift+U
        Export compiled Binary    Ctrl+Alt+S
void s
  // p    Show Sketch Folder       Ctrl+K
          Include Library              ▶
}         Add File...

void loop() {
  // put your main code here, to run repeatedly:

}
```

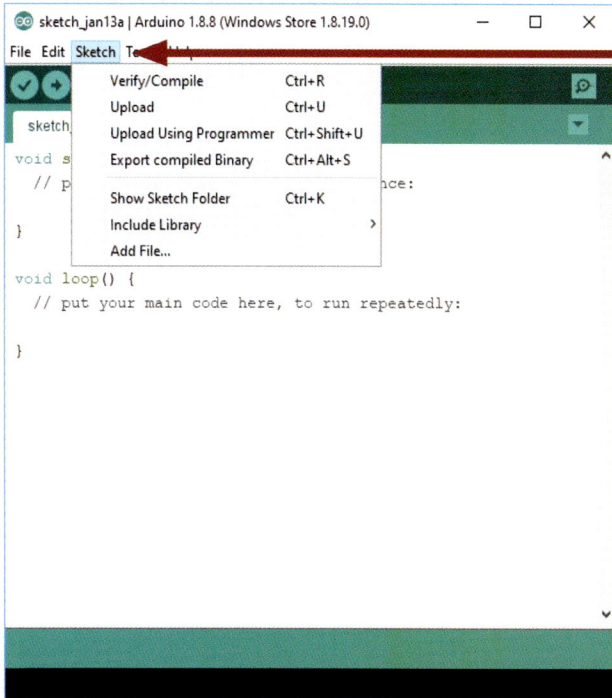

Sketch Menu - (Keyboard Shortcuts shown to the right)

Verify / Compile - Check a completed sketch that the code works. The IDE software will display issues with the code in the Task Details zone of the window (black).

Upload - Send the sketch to the board.

Uploading Using Programmer - This is an advanced feature that changes how the sketch is uploaded. When uploading, the sketch sends a small program first then the sketch. This feature changes the first program uploaded.

Export compiled Binary - Another advanced option that facilitates sending the sketch to another device.

Show Sketch Folder - Opens the file explorer to the location of the sketch under development.

Include Library - Facilitates adding a library (p. 62) to the sketch, manages current libraries, or allows for uploading a custom library.

Add file - Another advanced option that allows you to add a custom file to your sketch.

Tools Menu (Keyboard Shortcuts shown to the right)

Auto Format - Set the colors, tabs, and font of the text in the sketch to the default settings.

Archive Sketch - Make a .zip file of the sketch for archiving or sharing.

Fix Encoding and Reload - Fixes issues with the special characters that might be different between the IDE and the computer operating system.

Manage Libraries - Add, delete, rename, or update the libraries (supporting programs - p. 62) for sketches.

Serial Monitor - Send (Output) data from the sketch as it operates on the board to the computer screen. Requires specific lines of code in the sketch.

Serial Plotter - Graphs the data output of the sketch as it operates on the board. Requires specific lines of code in the sketch.

Firmware Updater - Updates firmware (board programming) if needed.

Board Manager - Select the version of board connecting to the IDE.

Port - Select the COM port.

Get Board Info - Verifies the version of board attached.

Programmer - Select the program to load with the sketch (do not change unless instructed).

Burn Bootloader - Select the program to load before your sketch to operate the board (do not change unless instructed).

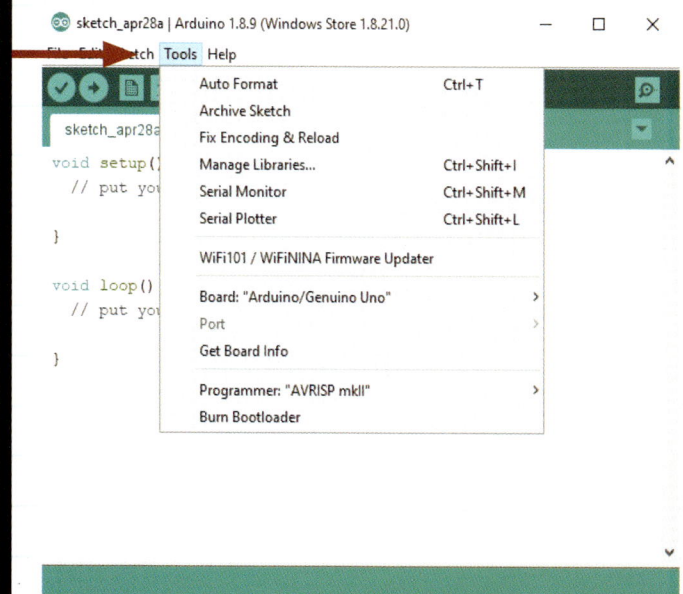

```
sketch_apr28a | Arduino 1.8.9 (Windows Store 1.8.21.0)         —   □   ×
     tch  Tools  Help
            Auto Format              Ctrl+T
            Archive Sketch
  sketch_apr28a  Fix Encoding & Reload
void setup()  Manage Libraries...      Ctrl+Shift+I
  // put yo  Serial Monitor           Ctrl+Shift+M
            Serial Plotter           Ctrl+Shift+L
}
            WiFi101 / WiFiNINA Firmware Updater
void loop()
  // put yo  Board: "Arduino/Genuino Uno"    ▶
            Port                           ▶
}           Get Board Info

            Programmer: "AVRISP mkII"      ▶
            Burn Bootloader
```

Take a video tour of the IDE software online.

SKILLS
Sketches and Coding

Install the Arduino® IDE software for your computer from https://www.arduino.cc/en/main/software

A typical project generally has three major sections.

1. Libraries
2. Setup →
3. Loop →

Libraries are always listed first in the sketch. For the projects in the first section of the book, libraries are not used. Libraries are covered in the second section of the book and explained on page 62.

The "Setup" section is required in all sketches. Despite not being required to run a sketch, the "Loop" section is the most powerful part of a sketch.

Each project in the book we will cover more code skills. Start by copying the code provided in each project and modify it to complete the extensions.

Follow along with these steps and you will write your first sketch for the UNO board.

Special Characters in Sketches

- Parentheses "( )" denote the possibility of values being set. You see this with the parentheses after pinMode, digitalWrite, and delay.
- Curvy brackets "{ }" indicate the start and stop of a section of the sketch
- A semi-colon ";" indicates the end of a line of instruction
- Two forward slashes "//" tell the program to ignore the text after them. They are used to put programming remarks and notes about a line or section of the sketch

If you miss one of these special characters the sketch will not compile and not be verified. You also will get the positive visual indicator of the command values turning colors that you are close.

Step-by-Step 1-3

1 After installing the IDE software, launch the program. A blank sketch will open or the last sketch opened on the computer.

Notice that the sketch has been assigned a name starting with "sketch_" and then the month and day. The last part of the name is a letter designation starting with the letter "a." In this case the sketch is called: sketch_jan23a

See that the sketch has the setup and loop sections defined.

2 Either copy and paste the code from the website or type the code on this page.

```
void setup()
{
  pinMode(10, OUTPUT);
}

void loop()
{
  digitalWrite(10, HIGH);
  delay(1000); // Wait for 1000 millisecond(s)
  digitalWrite(10, LOW);
  delay(1000); // Wait for 1000 millisecond(s)
}
```

Note: the color of the text will change as each entry is completed.

3 Click the verify button (check mark icon) in the upper left of the software to check the code. Errors will be highlighted for needed correction. Verify after each correction until the sketch works.

Access Tinkercad® at Tinkercad.com on any Internet connected computer.

There are two free programs for prototyping Fritzing and Tinkercad® that are useful in designing digital project prototypes. Tinkercad® prototypes have many advantages and will be the primary application for this book. Fritzing will be covered later (pages 64-66) and will be used for a few projects.

Tinkercad® is produced by Autodesk® (an industry giant in computer based design). Its "Circuits" feature has the functionality of both digital design and simulating the project with code.

Almost all projects start with the digital prototype. If the project uses Tinkercad® entering the code will simulate the project's actions.

REGISTERING for Tinkercad®

Tinkercad® requires registration and verification. A valid email address is required to register. This allows any work completed to be saved online and shared. Tinkercad® will send the email address associated with the account a confirmation note along with a response/confirmation link.

Using a classroom account or individual student accounts per your organization's policies may also be an option for you.

Tinkercad® does not allow individuals who are 12 or under to register without the consent of a supervising adult.

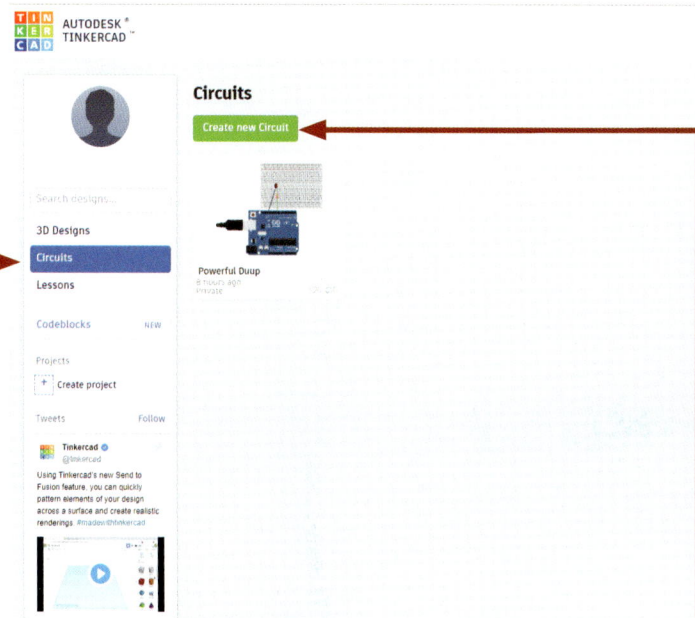

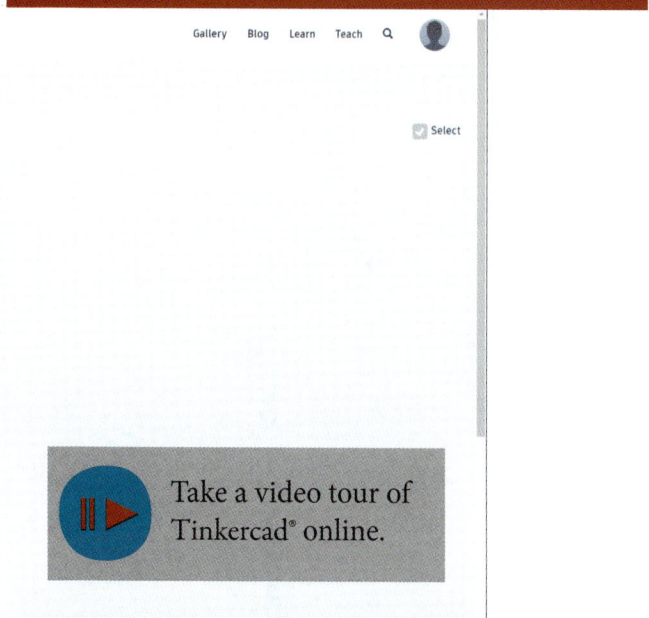

Take a video tour of Tinkercad® online.

Step-by-Step 1-15

1. Go to the Tinkercad® website, www.tinkercad.com

2. If needed, register for Tinkercad®.

3. Log in

4. Click on "Circuits" in the left side menu.

5. Click the green button "Create New Circuit."
Continued on next page.

Fritzing

Fritzing is a free open-source program which means it advances as the community of programmers add to it. Fritzing must be downloaded and installed on your computer from fritzing.org/download. Fritzing allows you to prototype your project and lay out its parts and design. Fritzing has a coding feature, but it does not integrate with projects to test them.

Fritzing can be installed on Mac, PC, and Chromebooks. Pages 64-66 cover how to use Fritzing.

6 Tinkercad® will open a new scene with a blank workspace.

7 A menu of images on the right side shows the components.

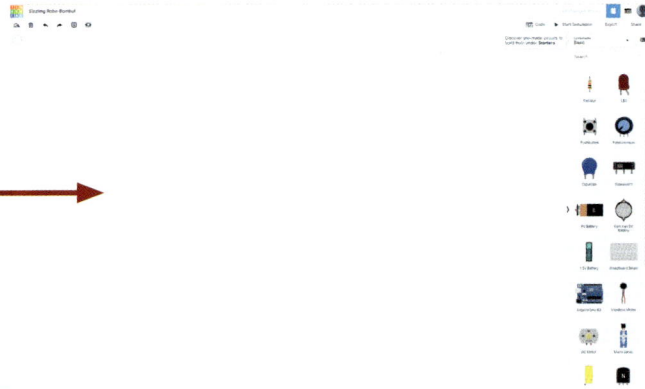

8 On the right side, find the UNO board. Drag the board to the workspace.

9 Drag a short breadboard to the workspace.

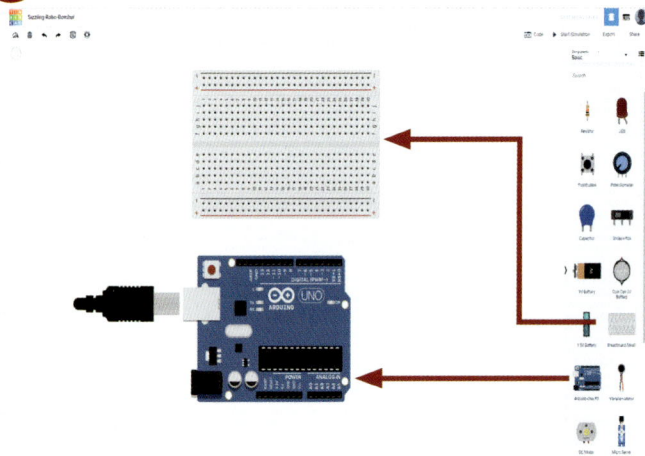

10 For this example, drag an LED and resistor to the workspace as shown. Use the search tool to find other parts if needed.

11 To add jumper wires, click on starting pin location, continue to hold the mouse button down and release the mouse button at the end location of the jumper wire and click again at the endpoint. Change the color of the wire as needed.

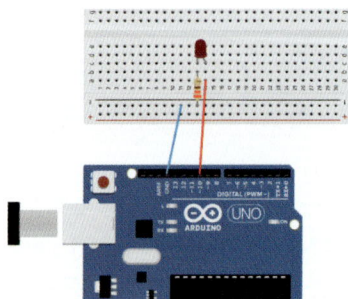

12 Next, add code to tell the prototype what to do. In the upper right of the screen locate the "Code" button and click it. Then click the drop down menu to change from "Blocks" to "Text."

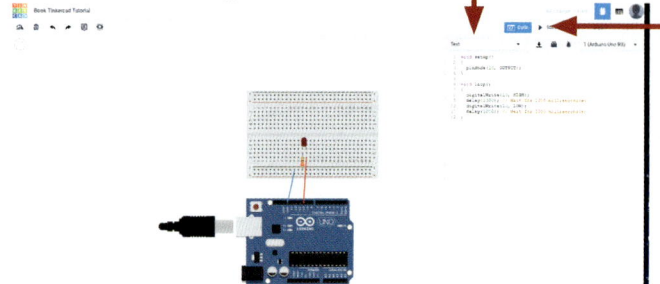

13 Starter code has already been entered. Change the pin number listed in all three locations from "13" to "10."

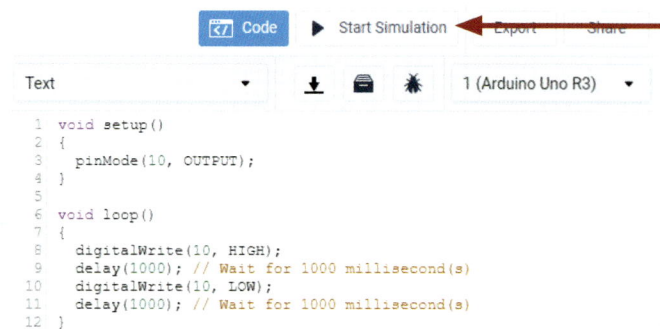

```
1  void setup()
2  {
3      pinMode(10, OUTPUT);
4  }
5
6  void loop()
7  {
8      digitalWrite(10, HIGH);
9      delay(1000); // Wait for 1000 millisecond(s)
10     digitalWrite(10, LOW);
11     delay(1000); // Wait for 1000 millisecond(s)
12 }
```

14 Click the "Start Simulation" button and see how things work.

15 If everything is connected correctly, the USB cable will plug in and the red LED will blink on for one second and off for one second.

Try This
- Change the resistor value.
- Move the jumper wires around (even if the wires are in the wrong place) or delete them.
- Change the timing referenced in the code to make the blinking rate change.
- Take the resistor out of the project.
- Flip the LED by 180 degrees so the anode and cathode leads are reversed.

See how the project responds to each change and fix to make the project work. The results of these changes will assist in future troubleshooting projects that may not work after the first build.

Assessing projects gives a more complete view of a student's growth and achievement because the assessment considers both end-product and process. To be valid, there should be multiple assessments points that reveal how the project was completed by the group and that understanding is deep across all team members.

We consider the best ultimate assessment to require learners to apply the new skills in the extensions and the solution of a real-world problem found at the end of each project. However, there are other places in the process that allow for assessment including project checkpoints and the project assessment rubric found on page 236.

Checkpoints:

1. **Whole Group Checkpoint** At any time, announce to the class that you are checking in with them. Have the learners raise their hand and indicate to you with the number of fingers held up to show you the their confidence in what they are doing.

    5 - complete confidence, no help needed
    4 - some challenges but no help needed
    3 - some challenges that may need help
    2 - significant challenges that need help soon
    1 - need help now

2. **Individual Group Checkpoint** At prescribed steps in each project have each group stop and ask for a

A project rubric can be found on page 236.

review what they have done so far.

Have the group do the following:
- Demonstrate what they have accomplished
- Confirm each member's participation
- Review their next steps
- Ask any questions

Once a group demonstrates an ability to self assess and if they have no questions, release them from doing these checkpoints.

Each checkpoint opportunity is marked with a green check-mark.

Group Member Assessment: Projects often require that one work in a cooperative group. As part of the assessment process, the group members should assess themselves and well as the other group members. This assessment should summarize and focus on effort, participation, cooperation, facilitation, and outcome as written in their journals. This should be done as a discussion and in their journals.

Project Assessment: Assess the learners based on the complete project including their knowledge and contributions. You will capture a complete picture if you use the self and group rubrics, the checkpoints, and your assessment of the project. A project assessment rubric can be found on page 236.

Presentation Assessment: Project Based Learning stresses one last element to form a complete assessment of the project. This assessment includes a report or presentation of the project to a group from outside the classroom. This may be other students from another class, parents or guardians, older students etc. This presentation could be in the form of a paper, poster, slide show, video, or other type of report or presentation. We have included links online to different reports and presentations that you may allow learners to select from.

One last note, we encourage you to change the project groups every three projects. For ideas on team formation, see the book website.

Project 1
UNO Heartbeat

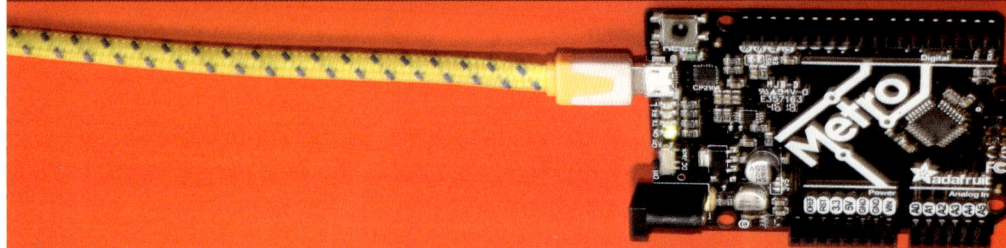

Getting Started

Lesson Integration:	Groupings:	Level:	Time to Complete:
Physics - Time (measurement), Anatomy of a Wave (Frequency) Biology - Circulatory System Mathematics - Cycles, Periodicity	1 - 2	Starter	15 min. for the project and 30 min. for extensions.

Objective:
- To control the UNO board indicator LED.

Prerequisite Skills:
- Time measurement (milliseconds)
- Arduino® IDE software (pages 11-13)
- Tinkercad® (pages 15-16)
- Ratios / Fractions for Extensions

Purpose and Skills:
- Introduce setting up the IDE software
- Connect the UNO board to the computer
- Produce a sketch
- Verify a sketch
- Upload a sketch
- Run a sketch
- Setting pin characteristics
- Loops

STEAM Connections:
Science - Metric Time, Heartbeat, Frequency
Technology - Code, Program Settings, Simulators, Digital Design, Electronic Components
Engineering - Designing, Building, and Using a Machine, Prototyping, Applied Physics
Allied Arts - Rhythm, Beats, Tempo
Math - Ratios, Conversions, Cycles

Key Vocabulary:
Analog - Describes data that is a number, letter, sound, or other information that is not limited in value or designation.
Code - The computer program in the form of line by line instructions.

Digital - One (1) or zero (0) logic value usually in a computer code. In a sketch or code the "HIGH" logic value is equal to a digital one (1) and "LOW" logic value is equal to a digital zero (0).
Diode - Electronic device that let a current pass through it in only one direction.
Engineering - A deliberate process based on math and science to create a technology, building, or form of transportation, or modify an environment.
Frequency - How often something happens.
LED - Light Emitting Diode. A device that gives off light when a small electrical current passes through its electronics. Usually can only emit one wavelength of light.
Sketch - Arduino®'s name for the code.
Prototype - The first operating version of a machine.
Technology - A tool, device, or machine to complete a task faster, more easily, or convert output.

Project Introduction:
-Introduce to the groups, the purpose of the project, the skills developed, the standards met, and the goal of the project.

Anticipatory Sets:
- Have students feel for their pulse.
- What is a heart beat?
- What is a pulse?
- What technologies can be used to detect a heart beat?
- What activities or professions rely on a heart beat or constant rate indicator?

Project 1
UNO Heartbeat

Educational Standards

ISTE Standards for Students
- Empowered Learner 1a, 1d
- Knowledge Constructor 3a, 3d
- Innovative Designer 4a, 4b, 4c, 4d
- Computational Thinker 5a, 5b, 5d
- Creative Communicator 6c, 6d
- Creative Communicator 6c, 6d
- Global Collaborator 7c, 7d

US Computer Science Standards
- Correlations can be found on the book's web page.

US NGSS - Middle School
DCIs
- MS-PS4-3 PS4.C
- MS-PS4-1 PS4.A
- MS-PS4-2 PS4.A PS4.B
- MS-LS1-3 LS1.A

Cross Cutting Concepts
- Cause and Effect
- Scale, Proportion and Quantity
- Structure and Function
- Patterns

Science and Engineering Practices
- Planning and Carrying Out Investigations
- Obtaining, Evaluating, and Communicating Information
- Analyzing and Interpreting Data
- Scientific Knowledge is Based on Empirical Evidence
- Using Mathematics and Computational Thinking

US NGSS - High School
- Correlations can be found on the book's web page.

US Common Core Language Arts and Mathematics
- Correlations can be found on the book's web page.

Step-by-Step 1-13

Materials List:
- Computer with Arduino® IDE software
- Connection to the Internet
- USB Cable
- UNO or UNO Compatible Microcontroller

Engineering Design - Digital Prototype

1. Launch and log into Tinkercad® (pages 15-16).

2. Start a new Circuits project.

3. Drag an UNO board to the workspace. Note: Use the UNO board even if you are using an UNO compatible board for the actual project.

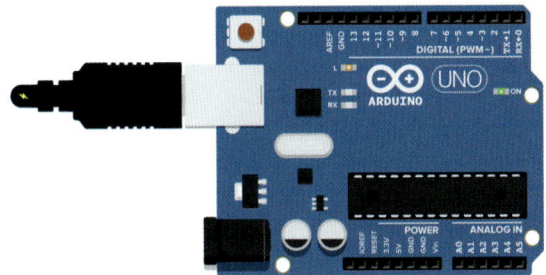

Coding - Digital Prototype

4. Enter the code in Tinkercad® to match the code on this page.

5. Start the simulation and observe the digital prototype.

```
void setup()
{
  pinMode(13, OUTPUT);
}

void loop()
{
  digitalWrite(13, HIGH);
  delay(300); // Wait for 300 milliseconds
  digitalWrite(13, LOW);
  delay(300); // Wait for 300 milliseconds
}
```

Engineering Design - Project Build

Some UNO boards may need computer drivers to connect to the computer. Visit the book web site for links to UNO drivers.

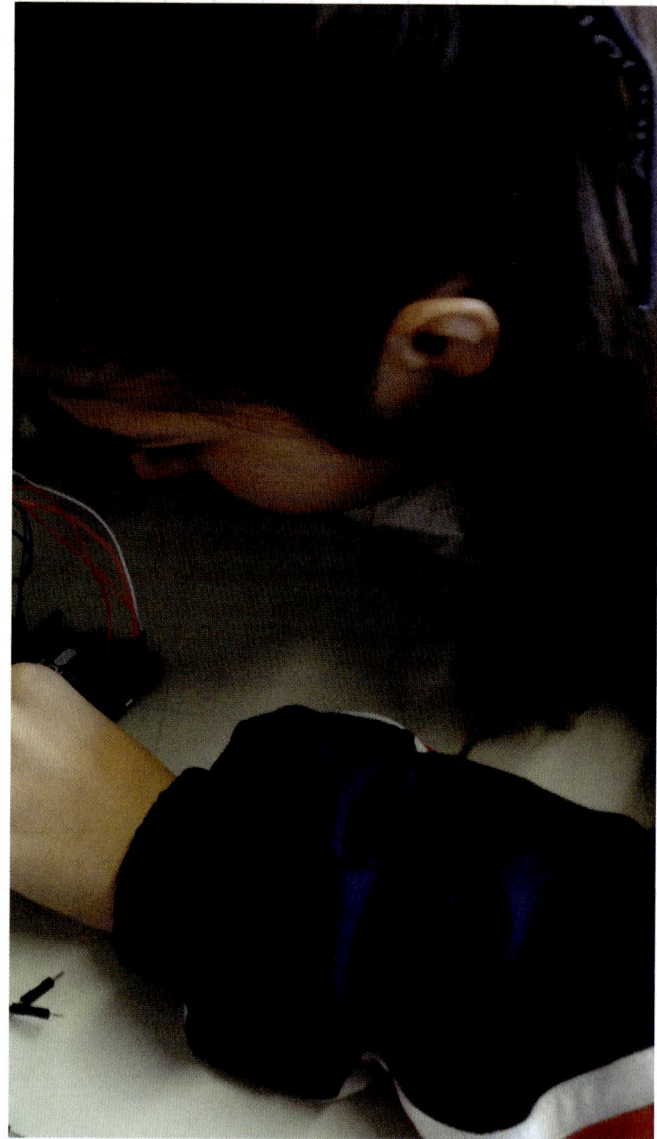

6 Connect the UNO board and computer using the USB cable.

7 Launch the Arduino® IDE software (pages 11-13) to make sure the board is communicating with the computer.

Go to the Tools menu and verify that the correct board is selected from the Board Manager menu. If not, select the correct board from the options listed.

Then, confirm that the right Port is selected. If not, select the port that lists the UNO board from the options listed.

Coding - Project Build

8 Copy and paste the code from Tinkercad® or type the code from page 19 into the Arduino® IDE software (pages 11-13).

```
File Edit Sketch Tools Help

TAC_UE_V1_P1 §

//Project 1
//Arduino Uno Board Blinking LED
//thearduinoclassroom.com
//Copyright 2019, Isabel Mendiola and Peter Haydock

void setup()
{
  pinMode(13, OUTPUT);
}

void loop()
{
  digitalWrite(13, HIGH);
  delay(300); // Wait for 300 millisecond(s)
  digitalWrite(13, LOW);
  delay(300); // Wait for 300millisecond(s)
}

Done Saving.
```

9 Save the sketch (rename as needed).

10 Verify the sketch.

11 Upload the sketch to the UNO board.

12 Observe the LED on the UNO board.

13 Document this project, discuss the project, and complete the extensions assigned from the next page.

Connection - According to the United Nations, 80% of the 50 million tons of the world's electronics thrown away end up in a landfill. Worldwide, the remaining 20% is recycled or refurbished, but not always in environmentally friendly ways.

In the United States, the recycling rate is a bit higher at 25%. Electronics recycling or refurbishing is now a $20 billion per year industry in the United States, both creating jobs and protecting the environment.

If an electronic device is turned in for recycling, it is first inspected to determine if it can be repaired and resold. If it cannot be repaired, the individual components in the device are assessed for resale. At the end of the process anything not repaired or resold is then scrapped for recycling.

The recycling industry can recover gold, silver, aluminum, lead, copper, glass, plastics, and steel from scrap electronics.

For example, more gold can be recovered from one ton of computer parts than 17 tons of gold ore. If all electronics in the United States were recycled, $60,000,000 in gold (like that found in the nugget shown on this page) would be recovered.

There are small amounts of recyclable metals in the UNO board you are using which can be recycled when the board no longer works.

Project 1
UNO Heartbeat

Project Reflection and Summative Activities, Discussion Starters, Extensions, and Problem

A = Essential | **B** = Recommended | **C** = Optional

Reflection/Summative Activities:

- Have each team member document in their project/classroom journal who was on their team, what went well, what they could improve upon, what they would do differently if they were to do the project again, and verify that every person in the group can do the project.

- In a classroom discussion led by the guide, have each team summarize their observations.

For Discussion:

- The code uses a timer / clock on the UNO board. Why would a board need a timer / clock?

- Where are timers and clocks effectively used to control the pace or sequence of events?

- As the sketch uploads, describe what you observe on the UNO Board and in the black window of the IDE software. What is happening?

Extension:

- Adjust the delay value(s) to make the LED blink faster or slower.

Professional Connection:

- Research and write a short description of a profession that relies on precise timing in your community. Be sure to include the education the profession requires, what devices are used, and what activities are performed.

Extensions:

- Add lines of code to the sketch to make a repeating pattern of blinking.

- Change the pinMode number to another digital pin number in the sketch. What happens when you verify the sketch? What happens when you upload and run the sketch?

- Build a project that would aid in keeping time or counting and evaluate its usability?

Extension:

- How might you use this project to keep time for music? Determine the delay values in the sketch to make the LED a metronome that blinks at 120 beats per minute, 90 beats per minute, 60 beats per minute, and 30 beats per minute? Build the project.

Problem:

- Identify a problem that requires precise timing and how this project could be a solution. Include in your solution the following:

Description of the problem.

Document how the problem is solved by this project.

What other resources are needed to solve the problem.

Build and demonstrate the solution.

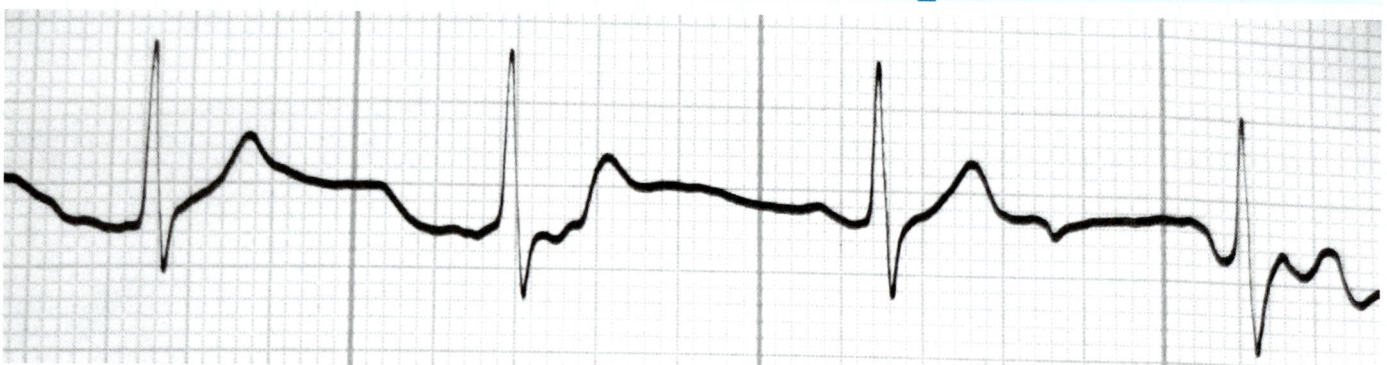

A project rubric can be found on page 236.

Anticipatory Sets
(from page 18)

- What is a heartbeat?
The pulse of a single contraction of the heart. Often used to determine if something is alive. Has contraction and relaxation phases.

- What is a pulse?
The rhythm of multiple heartbeats.

- What technologies can be used to detect a heartbeat?
An EKG machine, listening with a stethoscope, feeling a pulse on the neck or wrist.

- How is heartbeat rate indicated?
Beats per minute

- What activities or professions rely on a heartbeat or consonant rate indicator?
Musicians, nurses, trainers, dietitians, doctors.

A

Reflection/Summative Activities:
Responses Will Vary

For Discussion:
- The code uses a timer / clock on the UNO board. Why would a board need a timer / clock?
The clock on a microcontroller board ensures that instructions are carried out in the order and for the amount of time required. On other computer boards with more complex programs the clock plays a more important role. Think of playing a computer game. The game needs the timing of sound, video, interactions with other players, and controls to stay in sync as you play.

- Where are timers and clocks effectively used to control the pace or sequence of events?
Sports, GPS, Cooking, Music, Choreography

- As the sketch uploads, describe what you observe on the UNO Board and in the black window of the IDE software. What is happening?
The UNO board is indicating that it is receiving data from the computer.
The IDE software is translating the sketch into data and commands the UNO board will use. This is called compiling.

Extension:
- Adjust the delay value(s) to make the LED blink faster or slower.
Responses Will Vary

B

Professional Connection:
Responses Will Vary

Extensions:
- Add lines of code to the sketch to make a repeating pattern of blinking.

Responses Will Vary

- Change the pinMode number to another digital pin number in the sketch. What happens when you verify the sketch? What happens when you upload and run the sketch?
The sketch will verify and upload, but the LED will not blink.

- Build a project that would aid in keeping time or counting and evaluate its usability?
Responses Will Vary

C

Extension:
- How might you use this project to keep time for music?
The delay values could be adjusted to keep time with a prescribed rate.

- Determine the delay values for music that would need be in the sketch for:
120 beats per minute
(ex. 250 ms for both digitalWrite HIGH and LOW)
90 beats per minute
(ex. 333 ms for both digitalWrite HIGH and LOW)
60 beats per minute
(ex. 500 ms for both digitalWrite HIGH and LOW)
30 beats per minute
(ex. 1000 ms for both digitalWrite HIGH and LOW)
Build the project.
Responses will vary.

Problem:
Responses Will Vary

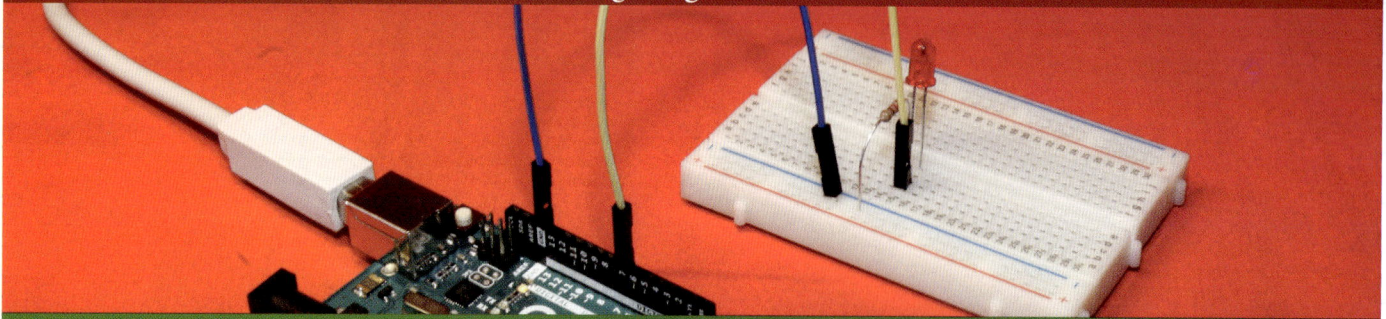

Project 2
Lighting Effects

Lesson Integration:		Groupings:	Level:	Time to Complete:
Physics - Time (measurement), Ohm's Law, Circuits Biology - Inter and Intra Species Communication Mathematics - Cycles, Periodicity		1 - 2	Starter	45 min. for the project and 45 min. for extensions.

Objectives:
- Investigate how circuits and electronic components interact with electrical energy.
- Control an LED in a prototype and machine.
- Control multiple LEDs in a machine.

Prerequisite Skills:
- Time measurement (milliseconds)
- Arduino® IDE software (pages 11-13)
- Tinkercad® (pages 15-16)
- Ratios / Fractions for Extensions

Purpose and Skills:
- Set up a circuit.
- Differentiate between anodes(+) and cathodes(-).
- Differentiate between series and parallel circuits.
- Introduce coding with subroutines.

STEAM Connections:
Science - Metric Time, Circuits, Biological Communication, Energy Discharge
Technology - Parts of Coding, Program settings, Resistors, Cathodes, Anodes, Communication Devices
Engineering - Digital Design, Wiring, Building and Using a Machine, Applied Math, Prototyping
Allied Arts - Rhythm, Patterns, Color, Communications
Math - Ratios, Conversions

Key Vocabulary:
Anode - The positive side of a circuit. The source of electrons. "+" Indicates an excess of electrons that will flow to the cathode(-) in the circuit.

Cathode - The negative side of a circuit. The electrons flow here after leaving the anode(+).
Ground - A part in an electrical circuit that is associated with the cathode(-).
Parallel - When two elements of an electrical circuit share electricity by dividing the flow of electrons.
Resistor - A part in an electrical circuit that slows (resists) the flow of electrons. Usually used to avoid burning out another part in the circuit.
Series - When two elements of an electrical circuit are in sequence such that the electricity flows directly from the first element to the second.
Subroutine - a portion of code that can be used many times over with inputs from many lines in the code.

Project Introduction:
- Introduce to the groups the purpose of the project, the skills developed, the standards met, and the goal of the project.
- Introduce how to read the color indicators on a resistor (inside back cover).
- Introduce how to read the breadboard rows and columns (page 10 and inside back cover).

Anticipatory Sets:
- Identify where blinking lights or multi-color light technologies are seen. List at least three different places or situations where blinking lights or multi-color light technologies are used, describe the lights, and explain the purpose of the lights.
- Create a class collage of images of LEDs in use.

Project 2
Lighting Effects

Educational Standards

ISTE Standards for Students
- Empowered Learner 1a, 1d
- Knowledge Constructor 3a, 3b, 3c, 3d
- Innovative Designer 4a, 4b, 4c, 4d
- Computational Thinker 5a, 5b, 5c, 5d
- Creative Communicator 6c, 6d
- Global Collaborator 7c, 7d

US Computer Science Standards
- Project correlations can be found on the book's web page.

US NGSS - Middle School
DCIs
- MS-PS4-1 PS4.A
- MS-PS4-2 PS4.A PS4.B
- MS-PS4-3 PS4.C
- MS-LS1-3 LS1.A

Cross Cutting Concepts
- Cause and Effect
- Scale, Proportion and Quantity
- Structure and Function

Science and Engineering Practices
- Planning and Carrying Out Investigations
- Obtaining, Evaluating, and Communicating Information
- Analyzing and Interpreting Data
- Scientific Knowledge is Based on Empirical Evidence
- Using Mathematics and Computational Thinking

US NGSS - High School
- Correlations can be found on the book's web page.

US Common Core Language Arts and Mathematics
- Correlations can be found on the book's web page.

Step-by-Step 1-25

Materials List:
- Computer with IDE software
- Connection to the Internet
- USB Cable
- UNO or UNO Compatible Microcontroller
- Short Breadboard
- Seven (7) Jumper Wires (Male to Male)
- Six (6) 220 Ohm Resistor
- Six (6) LEDs (any color)

Engineering Design - Digital Prototype

1. Launch and log into Tinkercad® (pages 15-16).

2. Start a new Circuits project.

3. Drag an UNO board to the workspace.

4. Drag a short breadboard to the workspace.

5. Drag a 220 Ohm resistor on to the breadboard. Place it in a row that spans a lettered column and the cathode(-) column.

6. Drag an LED on to the breadboard so that the cathode(-) leg of the LED is in the same row as the resistor.

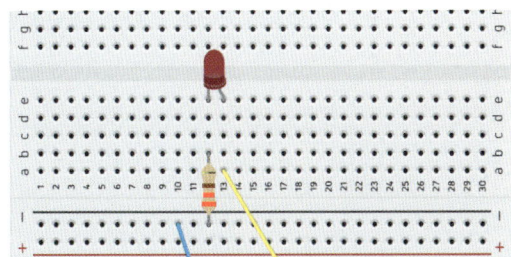

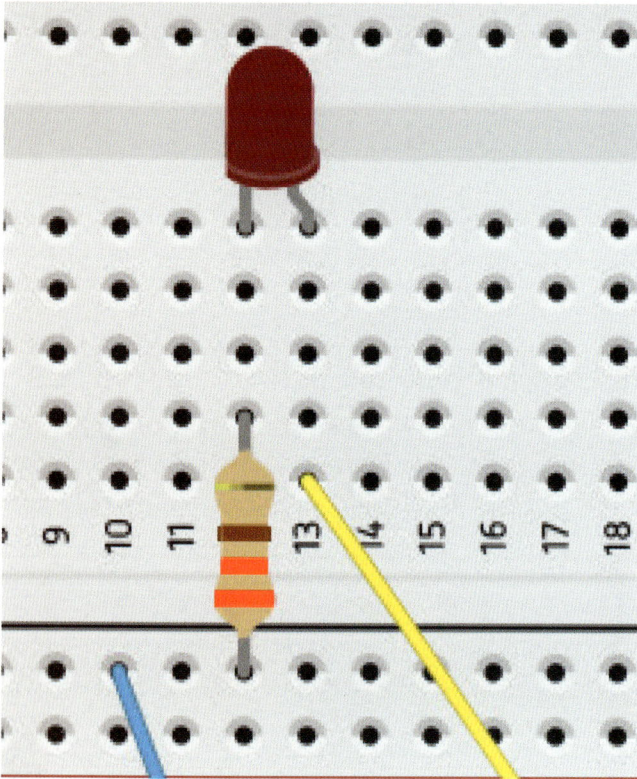

Connection - This long exposure photo shows male fireflies announcing their location to females on the ground as part of their mating display.

A biological chemical reaction called bioluminescence creates the light in a special organ in their abdomen. Other species of firefly use the light to attract prey that they catch and eat.

7 Connect with jumper wires the following:
- a ground (GND) pin on the UNO board to the cathode(-) column (the same that the resistor is connected to)
- pin 7 to the same row as the anode(+) leg of the LED

Coding - Digital Prototype

8 Enter the code in Tinkercad® to match the code on this page.

9 Start the simulation and observe the digital prototype.

```
1  //Project 2
2  //Arduino Uno Blinking LED
3  //thearduinoclassroom.com
4  //Copyright 2019, Isabel Mendiola and Peter Haydock
5
6
7  int led = 7; // LED connected to pin 7
8  void setup() {
9    pinMode(led, OUTPUT); // LED is declared as an output
10 }
11
12 void loop() {
13   digitalWrite(led, HIGH); // LED is on
14   delay(500);
15   digitalWrite(led, LOW); // LED is off
16   delay(500);
17 }
```

```
int led = 7; // LED connected to pin 7
void setup() {
  pinMode(led, OUTPUT); // LED is declared as
an output
}

void loop() {
  digitalWrite(led, HIGH); // LED is on
  delay(500);
  digitalWrite(led, LOW); // LED is off
  delay(500);
}
```

Engineering Design - Project Build Part 1

10 Place a 220 Ohm resistor on the breadboard such that it that spans a lettered column and the cathode(-) column.

11 Place an LED so that the cathode(-) leg of the LED is in the same row as the resistor.

12 Connect with jumper wires the following:
- a ground (GND) pin on the UNO board to the cathode(-) column
- pin 7 to the same row as the anode(+) leg of the LED

13 Study the circuit diagram on this page and compare it to your project.

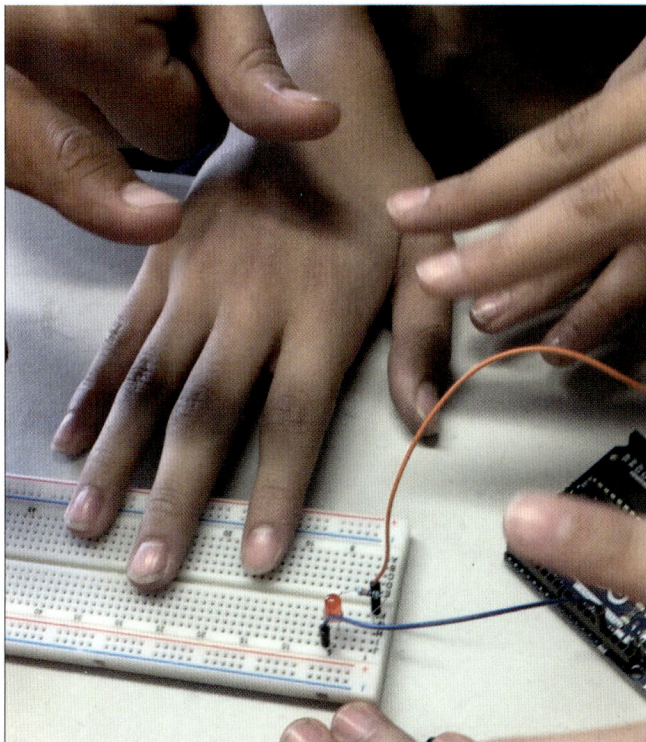

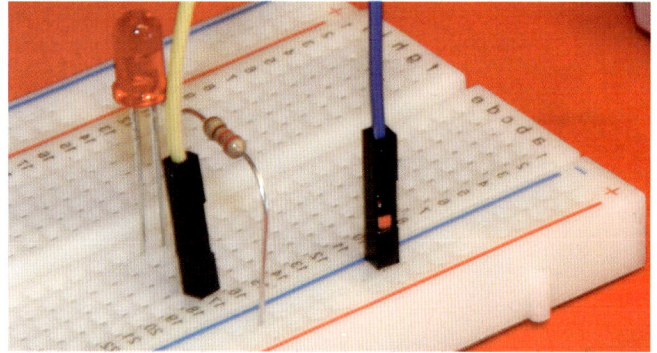

Connection - Blinking warning lights are required on tall structures around the world including towers and buildings. The lights must be red and blink at a steady state. Many of the lights are also shown on maps pilots use to aid navigation.

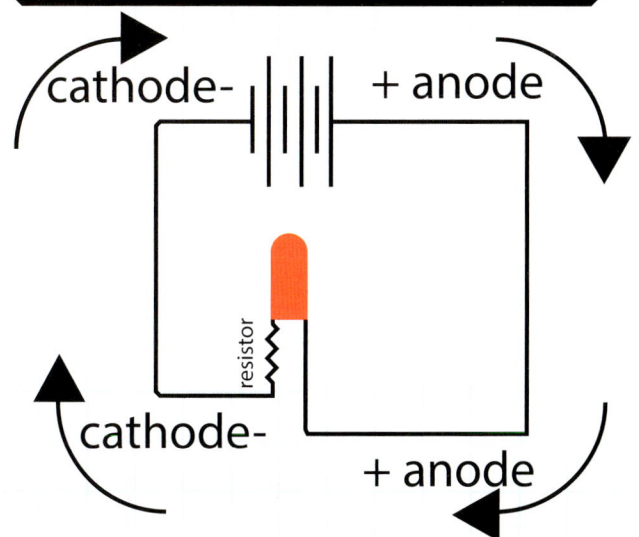

cathode- +anode

cathode-

resistor

+anode

14 Connect the UNO board and computer using the USB cable.

15 Launch the Arduino® IDE software (pages 11-13) to make sure the board is communicating with the computer.

Go to the Tools menu and verify that the correct board is selected from the Board Manager menu. If not, select the correct board from the options listed.

Then confirm that the right Port is selected. If not, select the port that lists the UNO board from the options listed.

Coding - Project Build Part 1

16 Either copy and paste the code from the Tinkercad® prototype or type the code on page 26 into the Arduino® IDE software (pages 11-13).

17 Save the sketch (rename as needed).

18 Verify the sketch.

19 Upload the sketch to the UNO board.

20 Observe the LED on the breadboard.

Engineering Design - Project Build Part 2

21 Use the diagrams and photos found on the next page to build the multi-light display. Place five (5) more 220 Ohm resistors on the breadboard such that each span a lettered column and the cathode(-) column.

22 Place five (5) more LEDs on the breadboard so that the cathode(-) leg of the LED is in the same row as the resistor.

Connection - Blinking warning lights are important safety and communication tools between law enforcement, traffic control, and construction workers and drivers. One can find white, red, blue, and amber colored lights that blink in a variety of patterns to warn others to pay attention. Each color conveys different information and importance. Laws and regulations by local, state, and federal governments determine which vehicles can use each color and at what time they should be used. Planes, helicopters, trains, some road signs, and even your vehicle have blinking lights to communicate potential hazards.

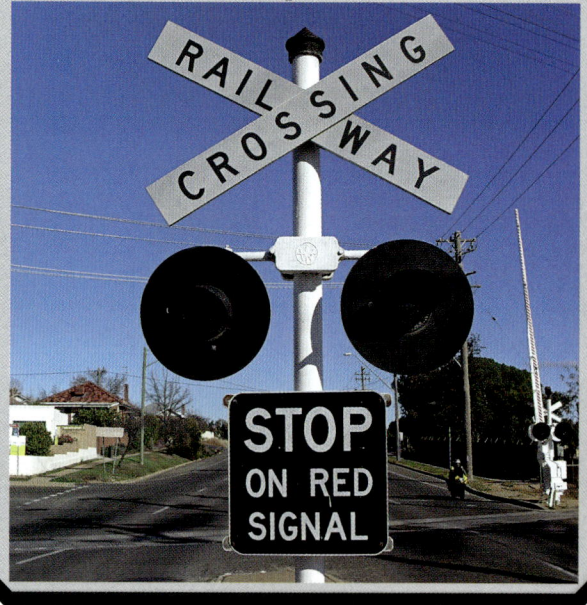

23 Connect with jumper wires the following:
- UNO pins eight (8) through 12 to the same row as the anode(+) leg of each LED

Coding - Project Build Part 2

24 Use the code found on the next page to complete the multi-light display.

25 Document this project, discuss the project, and complete the extensions assigned from page 30.

Multi-light notes.
- There are several ways to accomplish this part of the project. The code on this page shows the shortest solution using a subroutine.
- A longer way would be to introduce six (6) segments of code each with four (4) lines that sequentially turned on and off each LED inside the loop section. A segment of code would look like this.

```
digitalWrite(pin, HIGH);
delay(100);
digitalWrite(pin, LOW);
delay(100);
```

This would have placed 24 lines of code in the loop section for a total of 27 lines of code.
- The code shown on this page includes a subroutine which is seven lines of code called void flash.
- By doing this, 11 lines of code are saved. While in a small sketch like this, the savings does not have a great impact. In larger sketches this can significantly impact the efficiency of the code.
- Using a subroutine also allows for some very creative displays of the LEDs lighting up other than just in order.

```
void setup()
{
pinMode(12, OUTPUT);  //set pin 12 as output
pinMode(11, OUTPUT);  //set pin 11as output
pinMode(10, OUTPUT);  //set pin 10 as output
pinMode(9, OUTPUT); //set pin 9 as output
pinMode(8, OUTPUT); //set pin 8 as output
pinMode(7, OUTPUT);  //set pin 7 as output
}
void loop()
{
flash(12, 100); //set pin value, time value in ms
flash(11, 100); //set pin value, time value in ms
flash(10, 100); //set pin value, time value in ms
flash(9, 100); //set pin value, time value in ms
flash(8, 100); //set pin value, time value in ms
flash(7, 100); //set pin value, time value in ms
}
void flash(int pin, int wait) //sets pin and wait as integer variables.
{
 digitalWrite(pin, HIGH);
 delay(wait);
 digitalWrite(pin, LOW);
 delay(wait);
}
```

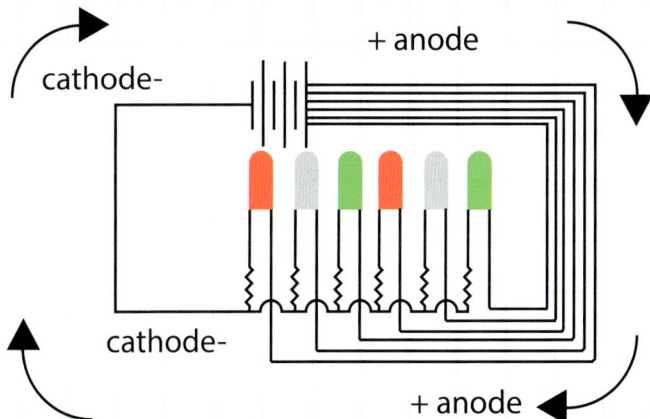

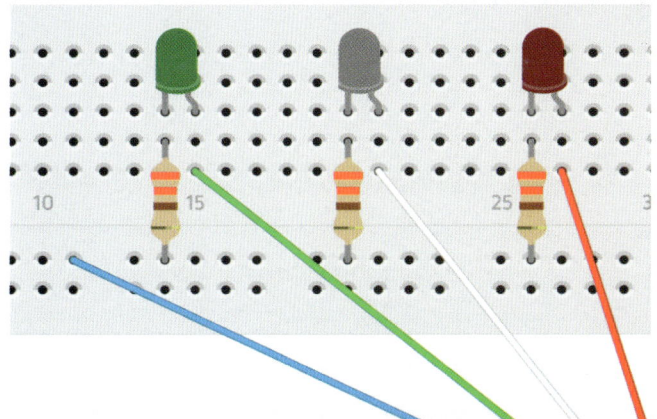

Project Reflection and Summative Activities, Discussion Starters, Extensions, and Problem

A = Essential

Reflection/Summative Activities:

- Have each team member document in their project/classroom journal who was on their team, what went well, what they could improve upon, what they would do differently if they were to do the project again, and verify that every person in the group can do the project.

- As a class, discuss the differences between this project and Project 1.

For Discussion:

- The first part of the project is to produce a digital prototype. What advantages and disadvantages can you identify with using this approach over building the actual project first?

- What is the purpose of a resistor?

- Generate a list of advantages and disadvantages of using multiple light colors in a display.

Extensions:

- Adjust the delay value(s) to make the LED(s) blink faster or slower. What is the fastest observable rate of blinking that can be coded?

- Research the chemical compounds used to make each color LED. Compare and contrast the compounds. Create a graphic design to illustrate.

B = Recommended

Reflection/Summative Activity:

- Collaboratively create signage for the classroom that documents the preparatory steps needed for each project.

Professional Connection:

- Research and write a short description of a profession that relies on blinking warning lights **or** a multi-color light display in your area. Be sure to include the education the profession requires, any laws or regulations that guide their use, and what activities are performed.

For Discussion:

- Why would programmers want to reduce the number of lines in a code? When would reducing the lines of code not be advantageous?

- Look at the second circuit diagram (page 29) for this project and explain why there are six wires on the anode(+) side of the circuit, but only one on the cathode(-) side.

Extension:

- Edit the sketch to create a different light pattern.

- Submit your project to the book page on the website. Register and Login to submit at: thearduinoclassroom.com

From the homepage:
Click
  Books
Click
  UNO Edition Vol. 1

C = Optional

Extensions:

- Investigate and design the delay values that seem to make the blinking LED(s) most visible as a warning device? As an added factor, investigate if the color of the LED a factor in visibility? Is the "on" time value more important than the "off" time value? Should the "on" and "off" values always be the same?

- The LEDs in modern TVs are similar to the LEDs used in this project, although smaller. Research how LED TVs create color images on the screen.

- Design a blinking light display that maximizes the number of LEDs controlled by the UNO board and sketch. What limitations do you have?

Problem:

- Identify a problem that requires a warning light or light display and create the solution. Include in your report the following:

Description of the problem.

Document how the problem is solved by this project.

What other resources are needed to solve the problem.

Design and produce a sign that would accompany the warning light **or** Design and produce the light display.

A project rubric can be found on page 236.

Anticipatory Sets
(from page 24)

Anticipatory Sets:
- Identify where blinking lights or multi-color light technologies are seen. List at least three different places or situations where blinking lights or multi-color light technologies are used, describe the lights, and explain the purpose of the lights.
Answers will vary, but may include the following: Indicators for a deaf person, traffic lights, proximity detectors, On/Off indicators, video game indicators.

- Create a class collage of LEDs in use.
Responses Will Vary

A

Reflection/Summative Activities:
Responses Will Vary

- As a class, discuss the differences between this project and Project 1.
The jumper wires, LED(s), resistor(s), breadboard, coding are different/new.
UNO board and IDE software are the same.

For Discussion:
- The first part of the project is to produce a digital prototype. What advantages and disadvantages can you identify with using this approach over building the actual project first?
Digital prototypes are often faster and easier to build. It also allows the testing of many different configurations quickly. Building a digital prototype often requires specialized and expensive programming first. Accept other answers as appropriate.

- What is the purpose of the resistor in this project?
The electricity in the circuit can overload the LED and damage it. The resistor reduces the amount of electrical current in the circuit. Recall V=IR.

- Generate a list of advantages and disadvantages of using multiple light colors in a display.
Changing light colors stimulate the brain and call attention to where the lights are.

Over-stimulation of the brain though the blinking lights can cause confusion. There are some combinations of color and blinking that can cause seizures.

Extensions:
- Adjust the delay value(s) to make the LEDs blink faster or slower. What is the fastest observable rate of blinking that can be coded?
Responses Will Vary

- Research the chemical compounds used to make each color LED. Compare and contrast the compounds. Create a graphic design to illustrate.
Responses Will Vary

B

Reflection/Summative Activity:
Responses Will Vary

Professional Connection:
Responses Will Vary

For Discussion:
- Why would programmers want to reduce the number of lines in a code? When would reducing the lines of code not be advantageous?
The code will run faster. There are less places for mistakes in the code. There are fewer opportunities to document in the code. Accept other answers as appropriate.

- Look at the second circuit diagram (page 29) for this project and explain why there are six wires on the anode(+) side of the circuit, but only one on the cathode(-) side.
The anode(+) side controls the signal to each LED separately where the cathode(-) only completes the circuit.

Extension:
Responses Will Vary

C

Extensions:
Responses Will Vary

Problem:
Responses Will Vary

Project 3
Buzz Me

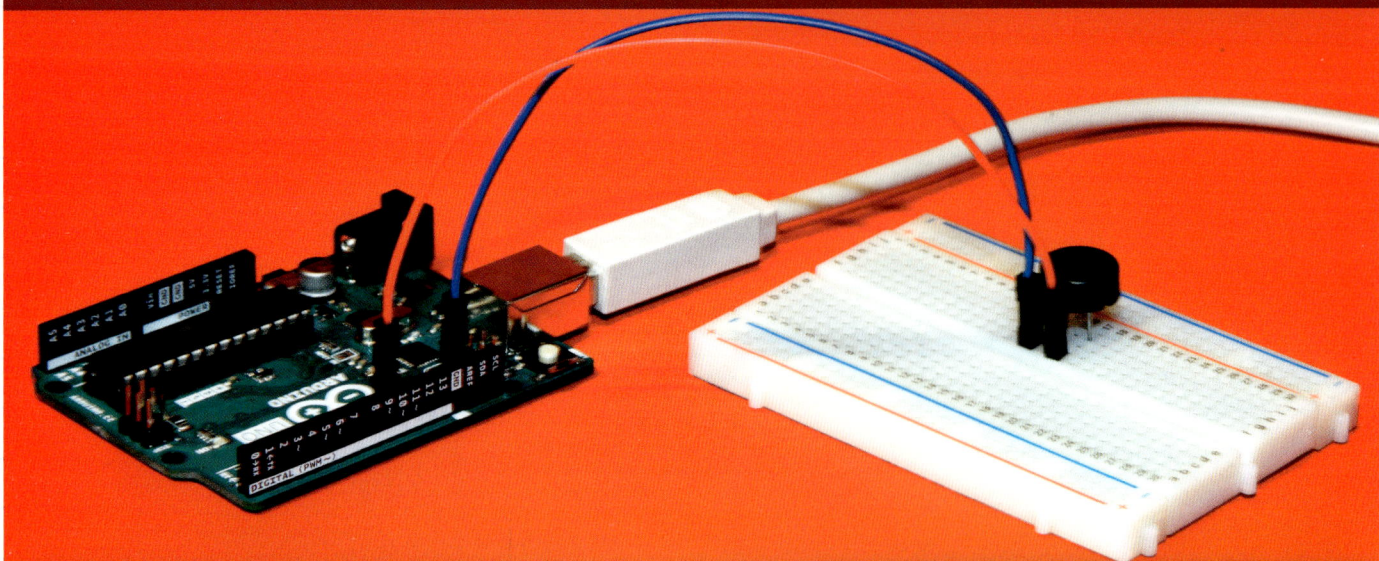

Getting Started

Lesson Integration:	Groupings:	Level:	Time to Complete:
Physics - Time (measurement), Ohm's Law, Circuits, Sound, Waves, Speed Biology - Inter and Intra Species Communication Mathematics - Cycles, Periodicity	1 - 2	Starter	30 min. for the project and 45 min. for extensions.

Objectives:
- Investigate how circuits and electronic components interact with electrical energy.
- Control a sound making device in a prototype and machine.

Prerequisite Skills:
- Time measurement (milliseconds)
- Understanding of sound frequencies
- Arduino® IDE software (pages 11-13)
- Tinkercad® (pages 15-16)

Purpose and Skills:
- Control a sound making device

STEAM Connections:
Science - Metric Time, Circuits, Sound, Ohm's Law, Biological Communication
Technology - Code, Program Settings, Simulators, Digital Design, Electronic Components, Communication Devices
Engineering - Designing, Building, and Using a Machine, Prototyping, Applied Physics
Allied Arts - Music, Communications
Math - Ratios, Conversions

Key Vocabulary:
Frequency - How often something happens. In the case of sound, frequency indicates the note or pitch of the sound. Higher notes have higher frequencies and lower notes have lower frequencies. The unit of frequency is Hertz (Hz) or occurrences per second s^{-1}.
Piezoelectric - a material that changes its characteristics (sound or color) when electricity is applied to it.

Project Introduction:
- Introduce the groups to the purpose of the project, the skills developed, the standards met, and the goal of the project.

Anticipatory Sets:
- Where can you find buzzers or horns? List at least three different places or situations where these technologies are used, describe the devices, and explain the purpose of the devices.
- Where in nature do animals use repeating sounds? For what purpose? How are they different than a buzzer or horn?

Project 3
Buzz Me

Educational Standards

ISTE Standards for Students
- Empowered Learner 1a, 1b, 1d
- Knowledge Constructor 3a, 3b, 3c, 3d
- Innovative Designer 4a, 4b, 4c, 4d
- Computational Thinker 5a, 5b, 5c, 5d
- Creative Communicator 6b, 6c, 6d
- Global Collaborator 7c, 7d

US Computer Science Standards
- Project correlations can be found on the book's web page.

US NGSS - Middle School
DCIs
- MS-PS4-1 PS4.A
- MS-PS4-2 PS4.A PS4.B
- MS-PS4-3 PS4.C

Cross Cutting Concepts
- Cause and Effect
- Scale, Proportion and Quantity
- Structure and Function

Science and Engineering Practices
- Planning and Carrying Out Investigations
- Obtaining, Evaluating, and Communicating Information
- Analyzing and Interpreting Data
- Scientific Knowledge is Based on Empirical Evidence
- Using Mathematics and Computational Thinking

US NGSS - High School
- Correlations can be found on the book's web page.

US Common Core Language Arts and Mathematics
- Correlations can be found on the book's web page.

Materials List:
- Computer with IDE software
- Connection to the Internet
- USB Cable
- UNO or UNO Compatible Microcontroller
- Short Breadboard
- Piezoelectric Buzzer (Piezo)
- Two (2) Jumper Wires (Male to Male)

Engineering Design - Digital Prototype

1. Launch and log into Tinkercad® (pages 15-16).

2. Start a new Circuits project.

3. Drag an UNO board to the workspace.

4. Drag a breadboard to the workspace.

5. Drag one (1) piezoelectric buzzer (piezo) to the breadboard and place the two leads in different rows in the same column.

Hover over the ends of the leads of the piezo to determine which is the cathode(-) and which is the anode(+) lead.

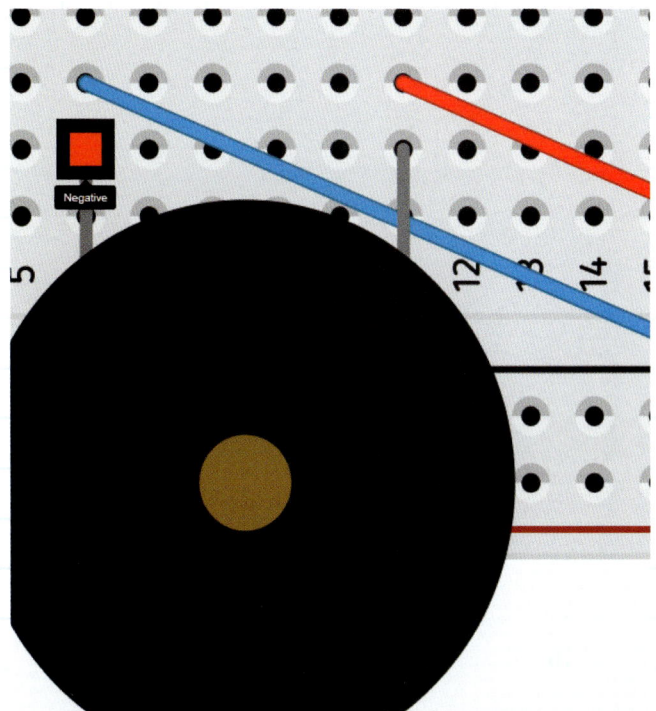

6 Connect with jumper wires the following:
- a ground pin on the UNO board to a pin in the same row as the cathode(-) lead of the piezo
- digital pin 9 to the same row as the anode(+) lead of the piezo

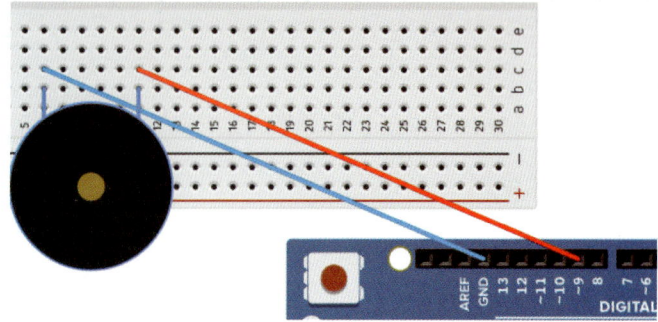

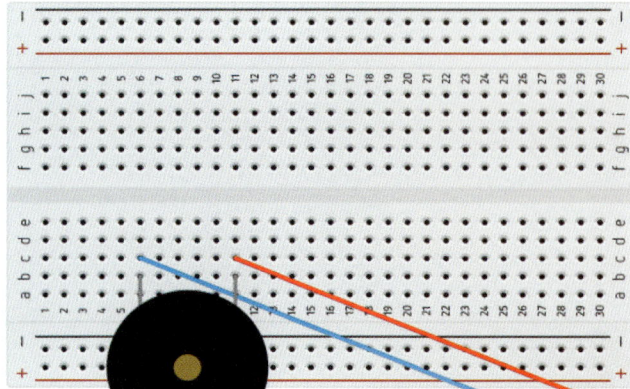

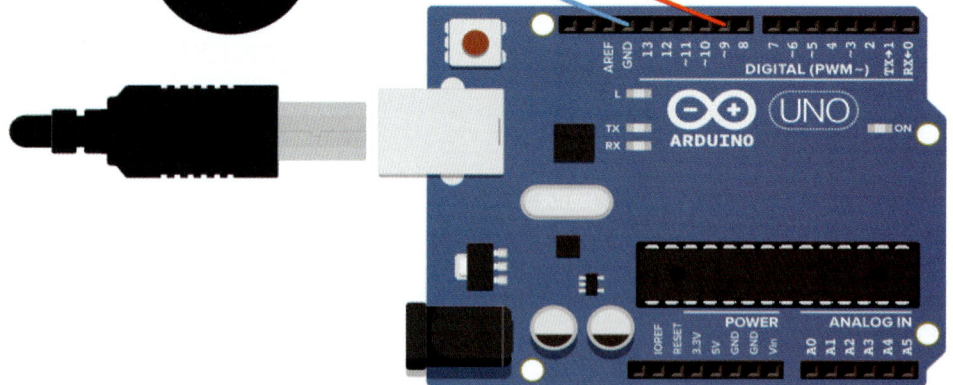

Connection - Warning devices often use a shrill or piercing continuous or repeating sound to announce an emergency. In this photo, a standard building fire alarm is mounted to the wall near the ceiling. It has both a strobe light and a horn for signaling to people that there is a fire in the building. The strobe can be seen by those in the room, but for those who are vision impaired or out of sight of the strobe, the warning tone communicates to them the danger.

Strict laws govern the operation and placement of fire alarms in public buildings and homes. Most schools have fire alarm drills once a month to teach students how to evacuate.

Project 3
Buzz Me

Coding - Digital Prototype

7 Enter the code in Tinkercad® to match the code on this page.

8 Start the simulation and observe the digital prototype.

```
const int buzz = 9; //buzzer to UNO pin 9

void setup()
{
pinMode(buzz, OUTPUT); //set pin 9 as output
}
void loop()
{
tone(buzz, 550);   // turn on pin
delay(300);        // delay for 300 milliseconds
noTone(buzz);      // turn off pin
delay(300);        //delay for 300 milliseconds
}
```

Engineering Design - Project Build

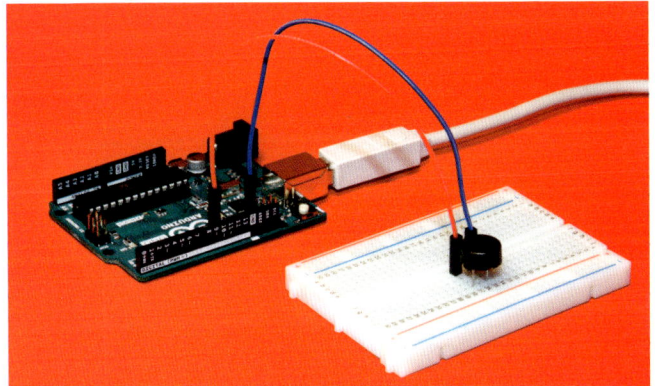

9 Place the piezo on the breadboard such that the two leads are in different numbered rows in the same column.

The piezo should be marked with a "+" to indicate the anode lead. If the piezo is not marked with a "+" the polarity does not matter and either lead can be connected to the anode(+) or cathode(-) sides of the circuit.

10 Connect with jumper wires the following:
- a ground pin on the UNO board to a pin in the same row as the cathode(-) lead of the piezo
- digital pin 9 to the same row as the anode(+) lead of the piezo

```
 1  //Project 3
 2  //Buzzer
 3  //thearduinoclassroom.com
 4  //Copyright 2019, Isabel Mendiola and Peter Haydock
 5
 6
 7  const int buzz = 9; //buzzer to UNO pin 9
 8
 9  void setup()
10  {
11
12    pinMode(buzz, OUTPUT); // Set pin 9 as output
13  }
14
15  void loop()
16  {
17    tone(buzz,550);       // 550 Hz sound
18    delay(300);           // delay for 300 milliseconds
19    noTone(buzz);         // stop sound
20    delay(300);           // no sound for 300 milliseconds
21
22  }
```

Text   1 (Arduino Uno R3)

11 Review the circuit diagram for this project.

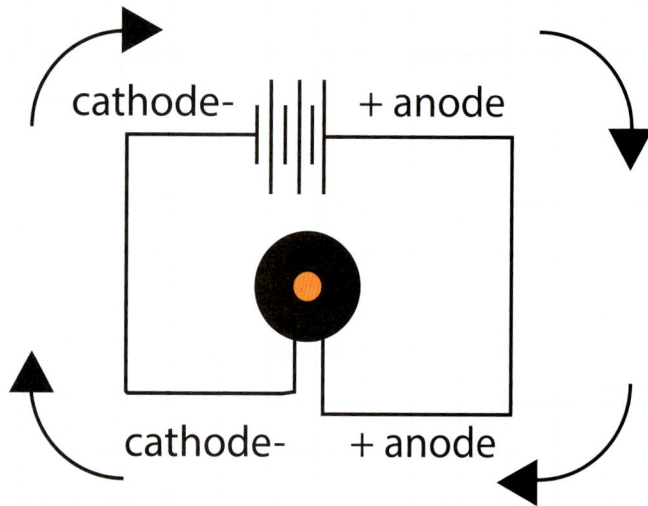

cathode-   + anode

cathode-   + anode

12 Connect the UNO board and computer using the USB cable.

13 Launch the Arduino® IDE software (pages 11-13) to make sure the board is communicating with the computer.

Go to the Tools menu and verify that the correct board is selected from the Board Manager menu. If not, select the correct board from the options listed.

Then confirm that the right Port is selected. If not, select the port that lists the UNO board from the options listed.

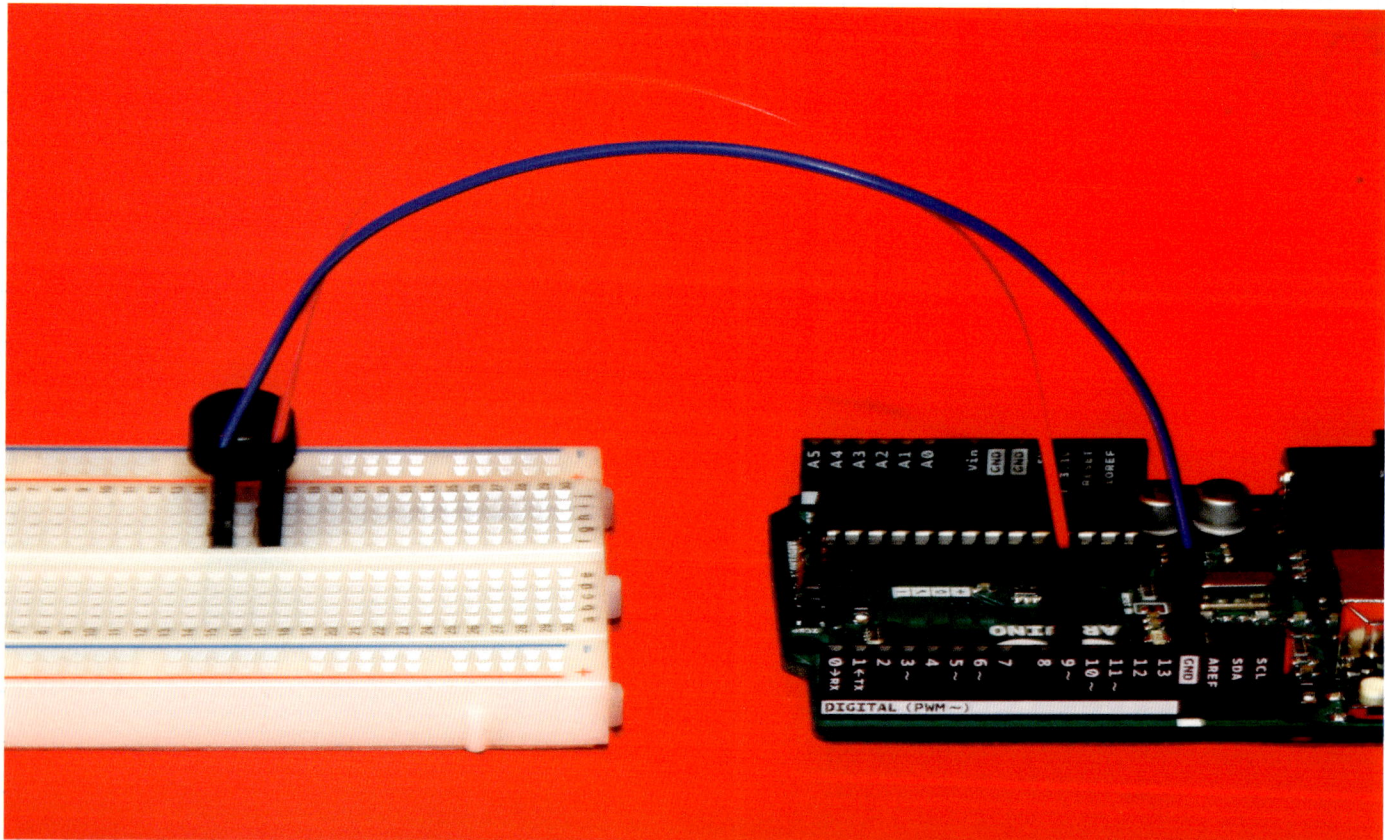

Connection - Many animals use sounds to indicate feelings, warnings, or location. Frogs, like this Northern Leopard Frog, can be identified by their croaking. This male frog's croaking in spring is loud and rhythmic to attract a mate. In a pond with dozens of frogs, females listen for the loudest croak to determine which male will be their mate. Once spring mating season is over, the frogs will croak less frequently and less loudly.

Project 3
Buzz Me

Coding - Project Build

14 Either copy and paste the code from the Tinkercad® prototype or type the code from page 35 into the Arduino® IDE software (pages 11-13).

15 Save the sketch (rename as needed).

16 Verify the sketch.

17 Upload the sketch to the UNO board.

18 Observe what happens with the piezo on the breadboard.

19 Document this project, discuss the project, and complete the extensions assigned from the next page.

```
TAC_UE_V1_P4a | Arduino 1.8.9 (Windows Store 1.8.21.0)
File Edit Sketch Tools Help

TAC_UE_V1_P4a
//Project 4
//Buzzer
//thearduinoclassroom.com
//Copyright 2019, Isabel Mendiola and Peter Haydock

const int buzz = 9; //buzzer to UNO pin 9

void setup()
{
  pinMode(buzz, OUTPUT); // Set buzzer - pin 9 as an output
}

void loop()
{
  tone(buzz, 550); // 550 Hz sound
  delay(300);       // delay for 300 milliseconds
  noTone(buzz);     // stop sound
  delay(300);       // no sound for 300 milliseconds

}
```

```
Done Saving.
Sketch uses 1664 bytes (5%) of program storage space. Maximum is 3
Global variables use 28 bytes (1%) of dynamic memory, leaving 2020
15                                    Arduino/Genuino Uno on COM4
```

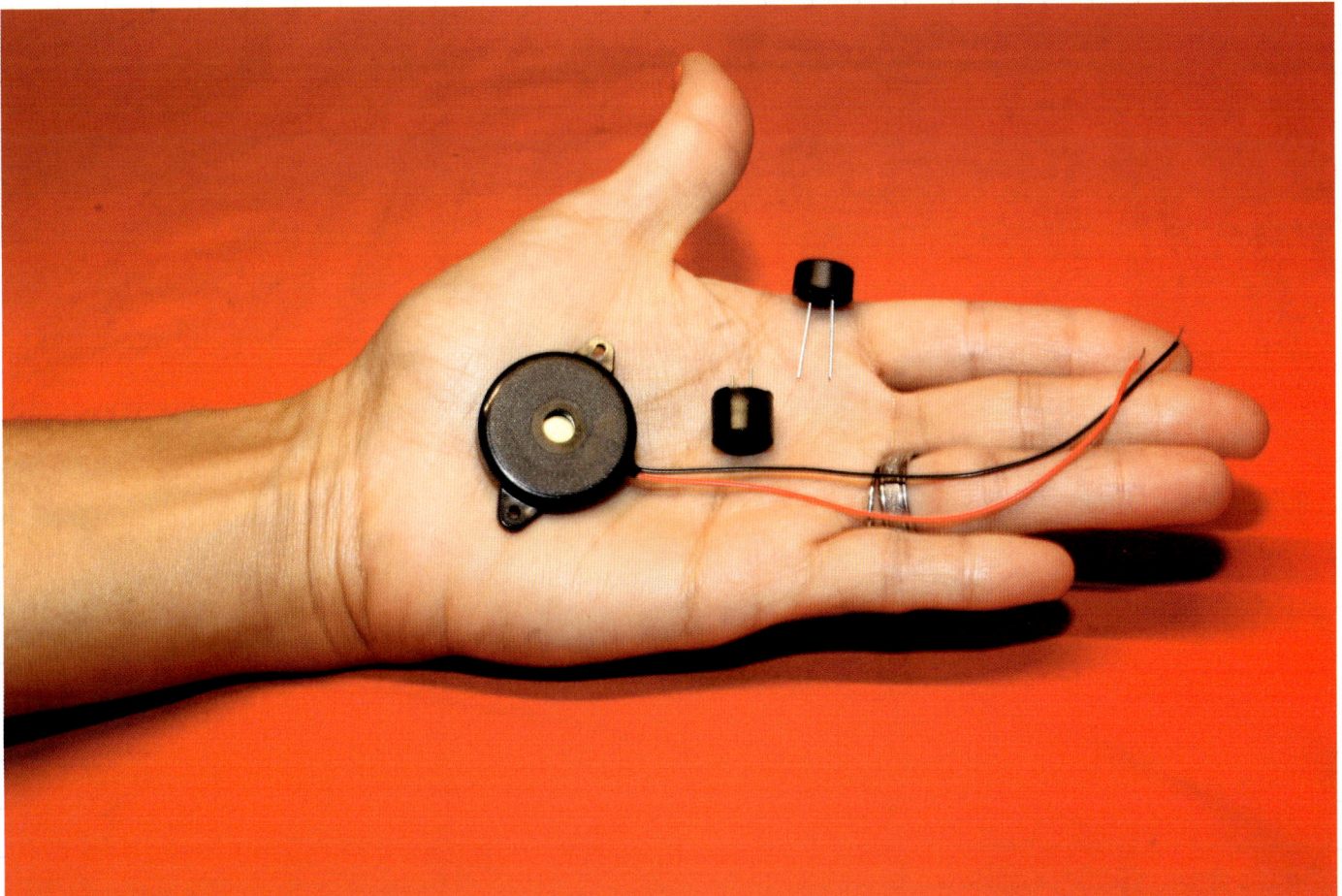

Project Reflection and Summative Activities, Discussion Starters, Extensions, and Problem

A = Essential	B = Recommended	C = Optional

A = Essential

Reflection/Summative Activityies:

- Have each team member document their project. In their journals create a **storyboard** presentation of the project including who was on their team, what parts they used, what they did, what went well, what they could improve upon, what they would do differently if they were to do the project again, and verify that every person in the group can do the project.

- Have each team share their storyboards with the class and solicit evaluations of their work.

For Discussion:

- What similarities and differences are there between coding for a light indicator and a sound indicator like the one in this project?

- How does the use of a single repeating tone impact a person's ability to hear the tone?

- Identify elements of sound that are not controlled in this project.

- Explain why you think a resistor is not used in this project.

Extensions:

- Adjust the delay value(s) to make the tone sound faster or slower. Demonstrate the solution.

- Change the frequency of the tone. Demonstrate the solution.

B = Recommended

Professional Connection:

- Research and write a short description of a profession that relies on warning tones in your area. Be sure to include the education the profession requires, any laws or regulations that guide their use, and what activities are performed. Include a list of professions that would not use a warning tone yet still need to alert a person of a warning.

Extensions:

- Evaluate Projects One and Two and determine if adding a buzzer enhances the value of the project.

- Add or edit lines of code to to make a different repeating pattern of tones.

- Add or edit lines of code to communicate the team member's names in Morse Code.

- Submit your project to the book page on the website. Register and Login to submit at: thearduinoclassroom.com

From the homepage:
Click
    Books
Click
    UNO Edition Vol. 1

C = Optional

Extensions:

- Investigate why does a stereo system found in a car or home generally have multiple speakers of differing sizes. Generate a short report on what you found.

- Assign the project teams the following challenge. Every team should build a tone generating device and synchronize them in a multi-device performance.

Problem:

- Identify a problem that requires a buzzer and create the solution. Include in your report the following:

Description of the problem.

Document how the problem is solved by this project.

What other resources are needed to solve the problem.

Design and produce the buzzer solution.

A project rubric can be found on page 236.

Anticipatory Sets
(from page 32)

Anticipatory Sets:
- Where can you find buzzers or horns? List at least three different places or situations where these technologies are used, describe the devices, and explain the purpose of the devices.
Fire alarms, alarm clocks, fire trucks, etc. Warnings and announcements.

- Where in nature do animals use repeating sounds? For what purpose?
Frogs croaking, sheep bleating, geese honking, etc.

- How are they different than a buzzer or horn?
Animal sounds often have multiple rhythms, tones and volumes.

A

Reflection/Summative Activities:
Responses Will Vary

For Discussion:
- What similarities and differences are there between coding for a light indicator and a sound indicator like the one in this project?
Similarities include coding the pin values, time values for the "On" and "Off "states. Differences include not needing

a resistor.

- How does the use of a single repeating tone impact a person's ability to hear the tone?
Single repeating tones are not often found in nature or even in man made constructs. This repetition cuts through background noise as those sounds vary and do not repeat making a single repeating tone more obvious.

- Identify elements of sound that are not controlled in this project.
Responses should include volume and frequency (the note) of the tone.

- Explain why you think a resistor is not used in this project.
The buzzer has sufficient resistance to function in the circuit.

Extensions:
Adjust the delay value(s) to make the tone sound faster or slower. Demonstrate the solution.
Responses Will Vary

Change the frequency of the tone. Demonstrate the solution.
Responses Will Vary

B

Professional Connection:
Responses Will Vary

Extensions:
- Evaluate Projects One and Two

and determine if adding a buzzer enhances the value of the project.
Responses Will Vary

- Add or edit lines of code to to make a different repeating pattern of tones.
Responses Will Vary

- Add or edit lines of code to communicate the team member's names in Morse Code.
Responses Will Vary

- Submit your project to the book page on the website.
Responses Will Vary

C

Extensions:
- Investigate why does a stereo system found in a car or home generally have multiple speakers of differing sizes. Generate a short report on what you found.
Responses Will Vary

- Assign the project teams the following challenge. Every team should build a tone generating device and synchronize them in a multi-device performance.
Responses Will Vary

Problem:
Responses Will Vary

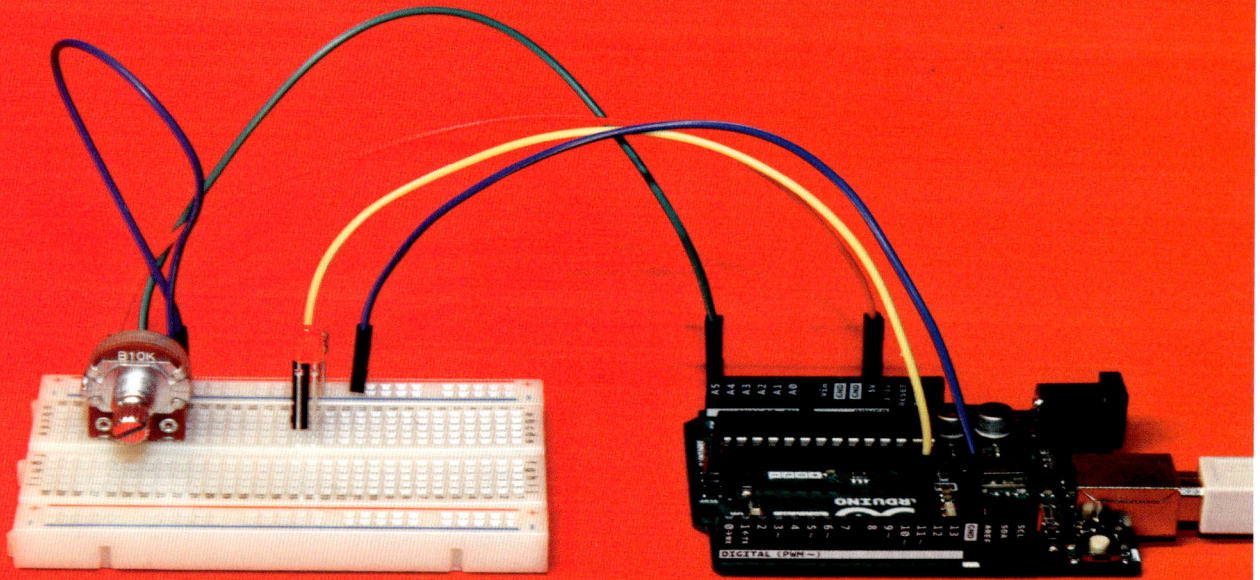

Project 4
Dimmer Switch

Lesson Integration:	Groupings:	Level:	Time to Complete:
Physics - Ohm's Law, Circuits, Voltage, Resistance Mathematics - Ratios, Algebra	1 - 2 Change groupings.	Intermediate	45 min. for the project and 30 min. for extensions.

Objectives:
- Investigate how circuits and electronic components interact with electrical energy.
- Control light intensity in a prototype and machine with a potentiometer.

Prerequisite Skills:
- Time measurement (milliseconds)
- Arduino® IDE software (pages 11-13)
- Tinkercad® (pages 15-16)

Purpose and Skills:
- Control an LED light using an analog input
- Control an LED light using a potentiometer

STEAM Connections:
Science - Circuits, Resistance, Ohm's Law, Light Intensity
Technology - Code, Program Settings, Simulators, Digital Design, Electronic Components, Potentiometer, Analog values
Engineering - Designing, Building, and Using a

Machine, Prototyping, Applied Physics
Allied Arts - Light Intensity
Math - Ratios, Conversions, Scale, Proportions, Fractions, Binary based numbers.

Key Vocabulary:
Potentiometer - A device that varies the amount of resistance in a circuit.

Project Introduction:
- Introduce the groups to the purpose of the project, the skills developed, the standards met, and the goal of the project.

Anticipatory Sets:
- Where would you see a potentiometer used? List at least three different places or situations where this technology is used, describe the devices, and explain the purpose of the devices.

- Why is the potentiometer connected to the analog pin in this project?

Project 4
Dimmer Switch

Educational Standards

ISTE Standards for Students
- Empowered Learner 1a, 1b, 1d
- Knowledge Constructor 3a, 3c, 3d
- Innovative Designer 4a, 4b, 4c, 4d
- Computational Thinker 5a, 5b, 5c, 5d
- Creative Communicator 6c, 6d
- Global Collaborator 7c, 7d

US Computer Science Standards
- Project correlations can be found on the book's web page.

US NGSS - Middle School
DCIs
- MS-PS4-1 PS4.A
- MS-PS4-2 PS4.A PS4.B
- MS-PS4-3 PS4.C

Cross Cutting Concepts
- Cause and Effect
- Scale, Proportion and Quantity
- Structure and Function

Science and Engineering Practices
- Planning and Carrying Out Investigations
- Obtaining, Evaluating, and Communicating Information
- Analyzing and Interpreting Data
- Scientific Knowledge is Based on Empirical Evidence
- Using Mathematics and Computational Thinking

US NGSS - High School
- Correlations can be found on the book's web page.

US Common Core Language Arts and Mathematics
- Correlations can be found on the book's web page.

Step-by-Step 1- 23

Materials List:
- Computer with IDE software
- Connection to the Internet
- USB Cable
- UNO or UNO Compatible Microcontroller
- Short Breadboard
- 10k Potentiometer
- LED
- 220 Ohm Resistor
- Five (5) Jumper Wires

Engineering Design - Digital Prototype

1. Launch and log into Tinkercad® (pages 15-16).

2. Start a new Circuits project.

3. Drag an UNO board to the workspace.

4. Drag a breadboard to the workspace.

5. Drag one 10k potentiometer to the breadboard such that the leads span three rows in a column.

6. Place an LED across two rows of a column.

7. Place a 220 Ohm resistor on the breadboard such that it connects the cathode(-) column and the row that has the cathode(-) lead of the LED.

8. Connect with jumper wires the following:
   - analog pin five (A5) to the same row as the input lead (center) of the potentiometer
   - the five volt (5v) pin to the same row as the anode(+) lead of the potentiometer
   - a ground (GND) pin on the UNO board to the cathode(-) column on the breadboard.
   - digital pin 11 to the same row as the anode(+) lead of the LED.
   - the cathode(-) lead of the potentiometer to the cathode(-) row

Coding - Digital Prototyping

9 Enter the code in Tinkercad® to match the code on this page.

10 Start the simulation. Rotate the potentiometer and observe the digital prototype.

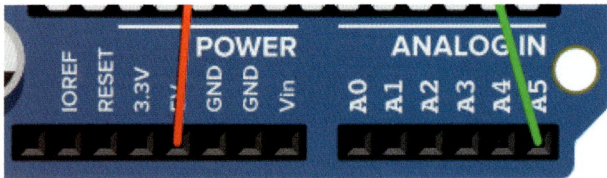

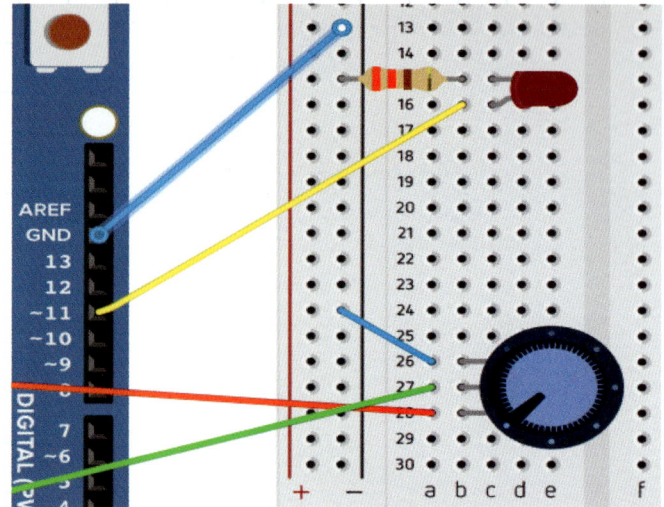

```
int input = A5; // Microcontroller Analog connection for the potentiometer
int LED = 11;  //  LED connected to digital pin 11
int value=0; // sets potentiometer input value at zero

void setup ()
{
 pinMode(LED ,OUTPUT);
}
void loop()
{
int  value = analogRead(input); // Reading the value of the potentiometer
analogWrite(LED, value/4); // The highest value of analog input is 1023 and PWM has
//a resolution of 256, so by dividing input by you scale input to PWM
}
```

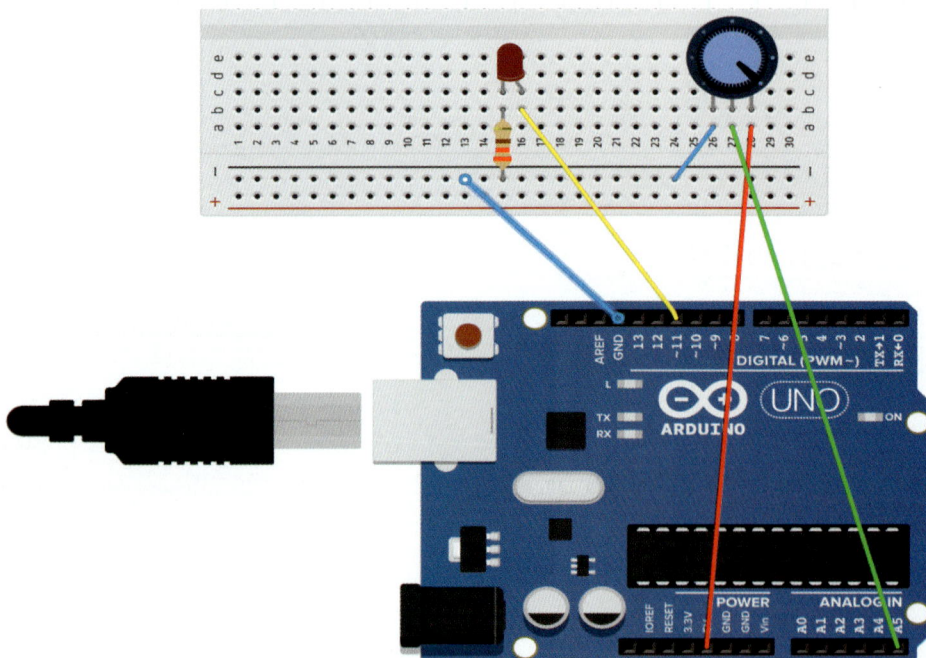

Text	▼	↓ 🗃 🐞	1 (Arduino Uno R3) ▼

```
1   //Project 7
2   //10k potentiometer with a LED
3   //thearduinoclassroom.com
4   //Copyright 2019, Isabel Mendiola and Peter Haydock
5
6   int input = A5; // Microcontroller Analog connection for the potentiometer
7   int LED = 11;  //  LED connected to digital pin 11
8   int value=0; // sets potentiometer input value at zero
9
10  void setup ()
11  {
12   pinMode(LED ,OUTPUT);
13  }
14  void loop()
15  {
16  int value = analogRead(input); // Reading the value of the potentiometer
17  analogWrite(LED, value/4); // The highest value of analog input is 1023 and PWM has
18  //a resolution of 256, so by dividing input by you scale input to PWM
19  }
20
```

Connection - The potentiometer is a special type of rheostat. A rheostat (pictured below on the right of the photo) is any device that controls an output of current. A standard rheostat only uses two leads and is designed for heavier currents.

The rheostat takes an incoming current and applies a variable resistance to the circuit based on the rotation of the rheostat's knob and then completes the circuit with the output current.

The potentiometer is a bit different and uses a ratio of the resistances from the cathode and anode sides of the circuit and then outputs a voltage through a third lead based on the rotation of the knob.

Engineering Design - Project Build

11 Place one 10k potentiometer to the breadboard such that the leads span three rows in a column.

12 Place an LED across two rows of a column

13 Place a 220 Ohm resistor on the breadboard such that it connects the cathode(-) column and the row that has the cathode(-) lead of the LED

14 Connect with jumper wires the following:
- analog pin five (A5) to the same row as the input lead (center) of the potentiometer
- the five volt (5v) pin to the same row as the anode(+) lead of the potentiometer
- a ground (GND) pin on the UNO board to the cathode(-) column on the breadboard.
- digital pin 11 to the same row as the anode(+) lead of the LED.
- the cathode(-) lead of the potentiometer to the cathode(-) row

15 Review the circuit diagram for this project.

16 Connect the UNO board and computer using the USB cable.

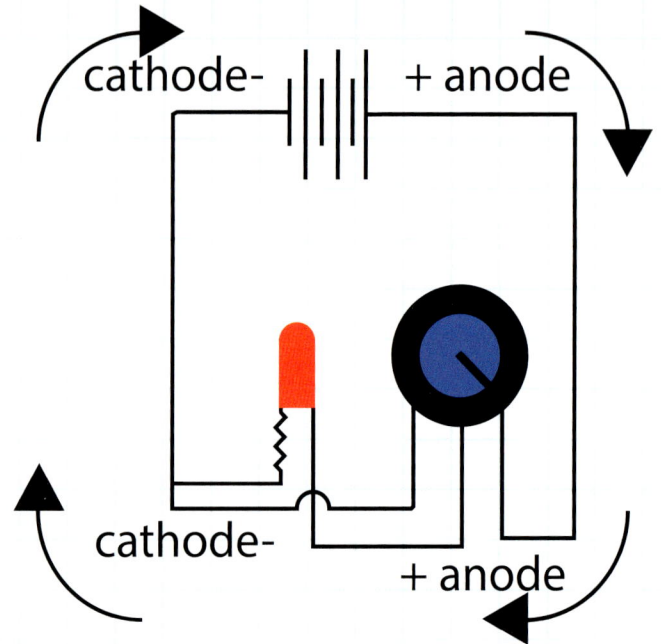

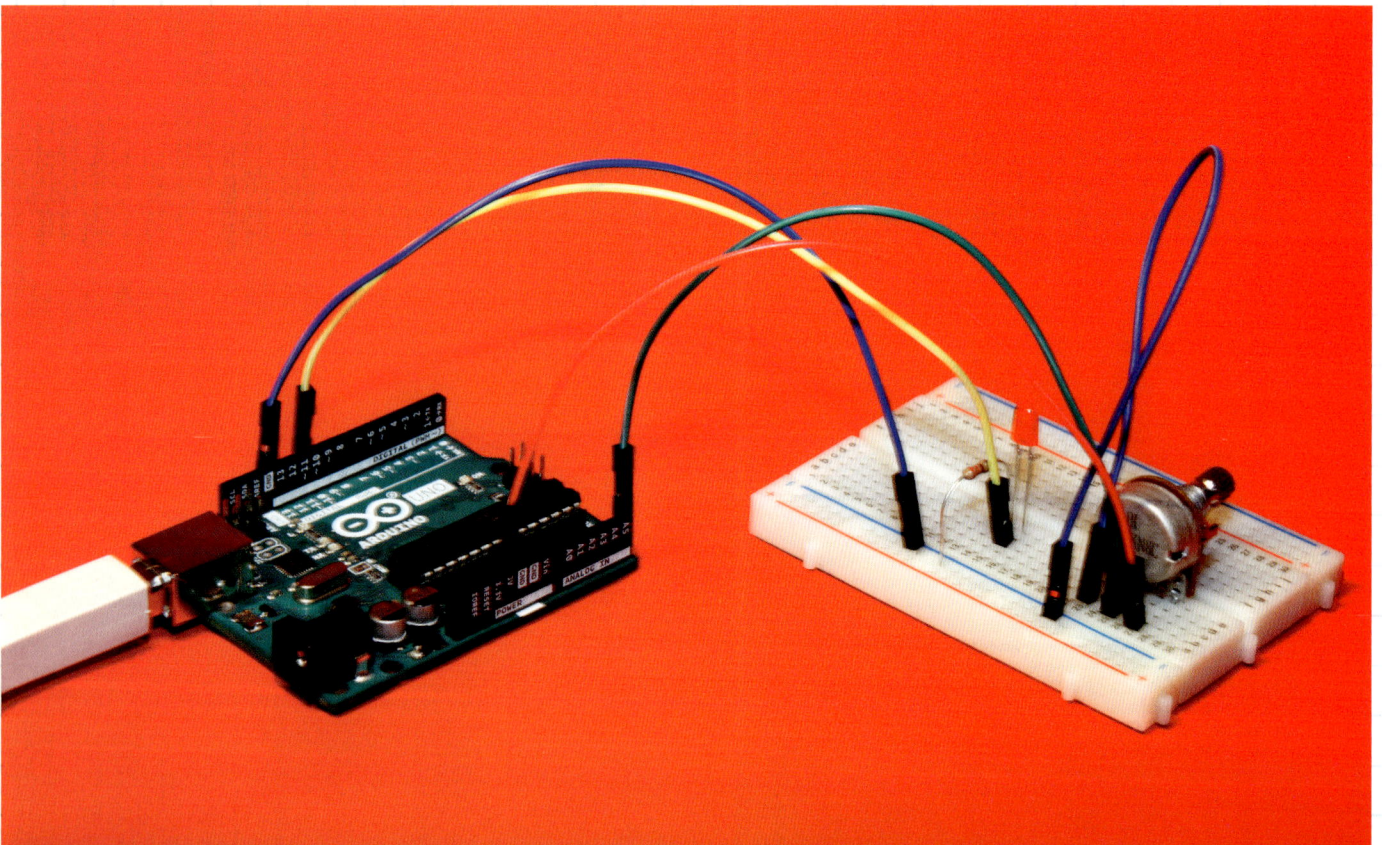

17 Launch the Arduino® IDE software (pages 11-13) to make sure the board is communicating with the computer.

Go to the Tools menu and verify that the correct board is selected from the Board Manager menu. If not, select the correct board from the options listed.

Then confirm that the right Port is selected. If not, select the port that lists the UNO board from the options listed.

Coding - Project Build

18 Either copy and paste the code from the Tinkercad® prototype or type the code from page 42 into the Arduino® IDE software (pages 11-13).

19 Save the sketch (rename as needed).

20 Verify the sketch.

21 Upload the sketch to the UNO board.

22 Adjust the potentiometer and observe the LED on the breadboard.

23 Document this project, discuss the project, and complete the extensions assigned from the next page.

Connection - Professional audio and video technicians control both sound and light levels for concerts, theaters, exhibits, and other public displays and performances. These technicians use mixing and control boards like the one below that bring together multiple inputs of sound and visuals to output for the audience. Important in that control process is the fine adjustment of the inputs and outputs to maximize the quality of experience for the audience.

Some controls are on sliders and others are on rotating knobs depending on the control needed. Each input and output is adjusted before the start of the performance to accommodate the venue and then constantly maintained as the circumstances require.

```
TAC_UE_V1_P7_NEW | Arduino 1.8.9 (Windows Store 1.8.21.0)      —   □   ×
File  Edit  Sketch  Tools  Help

TAC_UE_V1_P7_NEW

//Project 7
//10k potentiometer with a LED
//thearduinoclassroom.com
//Copyright 2019, Isabel Mendiola and Peter Haydock

int input = A5; // Microcontoller Analog connection for the potentiometer
int LED = 11;  //  LED connected to digital pin 11
int value=0; // sets potentiometer input value at zero

void setup ()
{
 pinMode(LED ,OUTPUT);
}
void loop()
{
int  value = analogRead(input); // Reading the value of the potentiometer
analogWrite(LED, value/4); // The highest value of analog input is 1023 and PW
//a resolution of 255, so by dividing input by you scale input to PWM
}

Done Saving.
```

Project 4
Dimmer Switch
Project Reflection and Summative Activities, Discussion Starters, Extensions, and Problem

A = Essential

Project Reflection/Summative Activities:

- Have each team member document in their project/classroom journal who was on their team, what went well, what they could improve upon, what they would do differently if they were to do the project again, and verify that every person in the group can do the project.

- As a team, create a single summary of their individual reflections.

For Discussion:

- What other output devices you have built would benefit from adding a potentiometer into the circuit?

- Compare and contrast the advantages and disadvantages of using a rotary control versus a slider control.

- What are the benefits of using "//" to make remarks and notes in the code?

Extension:

- Change the action of the potentiometer to increase or decrease the sensitivity of turning the knob thereby changing how the LED dims or brightens.

B = Recommended

Technical Writing:

- Write a description of the function of a potentiometer that would be understood by a six year old.

Extensions:

- Replace the LED with a buzzer and control the volume, tone, or repeat rate of the tone using the potentiometer.

- Add or edit lines of code to make the light brighten and dim but also blink faster or slower based on the input of the potentiometer.

- Submit your project to the book page on the website. Register and Login to submit at: thearduinoclassroom.com

From the homepage:
Click
  Books
Click
  UNO Edition Vol. 1

C = Optional

Extension:

- Add a buzzer to the project and control the volume or repeat rate of the tone along with the brightness of the LED using the potentiometer.

Problem:

- Identify a problem that would be solved by using the potentiometer. How could this project be a solution? Include in your report the following:

Description of the problem.

Description of how the problem is solved by this project.

What other resources you would need to solve the problem.

Design and produce the potentiometer solution.

A project rubric can be found on page 236.

Anticipatory Sets
(from page 40)

Anticipatory Sets:
 - Where would you see a potentiometer used? List at least three different places or situations where this technology is used, describe the devices, and explain the purpose of the devices.
Adjusting volumes on radios and other electronic equipment, light brightness, buzzers, volume, space heaters. All rely on a rotary action to change the ratio of resistances to increase or decrease the output current. The control allows for the raising of volume, heat, or brightness.

 - Why is the potentiometer connected to the analog pin in this project?
In order to read values other than digital values (1 or 0 - High or LOW). The potentiometer has at maximum 1024 values (0 to 1023).

A

Project Reflection/Summative Activities:
Responses Will Vary

For Discussion:
 - What other output devices you have built would benefit from adding a potentiometer into the circuit?
Buzzer volume, color value to an RGB LED, frequency of blinking rate of an LED, speed of text appearing on a screen, water flow, speed at which a senor operates.

 - Compare and contrast the advantages and disadvantages of using a rotary control versus a slider control.
A rotary control take less room, and can be integrated with a push button switch. Sliders can be more easily calibrated for comparison across several sliders.

 - What are the benefits of using "//" to make remarks and notes in the code?
Coders that come back to the original code will have guidance on the purpose of the line of code.

Extensions:
 - Change the action of the potentiometer to increase or decrease the sensitivity of turning the knob thereby changing how the LED dims or brightens.
Responses Will Vary

B

Technical Writing:
 - Write a description of the function of a potentiometer that would be understood by a six year old.
Responses Will Vary

Refer to the "Writing for Engineers" link on the book web page.

Extensions:
 - Replace the LED with a buzzer and control the volume, tone, or repeat rate of the tone using the potentiometer.
Responses Will Vary

 - Add or edit lines of code to make the light brighten and dim but also blink faster or slower based on the input of the potentiometer.
Responses Will Vary

 - Submit your project to the book page on the website.
Responses Will Vary

C

Extension:
 - Add a buzzer to the project and control the volume or repeat rate of the tone along with the brightness of the LED using the potentiometer.
Responses Will Vary

Problem:
Responses Will Vary

Project 5
Rainbow Light

Lesson Integration:	Groupings:	Level:	Time to Complete:
Physics - Ohm's Law, Circuits, Light, Color, Wavelengths, Spectrum Chemistry - Compounds, Pigments Biology - Inter and Intra Species Communication, Biology of Vision Mathematics - Cycles, Periodicity, Scale, Algebra, Formulas	1 - 2	Starter	30 min. for the project and 45 min. for extensions.

Objectives:
- Investigate how circuits and electronic components interact with electrical energy.
- Control light and color in a prototype and machine.

Prerequisite Skills:
- Time measurement (milliseconds)
- Understanding of color spectrum
- Arduino® IDE software (pages 11-13)
- Tinkercad® (pages 15-16)

Purpose and Skills:
- Control an RGB LED light to generate many colors
- Introduce coding non-binary data.

STEAM Connections:
Science - Metric time, Circuits, Color Spectrum, Additive Color Formation of Light
Technology - Code, Program Settings, Simulators, Digital Design, Electronic Components, Potentiometer, Analog values
Engineering - Designing, Building, and Using a Machine, Prototyping, Applied Physics
Allied Arts - Colors, Light, Pigments
Math - Ratios, Conversions, Formulas

Key Vocabulary:
Color Spectrum - Visible light can be displayed as discrete colors from red to violet. Each color has its own wavelength.
Wavelength - Colors of light are determined by wavelength. The units for wavelengths of light are nanometer (nm) or billionths of a meter. There are one billion nanometers in a meter.

Project Introduction:
- Introduce the groups to the purpose of the project, the skills developed, the standards met, and the goal of the project.

Anticipatory Sets:
- Where can you find lights that change color? List at least three different places or situations where this technology is used, describe the devices, and explain the purpose of the devices.
- Why is this LED called an RGB LED?

Educational Standards

ISTE Standards for Students
- Empowered Learner 1a, 1b, 1d
- Knowledge Constructor 3a, 3b, 3c, 3d
- Innovative Designer 4a, 4b, 4c, 4d
- Computational Thinker 5a, 5b, 5c, 5d
- Creative Communicator 6b, 6c, 6d
- Global Collaborator 7c, 7d

US Computer Science Standards
- Project correlations can be found on the book's web page.

US NGSS - Middle School
DCIs
- MS-PS4-1 PS4.A
- MS-PS4-2 PS4.A PS4.B
- MS-PS4-3 PS4.C

Cross Cutting Concepts
- Cause and Effect
- Scale, Proportion and Quantity
- Structure and Function

Science and Engineering Practices
- Planning and Carrying Out Investigations
- Obtaining, Evaluating, and Communicating Information
- Analyzing and Interpreting Data
- Scientific Knowledge is Based on Empirical Evidence
- Using Mathematics and Computational Thinking

US NGSS - High School
- Correlations can be found on the book's web page.

US Common Core Language Arts and Mathematics
- Correlations can be found on the book's web page.

Materials List:
- Computer with IDE software
- Connection to the Internet
- USB Cable
- UNO or UNO Compatible Microcontroller
- Short Breadboard
- RGB LED (Anode +)
- Four (4) Jumper Wires
- Three (3) 220 Ohm resistors

Engineering Design - Digital Prototype

1. Launch and log into Tinkercad® (pages 15-16).

2. Start a new Circuits project.

3. Drag an UNO board to the workspace.

4. Drag a breadboard to the workspace.

5. Drag one RGB LED to the breadboard. The leads span four rows in a column. Make sure to note the longest lead.

The only RGB LED available in Tinkercad® is the cathode(-). The code on the next page is for the cathode(-) RGB LED.

The materials list for this project specifies an anode(+) RGB LED. The anode(+) RGB LED will require two changes from the digital prototype.

1. Connect the anode(+) lead of the RGB LED to the 5v pin on the UNO board.
2. Include this code at the start of the sketch subroutine (SetColor) to correct the color value sent to the LED.

```
red = 255 - red;
green = 255 - green;
blue = 255 - blue;
```

Project 5
Rainbow Light

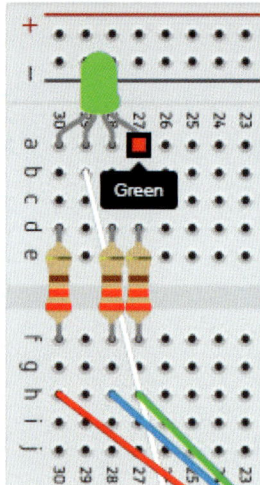

Lead	Function
1	Red Diode
2	GND for cathode(-) RGB LED
3	Blue Diode
4	Green Diode

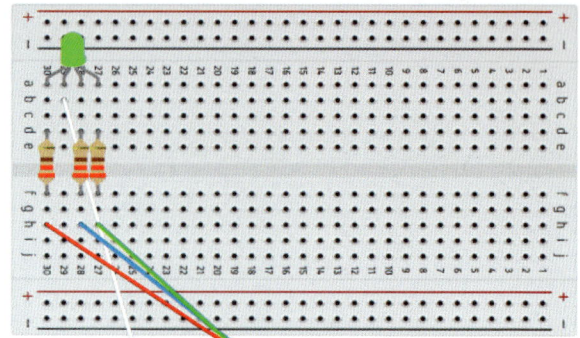

6. Place three (3) 220 Ohm resistors on the breadboard connecting the rows with color lead of the RGB LED (not the cathode(-) lead) with the same row but on the other half of the breadboard

7. Connect with jumper wires the following:
   - a ground (GND) pin on the UNO board to a pin in the same row as the cathode(-) lead of the RGB LED. The cathode(-) lead is the longest lead on the RGB LED.
   - digital pin two (2) to the same row as the resistor that connects to red lead of the RGB LED
   - digital pin four (4) to the same row as the resistor that connects to green lead of the RGB LED
   - digital pin three (3) to the same row as the resistor that connects to blue lead of the RGB LED. Hover over the RGB LED leads to verify the color connected to each lead.

Coding - Digital Prototype

8. Enter the code in Tinkercad® to match the code on this page.
   The code setup for this digital prototype is much like the second part of project two (2) with a subroutine. In this case the subroutine sends the red, green and blue values for 1.5 seconds (1500 ms).

```
int redPin = 2;
int greenPin = 4;
int bluePin = 3;

void setup()
{
 pinMode(redPin, OUTPUT); // set redPin as output
 pinMode(greenPin, OUTPUT); // set greenPin as output
 pinMode(bluePin, OUTPUT); // set bluePin as output
}
void loop()
{
 setColor(255, 0, 0); // red color
 setColor(0, 255, 0); // green color
 setColor(0, 0, 255); // blue color
}
void setColor(int red, int green, int blue) //Subroutine
{
//red = 255 - red; //remove "//" for anode(+) RGB LED
//green = 255 - green;  //remove "//" for anode(+) RGB LED
//blue = 255 - blue; //remove "//" for anode(+) RGB LED
analogWrite(redPin, red); // connection to pin and set red value
analogWrite(greenPin, green); // connection to pin and set green value
analogWrite(bluePin, blue); // connection to pin and set blue value
delay(1500);
}
```

9 Start the simulation and observe the digital prototype.

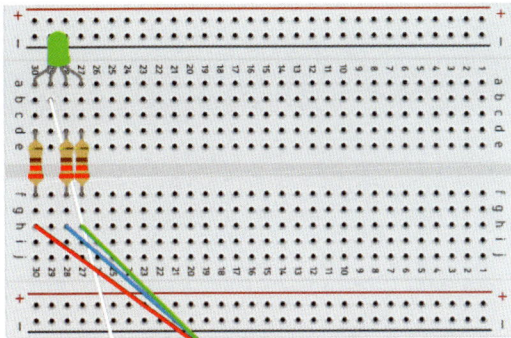

Engineering Design - Project Build

10 Place the anode(+) RGB LED on the breadboard such that the leads span four rows in a column. Make sure to note the longest lead.

11 Place three (3) 220 Ohm resistors connecting the rows with color lead of the RGB LED (not the anode(+) lead) with the same row but on the other half of the breadboard

12 Connect with jumper wires the following:
- the 5v pin on the UNO board to a pin in the same row as the anode(+) lead of the RGB LED. The anode(+) lead is the longest lead on the anode(+) RGB LED.
- digital pin two (2) to the same row as the resistor that connects to red lead of the anode(+) RGB LED
- digital pin four (4) to the same row as the resistor that connects to green lead of the anode(+) RGB LED
- digital pin three (3) to the same row as the resistor that connects to blue lead of the anode(+) RGB LED

Connection - Color mixing is governed by exact rules. The rules are different for pigments (like paints and crayons) as compared to light.

For pigments we can see that color mixing is additive going to black and subtractive to get white. Black pigment is made by the mixing of all colors. Whereas white pigment is made by removing all colors.

For light the reverse is true, white light is additive being made of all colors and black is subtractive being made of no color. Black in terms of light is the absence of any color light.

White light, as shown in this photo, is made by mixing Red, Blue and Green light (see the center overlap in the middle).

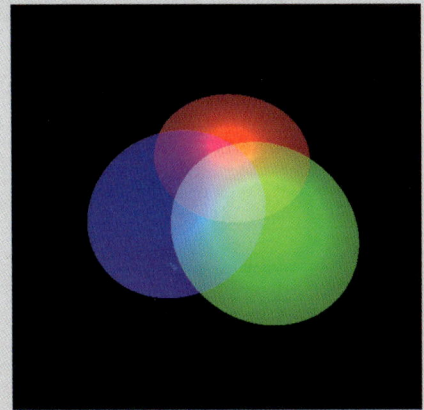

13 Review the anode(+) circuit diagram for this project.

14 Connect the UNO board and computer using the USB cable.

15 Launch the Arduino® IDE software (pages 11-13) software to make sure the board is communicating with the computer.

Go to the Tools menu and verify that the correct board is selected from the Board Manager menu. If not then select the correct board from the options listed.

Then confirm that the right Port is selected. If not then select the port that lists the UNO board from the options listed.

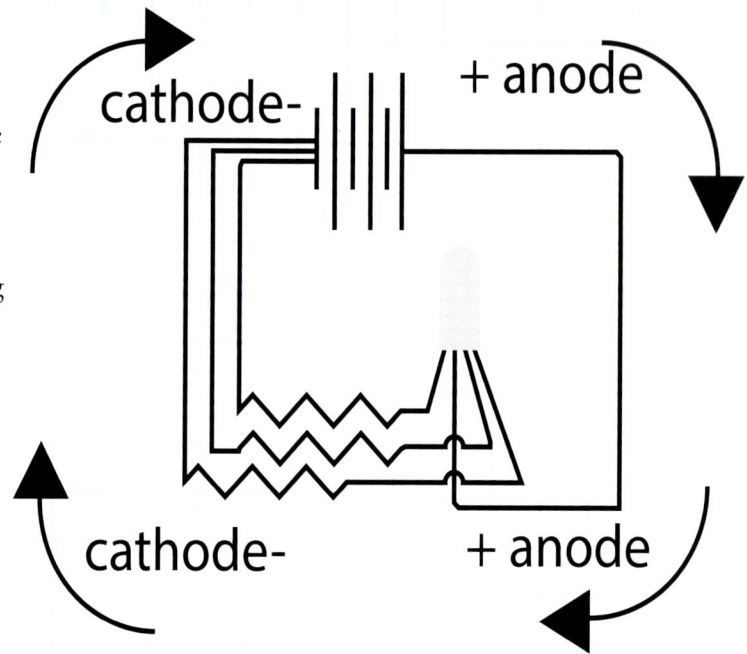

cathode-   + anode

cathode-   + anode

Project 5
Rainbow Light

```
TAC_UE_V1_P5_Anode2

//Project 5
//Rainbow Light
//thearduinoclassroom.com
//Copyright 2019, Isabel Mendiola and Peter Haydock

int redPin = 2;
int greenPin = 4;
int bluePin = 3;
void setup()
{
pinMode(redPin, OUTPUT); // set redPin as output
pinMode(greenPin, OUTPUT); // set greenPin as output
pinMode(bluePin, OUTPUT); // set bluePin as output
}
void loop()
{
 setColor(255, 0, 0); // red color
 setColor(0, 255, 0); // green color
 setColor(0, 0, 255); // blue color
}
void setColor(int red, int green, int blue)
{
  red = 255-red;
  green = 255-green;
  blue = 255-blue;
analogWrite(redPin, red); // connection to pin and set red value
analogWrite(greenPin, green); // connection to pin and set green value
analogWrite(bluePin, blue); // connection to pin and set blue value
delay(1500);
}
```

Done compiling.

RGB LED leads may have the Green and Blue leads switched. Changing the code and jumper wire colors to reflect this variation is recommended.

Coding - Project Build

16 Either copy and paste the code from the Tinkercad® prototype or type the code from page 50 into the Arduino® IDE software (pages 11-13). For the anode(+) RGB LED make the adjustments to the wiring and code noted on page 49.

17 Save the sketch (rename as needed).

18 Verify the sketch.

19 Upload the sketch to the UNO board.

20 Observe the RGB LED on the breadboard.

21 Document this project, discuss the project, and complete the extensions assigned from the next page.

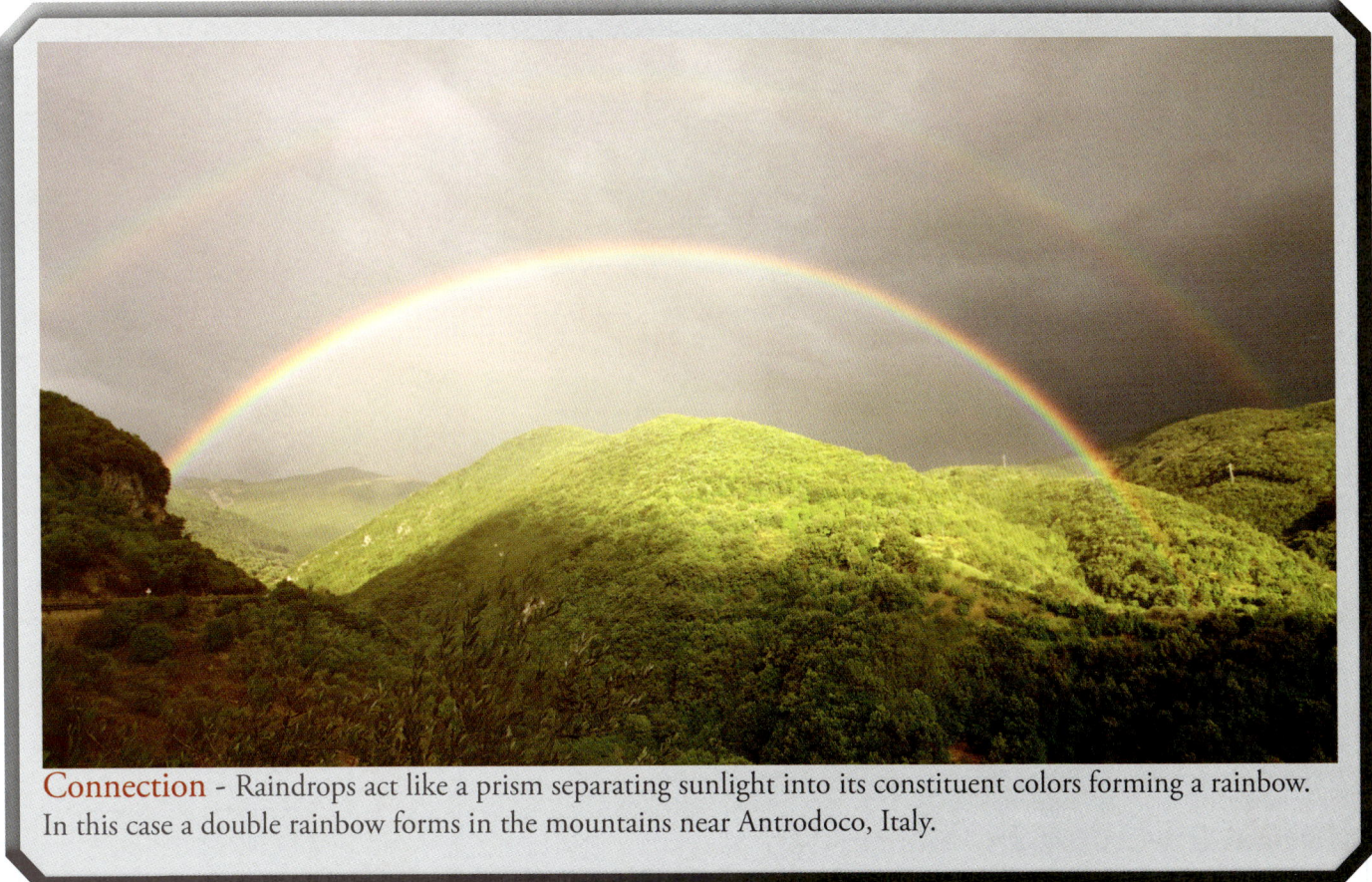

Connection - Raindrops act like a prism separating sunlight into its constituent colors forming a rainbow. In this case a double rainbow forms in the mountains near Antrodoco, Italy.

Project 5
Rainbow Light

Project Reflection and Summative Activities, Discussion Starters, Extensions, and Problem

A = Essential

Reflection/Summative Activities:

- Have each team member document in their project/ classroom journal who was on their team, what went well, what they could improve upon, what they would do differently if they were to do the project again, and verify that every person in the group can do the project.

- Produce a color chart or wheel that demonstrates color mixing of light in the RGB LED (see page 51). Recreate the chart/wheel on wax paper for displaying in the classroom.

For Discussion:

- Discuss the advantages and disadvantages to using an RGB LED for lighting. List the color(s) you are unable to make with this code.

- Discuss the code and wiring changes needed to use the anode(+) RGB LED. Why are these changes needed? Create a circuit diagram for the cathode(-) RGB LED.

Extensions:

- Change the code to add a white light to the pattern.

- Investigate how pigments, like those in paint, make colors. Compare and contrast color mixing with light versus mixing of pigments to produce different colors. Demonstrate the mixing of pigments with a color chart or wheel.

B = Recommended

Section Reflection/Summative Activities:

- Have the team create a graphic organizer summarizing the skills and knowledge they obtained in doing Projects One through Seven.

- Document in your journal key vocabulary, devices used, and skills developed in this section.

Artistic Connection:

- Demonstrate using a flashlight and colored gel paper in a darkened room how different lighting changes the appearance of everyday objects (food, clothing, etc..).

Extensions:

- Change the pattern of the colors and/or produce new colors by editing the code.

- Create a small box with a black interior that will house the RGB LED and a small object to observe how colors made by the RGB LED effect the appearance of the object.

- Submit your project to the book page on the website. Register and Login to submit at: thearduinoclassroom.com

From the homepage:
Click
  Books
Click
  UNO Edition Vol. 1

C = Optional

For Discussion:

- Discuss how a rainbow forms.

- As a class, research and discuss the best color lights for plants to grown under.

Extensions:

- Research and write an explanation of why number ranges do not end at the end of a place value (ex. ranges do not end with 9, 99, 999, etc..) but range from 0 to 1, 0 to 3, 0 to 7, 0 to 255, and 0 to 1023 (ending with 1, 3, 7, 255, 1023, etc.).

- Change the pattern of the colors to reproduce the color pattern of a rainbow in sequence.

Problem:

- Identify a problem that would be solved by light that varies in color like the one you have been working with in this project. How could this project be a solution? Include in your report the following:

Description of the problem.

Description of how the problem is solved by this project.

What other resources you would need to solve the problem.

Design and produce the RGB LED solution.

Project 5
Rainbow Light

Response Starters in Blue Italic

A project rubric can be found on page 236.

Anticipatory Sets
(from page 48)

Anticipatory Sets:
- Where can you find lights that change color? List at least three different places or situations where this technology is used, describe the devices, and explain the purpose of the devices.
In entertainment (theatre, shows), events (parties, exhibits), signage (streets signs and ads, businesses display), home (decoration, outdoor indoor lighting), Business (Stair cases, elevators), Automobiles, vehicles (limousine interiors, boats, cars) and artwork (frames and gallery lighting).

- Why is this LED called an RGB LED?
RGB stands for Red, Green, and Blue the three colors of light that can be easily mixed to make all the other colors save black. There are three diodes in the LED each making one color.

A
Reflection/Summative Activities:
Responses Will Vary

- Produce a color chart or wheel that demonstrates color mixing of light in the RGB LED (see page 51). Recreate the chart/wheel on wax paper for displaying in the classroom.

Responses Will Vary

For Discussion:
- Discuss the advantages and disadvantages to using an RGB LED for lighting. List the color(s) you are unable to make with this code.
The RGB LED can generate literally thousands of colors of light from one LED. If the LED goes out, no colors are generated where color device that uses three colors of LED to make a color might not be right but at least some color would be generated.

Black

- Discuss the code and wiring changes needed to use the anode(+) RGB LED, Why are these changes needed? Create a circuit diagram for the cathode(-) RGB LED.
The color values sent to the RGB LED will need to be adjusted in the code and the power wire will need to change to a 5v connection if an anode(+) RGB LED.

Extensions:
- Change the code to add a white light to the pattern.
Responses Will Vary

- Investigate how pigments, like those in paint, make colors. Compare and contrast color mixing with light versus mixing of pigments to produce different colors. Demonstrate the mixing of pigments with a color chart or wheel.

Responses will vary, but should include how black and white light and pigments are made as well as reflection and absorption of lights from a surface.

B
Section Reflection/Summative Activities:
Responses Will Vary

Artistic Connection:
Responses Will Vary

Extensions:
- Change the pattern of the colors and/or produce new colors by editing the code.
Responses Will Vary

- Create a small box with a black interior that will house the RGB LED and a small object to observe how colors made by the RGB LED effect the appearance of the object.
Responses Will Vary

- Submit your project to the book page on the website.
Responses Will Vary

C
Extensions:
- Number ranges research.
Responses Will Vary

- Change the pattern of the colors to reproduce the color pattern of a rainbow in sequence.
Responses Will Vary

Problem:
Responses Will Vary

The Arduino Classroom § 55

Section 2
Displays, Inputs, and Controls

SKILLS

The Serial Monitor and Serial Plotter are two powerful tools found in the Arduino® IDE that are especially useful for STEAM projects. They both are used to output data. Serial Monitor can output raw numbers and text while Plotter can create simple graphs.

To use Serial Monitor, two lines of code must be added and one setting change made in the IDE. A Serial.begin line needs to be added inside the void setup() section. Serial.begin must include a data value which indicates the speed at which the code will relay data to the Serial Monitor. It will look like this:

Serial.begin(9600);

The number inside is usually determined by the devices connected to the UNO board. The technical details of your devices will determine the setting.

The second line of code needed is at least one Serial.print line. The parameter will appear inside the parentheses. The code will print a variable value, some text, or both.

Serial.print("Hello");

This example line of code will print on the screen the word "Hello" with no line returns. Note the parentheses used.

Serial.println(x);

In this case, adding the "ln" after Serial.print forces a line change after the value of "x" is printed on the screen.

The last thing to setup is the connection speed. Launch the Serial Monitor and look to the bottom of the window and select the connection speed from the drop down menu second from the right.

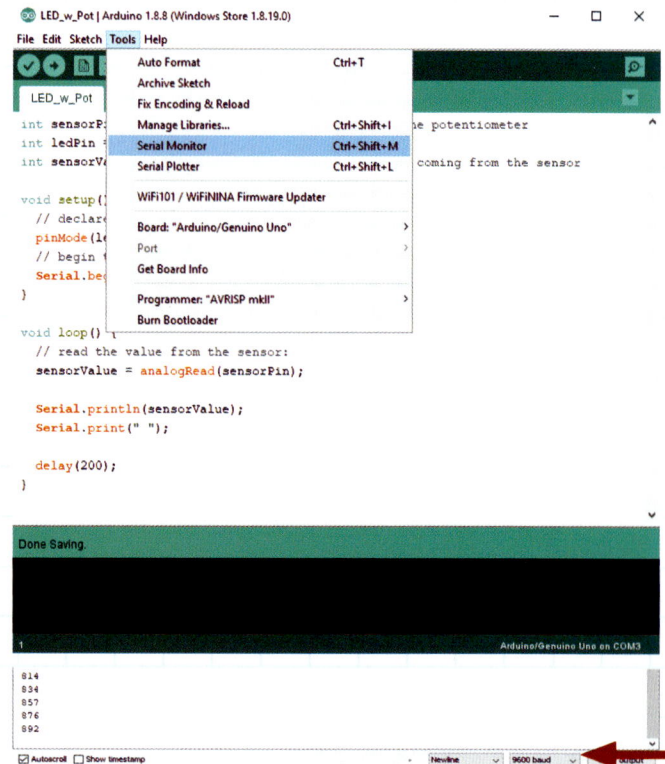

The code on this page is the same code from Project Four(4) but with four lines added. Serial.begin, Serial.print, and Serial.println lines and a small delay to make seeing the output easier have been added.

```
int input = A5; // Microcontroller Analog connection for the potentiometer
int LED = 11;  //  LED connected to digital pin 11
int value=0; // sets potentiometer input value at zero
void setup ()
{
 pinMode(LED ,OUTPUT);
 Serial.begin(9600); // begin the serial monitor @ 9600 baud
}
void loop()
{
int  value = analogRead(input); // Reading the value of the potentiometer
analogWrite(LED, value/4); // The highest value of analog input is 1023 and PWM has
//a resolution of 256, so by dividing input by you scale input to PWM
Serial.print(Potentiometer = ");
Serial.println(value);
 delay(200);
}
```

Open Project Four (4) and add the four (4) lines of code in Tinkercad® or to the actual project. Then launch the Serial Monitor. Check the connection speed. Watch the values change as the potentiometer is dialed from zero (0) to 1023. The LED will dim and brighten as the knob is turned. Notice the placement of Serial.print within the void loop.

For the Serial Plotter, follow the same steps. Write the code to include a Serial.begin line and then add a Serial.print line. Because the Serial Plotter does not accept words, you will have to remove any code with quotes. Once you have added the code launch the Plotter, set the connection speed and watch your data be graphed. You can see the graph we generated from Project 4 on this page.

COM3 (Arduino/Genuino Uno)

```
Potentiometer = 1023
Potentiometer = 1023
Potentiometer = 1023
Potentiometer = 1023
Potentiometer = 850
Potentiometer = 698
Potentiometer = 666
Potentiometer = 461
Potentiometer = 220
Potentiometer = 0
Potentiometer = 0
Potentiometer = 0
Potentiometer = 0
```

COM3 (Arduino/Genuino Uno)   — ☐ ✕

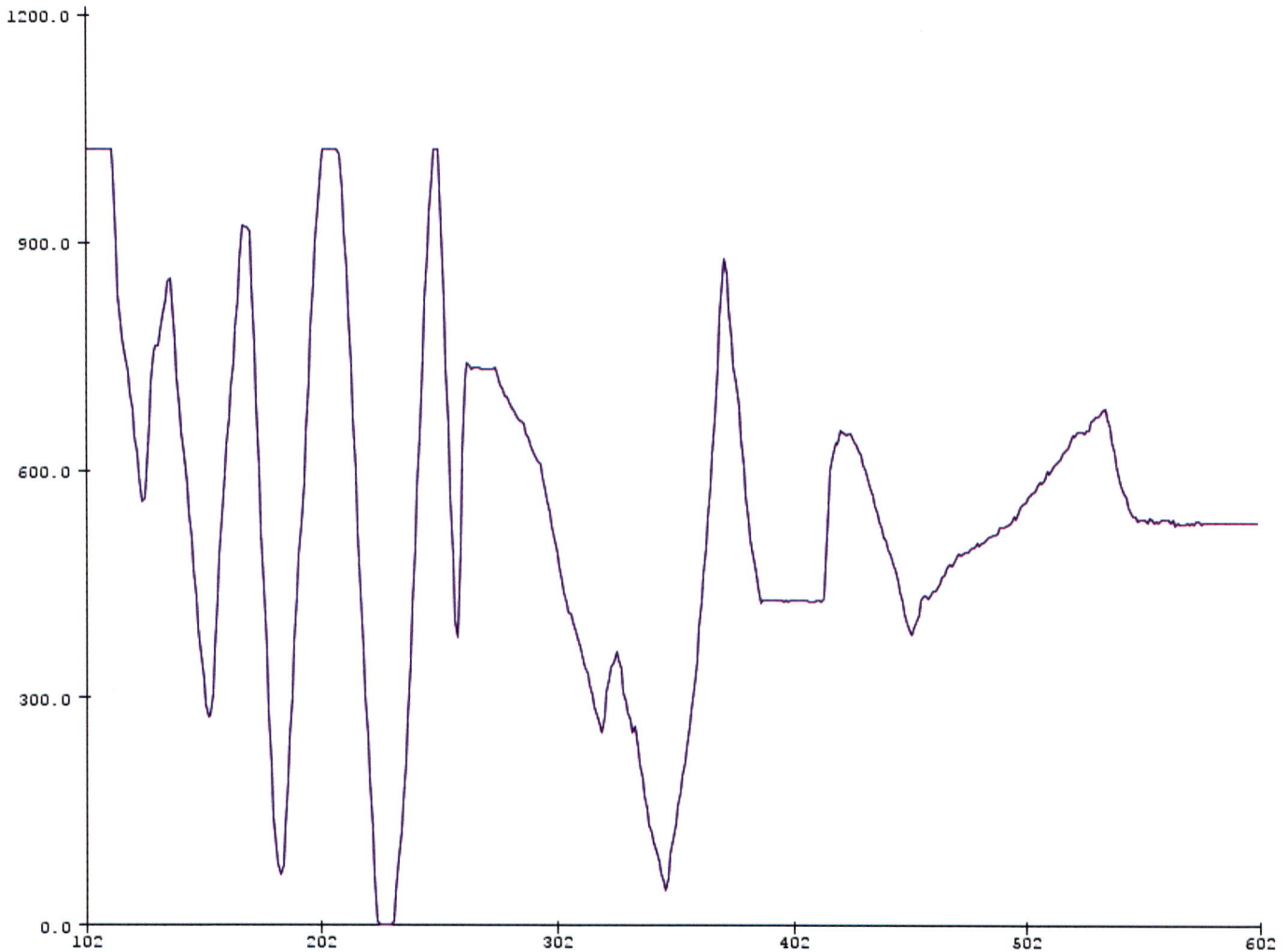

9600 baud

Project complexity will start to increase with the next projects and with that the need to manage time, resources, and work.

To be successful with these upcoming projects, learners should take time to plan and manage their work. Using a project management framework will increase the number of successfully completed projects.

Project management is now a skill that has been deemed so important to the success of businesses that there are hundreds of classes and dozens of programs that teach people how to manage a project. As much as a Bachelors, Masters, or Ph.D. signifies an educational accomplishment, professionally certified project managers are a highly valued asset to a business.

Take a look at the list of reasons projects fail in the purple box. When a project fails it costs the organization time, competitive advantages, and money.

Reasons Projects Fail

1. Poor Preparation (Planning, Assignments, Budget, and Calendar)
2. Poor Documentation and Tracking
3. Bad or Inexperienced Project Managers
4. Failure to Define Parameters and Enforce Them
5. Underestimated Cost or Time Predictions
6. Lack of Communication
7. Competing Priorities (Time, Resources)
8. Disregarding (Early) Project Warning Signs

The skills developed here will help build a framework that will increase the quality of projects and help everyone finish on time.

There are five steps to a project

1. Conception and Initiation
2. Definition and Planning
3. Launch and Execution
4. Performance and Control
5. Close

Project Conception and Initiation

At the start define the project, research the project's feasibility, and influence the stakeholders to greenlight the endeavor. This phase also estimates the likelihood of success.

In a classroom setting, this will be more than likely defined by the teacher or specialist.

Project Definition and Planning

In this phase, all of the resources and constraints are identified. The team, time lines, budgets, intermediate deliverables, and location(s) are identified. The terms of success are defined and codified. Tools for tracking, communication, collaboration, and reporting are initiated. The project should also identify how the group will track and notify when intermediate goals are not being met (risks) that might threaten the completion of the project on time. Also at this point determine how issues and problems will be resolved.

Project Launch and Execution

This phase is about doing. Here all of the resources are assigned, the work delegated, and the tracking and communications systems used. Those familiar with Engineering Design will see overlap in the processes starting here.

Project Performance and Control

Regular meetings or check-ins are held. Longer term projects may have intermediate goals set. Budgets and materials are tracked. As issues are identified, solutions are implemented with adjustments to team, time, and other resources made. Progress is documented and stakeholders are notified of the progress until completion

Project Close

A review of the completed project is done. This can be a formal written report, presentation, or a closure meeting. All documentation, communications, and products are archived for future reference.

We recommend that you introduce this framework to your learners and help them look at their projects in this context.

Mathematical Operations

Doing math within the code is fairly straightforward. There are two parts to each mathematical operation. The first step is to declare the variable or variables intended for use and set their initial value. For integers, the "int" command is used. Then the variable is declared along with its initial value.

```
int x  =5;
```

Once this is done, the variable can play a role in function. The correct programming is to declare the variable that will be changed and then carry out the function.

```
x= x + 1;
```

Depending on the context it may be better to set a new variable for each step in a function.

```
int x  =5;
int y = 0;
y= x + 1;
```

In this example, the "y" starts as "0" and ends as "6." All the arithmetic functions are available (+ , -, *, and /). There are many other functions to act on numbers.

To use numbers with decimals substitute the "float" command for "int."

```
float x=5.5;
```

Conditional Statements

Conditional statements allow programs to evaluate input and perform an action. The conditional statement has at least one line of code but almost always multiple lines for at least two or more actions.

The conditional statement starts with the command "if" placed in the code to signal that some condition is being tested. Generally the condition being tested is comparing a constant and a variable, but it can also be two variables.

When you have two numbers you can compare them in many ways. For example a statement like "5 is bigger than 2." Other comparisons of 5 and 2 include:

2 is smaller than 5 or 2 < 5

5 is not equal to 2 or 5 <> 2

It is possible to substitute a variable for one of the numbers. Examples of this would be

x is smaller than 5 or x < 5

x is not equal to 5 or x <> 5

In the code these statements would be written

```
x  <  5
x  != 5
```

Now combine this with an "if" and the code now has a full conditional statement.

```
if (x <  5);
if (x != 5);
```

The last step is to add an action to the conditional statement. It can be any action you can program like turning on an LED or displaying text. For example:

```
if (x < 5) digitalWrite(LEDpin1, HIGH);
if (x != 5) digitalWrite(LEDpin2, HIGH);
```

Multiple actions can be placed after the "if" comparison. Also, multiple "if" statements can be strung together in the code to do multiple comparisons.

The second part of a conditional statement can be the "else" command. "else" must be preceded by an "if." This allows the code to not only make a comparison but vary the outcome based on the comparison and complete different actions.

```
if (x < 5) {digitalWrite(LEDpin1, HIGH);}
else {digitalWrite(LEDpin2, HIGH);}
```

An "else" statement can contain an "if" statement to make multiple comparisons as well.

```
if (x < 5) {digitalWrite(LEDpin1, HIGH);}
else if (x >5) {digitalWrite(LEDpin2, HIGH);}
else {digitalWrite(LEDpin1, HIGH);
digitalWrite(LEDpin2, HIGH);}
```

Pay attention to the use of end line semicolons and curly brackets when writing "if else" statements.

The only remaining thing to cover here is the variable. The variable will come from either a setting at the beginning of the code (int x =5;), an input from a sensor (introduced later), or function within the code.

One of the best features of the Arduino® platform is the Library. Libraries are very similar to the subroutine covered in project two but are only one line of code at the start of a sketch and are usually designed to work with a specific piece of hardware.

The hardware documentation and certainly our project documentation will tell you if a library is used. You will need to know if your sensor, output, or other input piece of hardware has a library already installed. To do this:

1. Open the Arduino® IDE software (pages 11-13).

2. Click on the "Sketch" menu.

3. Click on the "Include Library" option.

4. You can scroll through the list and find your hardware library file listed.

5. Click on the library you will use and it will add the correct code to your sketch.

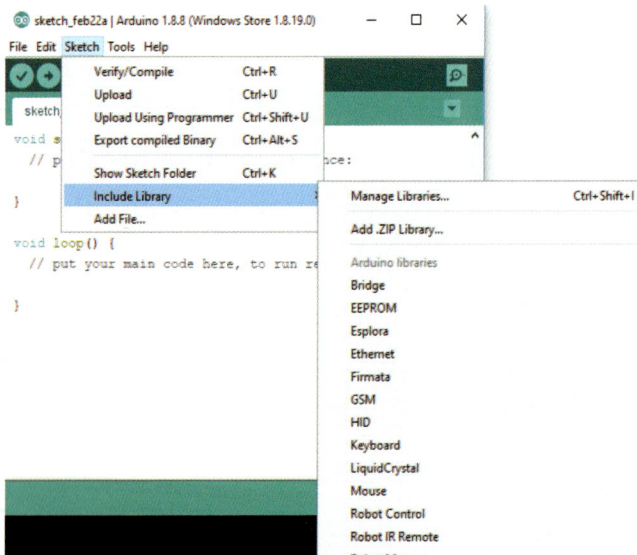

6. The code will look like this:
   `#include <LiquidCrystal.h>`.

If the library is not already installed, the book's project documentation will explain how to acquire the library and install it. To do this:

1. Go to the book web page for the file indicated in the documentation.

2. Download the file (usually a .zip file).

3. Install the file by first clicking on the "Sketch" menu.

4. Click on the "Include Library" option.

5. Click on the "Add .ZIP Library" option.

6. Locate the downloaded file and click on it.

7. Click "Open" and the library will be added to your files.

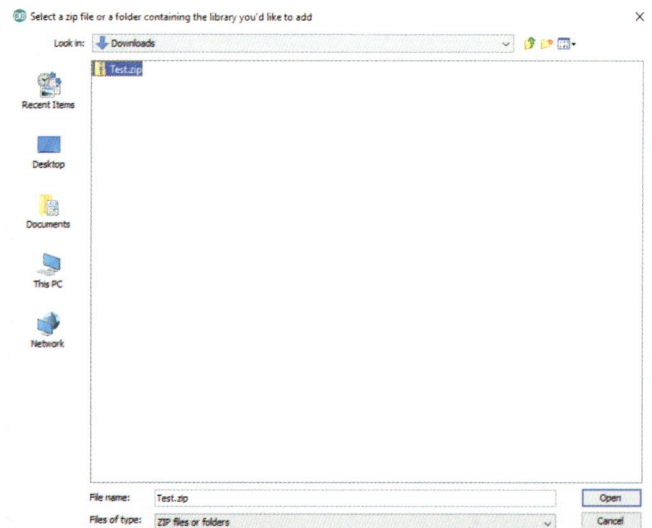

From time to time, the libraries need to be updated. The IDE software will place a notification at the bottom of the software.

Just click on the link in the message and the update will happen automatically.

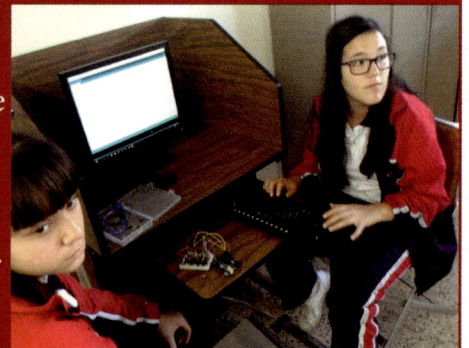

GitHub is one of the most powerful tools on the Internet for coders. According to the statistics listed on the site, 31 million coders use GitHub with over 100 million resources uploaded.

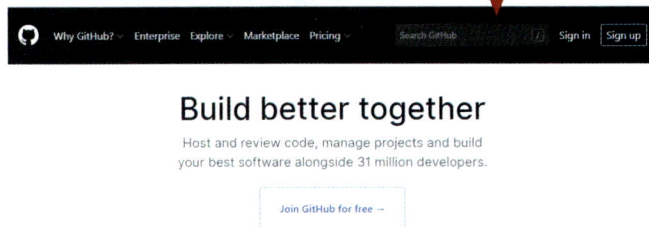

Build better together

Host and review code, manage projects and build your best software alongside 31 million developers.

Join GitHub for free →

GitHub allows for storing, sharing, and editing of resources (called repositories) for open source projects as well as closed or privately shared projects. Accessing open source resources on GitHub is free. No account is needed either. Simply go to GitHub.com use the search tool at the top of the page and resources matching the criteria entered will be listed.

The following can be found on GitHub:

1. Open Source Code for Arduino® projects
2. Fritzing parts for the digital prototypes
3. Arduino® libraries
4. Resources for other open source projects

1 Use the search at the top of the main page of GitHub to start looking for the resource needed. For example look for the liquid crystal display.

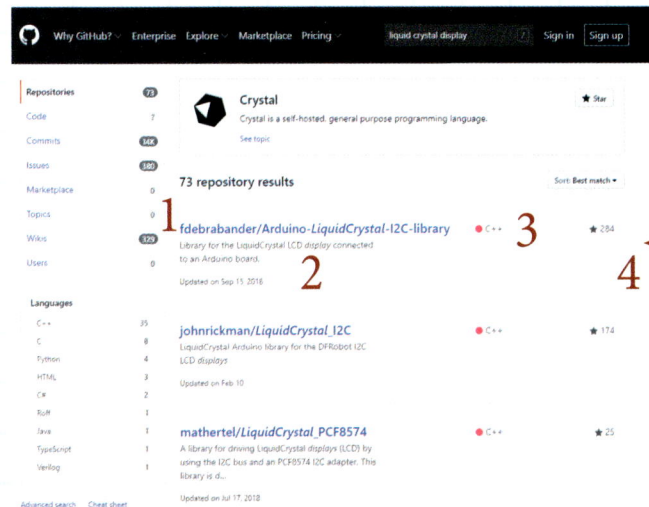

2 The search will return:

1. Links to the repositories.
2. A short description of the contents of each repository.
3. The code language of each repository.
4. The number of stars users of GitHub have assigned each repository.

3 Click on the link to access the repository.

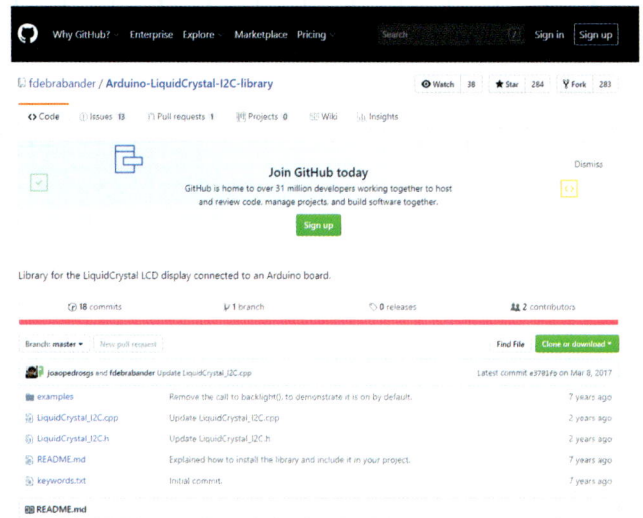

4 If the repository meets the needs of the project, click the green button on the page.

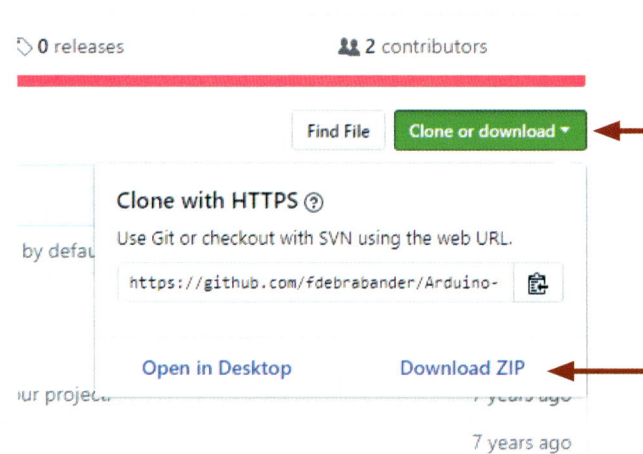

5 Select "Download ZIP."

6 Follow the instructions for adding the repository files to the program being used.

Fritzing is another powerful digital prototyping tool for setting up Arduino® UNO projects. The table on this page highlights the similarities and differences between Tinkercad® and Fritzing.

Tinkercad®	Fritzing
Online	Download
Free	Free
Drag and Drop Prototyping	Drag and Drop Prototyping
Runs project code	Does not run project code
Project Simulation	No Project Simulation
Fewer devices	More devices
Cannot add new devices	Can add new devices

Fritzing is an important tool in working with Arduino® projects especially when Tinkercad® does not have the devices in its inventory to build the prototype.

Step-by-Step 1-22

1. Go to the Fritzing website and download the application appropriate for your computer.

2. Install Fritzing.

3. Run Fritzing and the program will open to the landing page shown on this page.

4. For the projects covered in this book, we will look at the functions shown in the Breadboard section and how to add Parts.

5. The landing page also includes:
   - news announcements about Fritzing
   - a list of recent projects
   - access to the schematic (circuit) diagram building tools
   - access to the single layer printed circuit boards building tool
   - access to the code building tool

6. Click on the Breadboard button

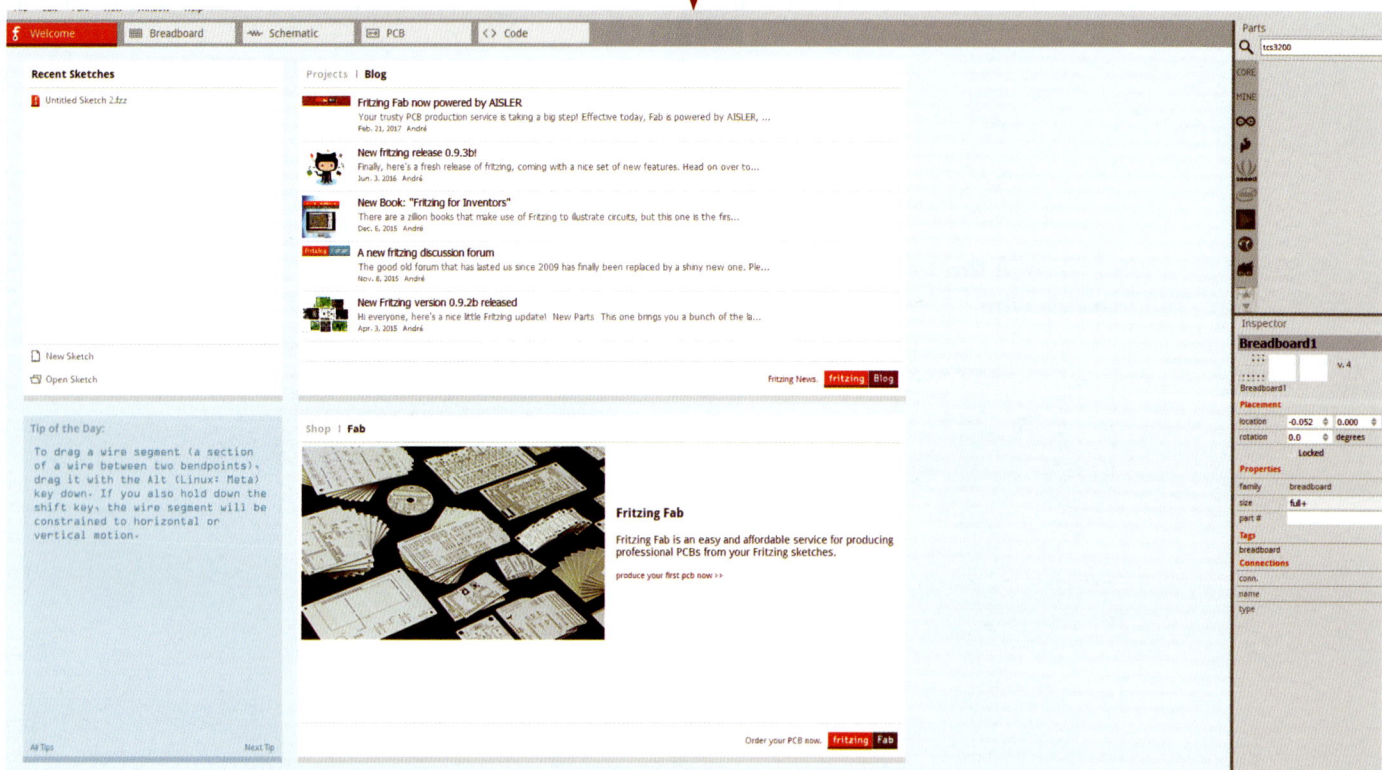

7 The Breadboard stage will launch with a breadboard in place

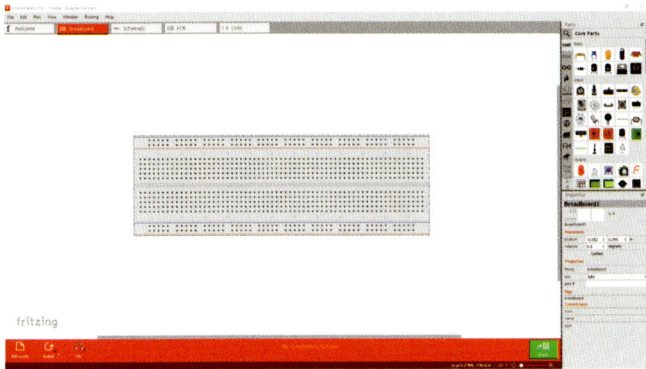

8 At the bottom right of the stage is the zoom feature to control the size of the stage size.

9 To draw a jumper wire, click and hold the mouse button down on the starting location of the jumper wire and drag to the ending location.

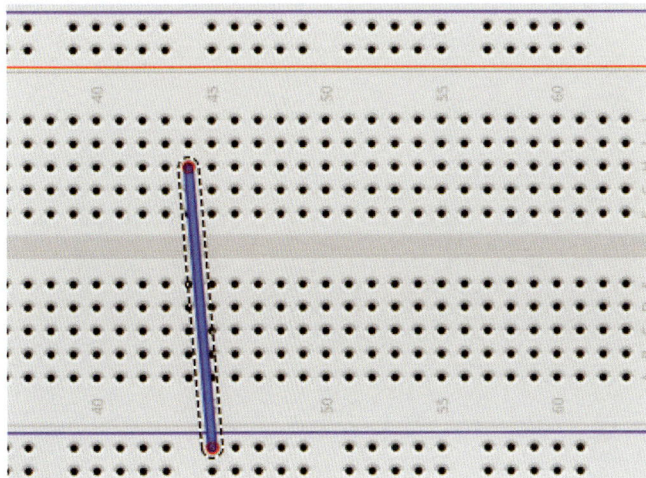

10 To introduce a bend point in the jumper wire, click in the middle of the wire.

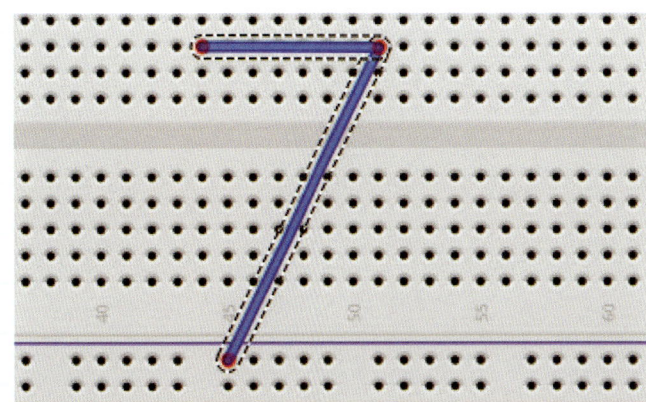

11 To introduce a part on to the stage, drag click through the parts lists on the right side of the stage.

The UNO board is found in the Arduino® parts list (the infinity sign with the "-" and "+").

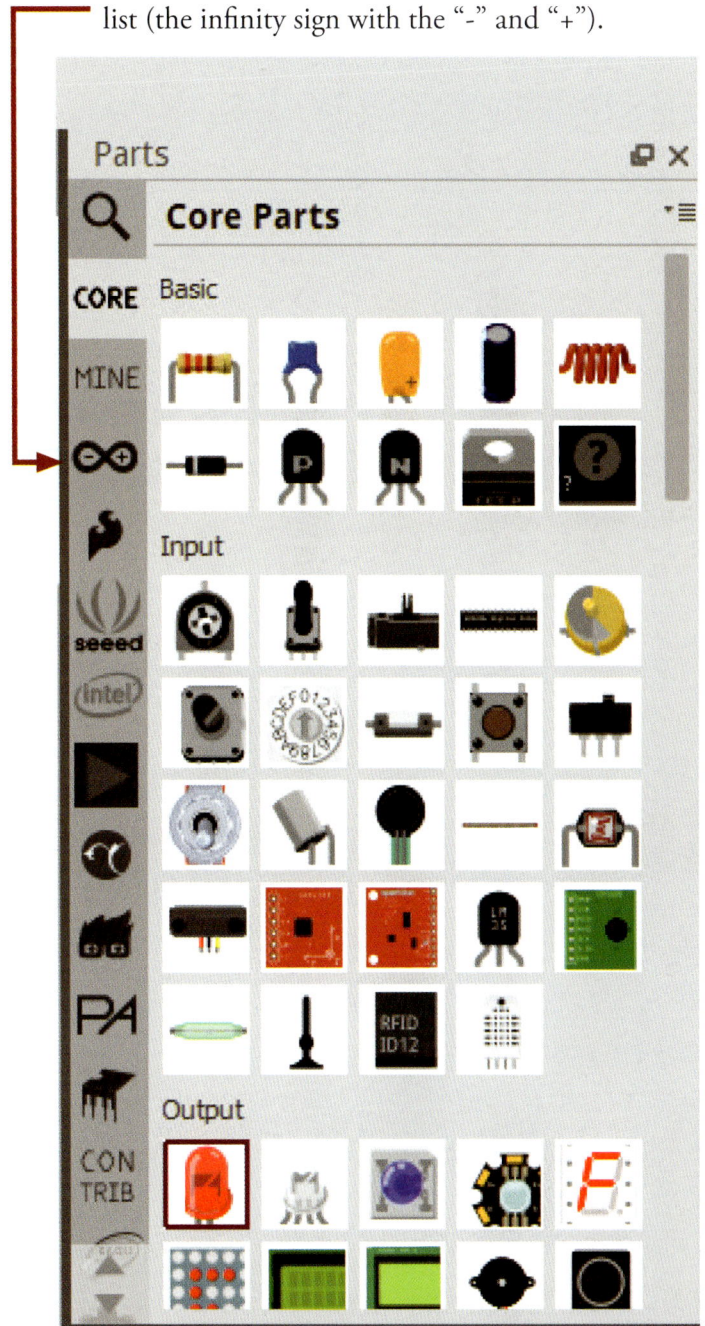

Use the search box to look for parts that might not be immediately seen.

12 Once the part is selected, look below the parts list to see the parts inspector.

This part of the stage allows you to change the properties of a part. For example, if a resistor is placed on the breadboard the resistance can be changed along with other properties.

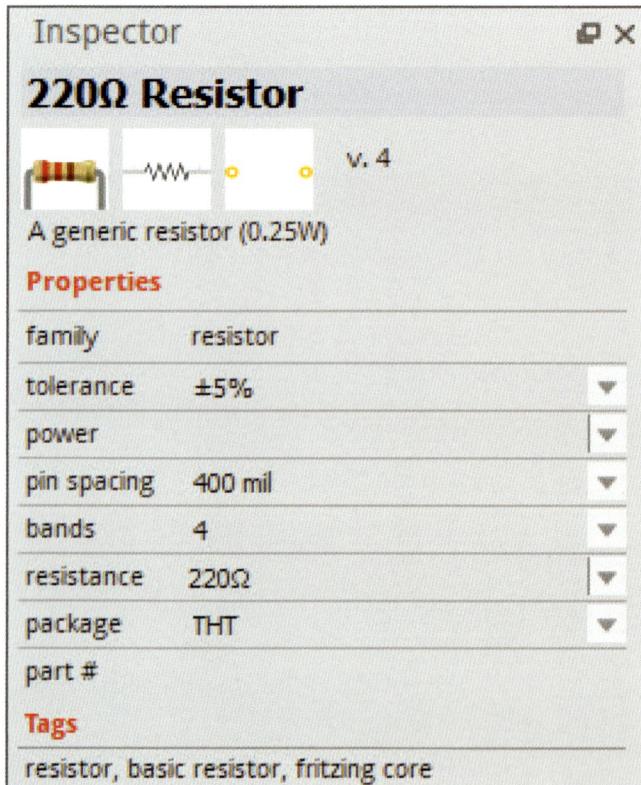

Inspector

220Ω Resistor

v. 4

A generic resistor (0.25W)

Properties

family	resistor
tolerance	±5%
power	
pin spacing	400 mil
bands	4
resistance	220Ω
package	THT
part #	

Tags

resistor, basic resistor, fritzing core

13 The Fritzing parts list is extensive, but if you do not find the part you want adding a part can be done in a few steps.

14 First search the Internet for the part, looking for a Fritzing compatible file which is designated with a file extensions of .fzpz, .fzbz, or .fzb. A good place to look is GitHub

15 Download the file to your computer and extract the file if the file or files are in a .zip file.

16 Import the file by clicking the "hamburger" in the upper right of the parts section of the stage.

Parts

17 Click on the Import option and navigate to the downloaded Fritzing file.

18 The part will show up in the Parts section under the button labeled as "MINE."

Parts

My Parts

CORE

MINE

19 Drag the part(s) to the stage to use.

20 Build your project as needed.

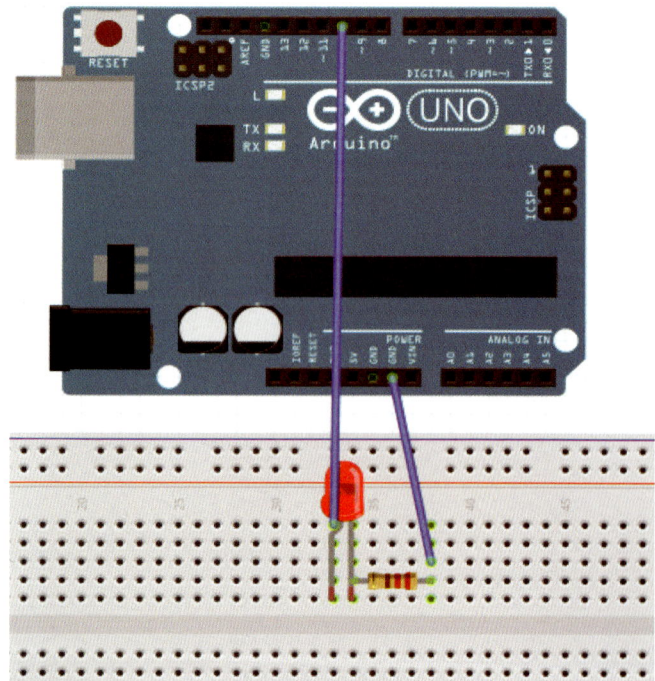

21 Save your sketch and confirm saving any downloaded parts for use later.

22 Recall that Fritzing does not have a code simulator so the project cannot be tested as it can in Tinkercad®.

SKILLS
Troubleshooting Projects

On this page find some ideas on how to fix projects that are not working as you expected.

Build in Tinkercad®
- First, build your project and code in Tinkercad®
- Run "Start Simulation" in Tinkercad®

Arduino® IDE Code
- Write the code in the Arduino® IDE program and verify the code. If there is something wrong with the code the line or lines with issues will be highlighted. You may need to verify the code several times until all the issues are resolved. The verification process will indicate what issue needs resolving.

- Look for the colors of the characters, words, and numbers. Make sure each are colored correctly. Most discrepancies in the code will show up as indicated by incorrectly colored words or numbers.
- Capital letters, spaces, and spelling matter.
- Read each line and look for missing semicolons, curly brackets, or parentheses.
- Visualize and trace the values through the code one step at a time. Printing out the code and writing down values as the code progresses can be helpful.

- Add a Serial Monitor write statement to output values as the code runs. By adding the write statement at critical parts of the code to show how variables are being calculated or what values a sensor is reading.
- If a "302" error is given, retype the line of code from scratch. There may be hidden character in the original line of code.

Hardware
Eventually you will want to build your project and issues will arise. Here are some things to try.

- Use red jumper wires for anode(+) connections.
- Use blue jumper wires for cathode(-) connections.
- Use different color jumper wires for other connections.
- Check the jumper wire connections.
- Reseat jumper wires.
- Verify the pins that each jumper should be connected to both on the breadboard and on UNO board.
- Cross-reference the jumper wires with the code.
- Verify jumper wires are in the right columns and rows to make the correct connections or avoid short circuits or inadvertent connections with other devices.
- Reseat the other devices.
- Verify the rows and columns the device should be in.
- Replace the device or devices.

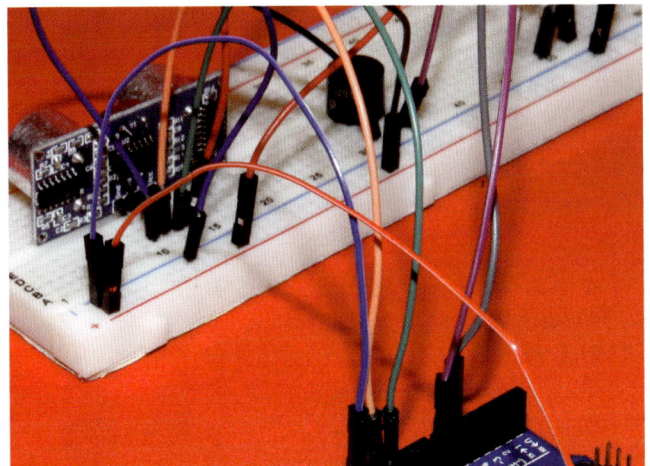

Project 6
Spin It Up

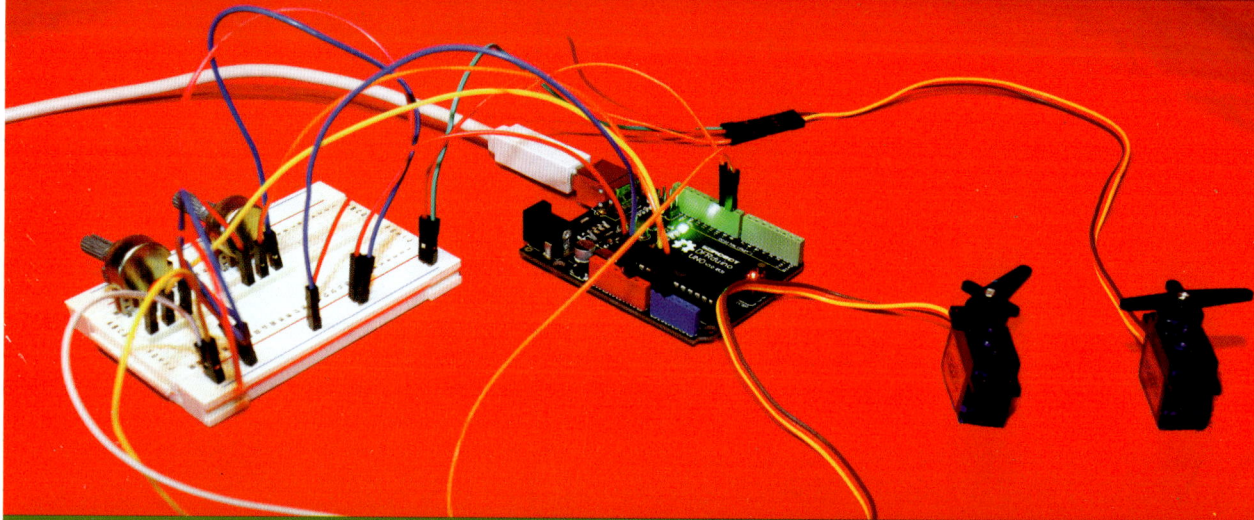

Lesson Integration:	Groupings:	Level:	Time to Complete:
Physics - Time (measurement), Ohm's Law, Circuits, Motion, Forces, Work, Torque, Gears, Motors, Mechanics Biology - Musculoskeletal System Mathematics - Cycles, Periodicity, Degrees, Rotation, Ratios, Angles, Radial Position	1 - 2	Intermediate	45 min. for the project and 45 min. for extensions.

Objectives:
- Investigate how circuits and electronic components interact with electrical energy.
- Control motion

Prerequisite Skills:
- Time measurement (milliseconds)
- Arduino® IDE software (pages 11-13)
- Tinkercad® (pages 15-16)
- Libraries (page 62)
- Angles

Purpose and Skills:
- To build a functioning servo
- Control movements in prescribed arcs

STEAM Connections:
Science - Metric time, Circuits, Ohm's Law, Mechanics, Work, Torque, Joints
Technology - Code, Program Settings, Simulators, Digital Design, Electronic Components, Potentiometer, Analog Values, Servos
Engineering - Designing, Building, and Using a Machine, Prototyping, Applied Physics, Biomedical Engineering

Allied Arts - Motion
Math - Ratios, Conversions, Degrees, Rotation, Angles, Radial Position

Key Vocabulary:
Library - A program designed for a specific device that standardizes inputs and outputs for a sketch or code.
Servo - A motor that can only spin with a limited range of motion. Usually less than 360 degrees.

Project Introduction:
- Introduce to the groups, the purpose of the project, the skills developed, the standards met, and the goal of the project.
- Introduce how to read the color indicators on a resistor.

Anticipatory Sets:
- Where might you see servos used? List at least three different places or situations that use this technology, describe each device that contains the servos, and explain the purpose of the servos.
- What are the advantages and disadvantages of designing a machine with a servo?

Project 6
Spin It Up

Educational Standards

ISTE Standards for Students
- Empowered Learner 1a, 1d
- Knowledge Constructor 3a, 3b, 3c, 3d
- Innovative Designer 4a, 4b, 4c, 4d
- Computational Thinker 5a, 5b, 5c, 5d
- Creative Communicator 6b, 6c, 6d
- Global Collaborator 7c, 7d

US Computer Science Standards
- Project correlations can be found on the book's web page.

US NGSS - Middle School
DCIs
- MS-PS4-1 PS4.A
- MS-PS4-2 PS4.A PS4.B
- MS-PS4-3 PS4.C

Cross Cutting Concepts
- Cause and Effect
- Scale, Proportion and Quantity
- Structure and Function

Science and Engineering Practices
- Planning and Carrying Out Investigations
- Obtaining, Evaluating, and Communicating Information
- Analyzing and Interpreting Data
- Scientific Knowledge is Based on Empirical Evidence
- Using Mathematics and Computational Thinking

US NGSS - High School
- Correlations can be found on the book's web page.

US Common Core Language Arts and Mathematics
- Correlations can be found on the book's web page.

Materials List:
- Computer with IDE software
- Connection to the Internet
- USB Cable
- UNO or UNO Compatible Microcontroller
- Short Breadboard
- 14 Jumper Wires (Male to Male)
- Two (2) 10k Potentiometers
- Two (2) Mini Servos
- Phillips head screwdriver

Engineering Design - Digital Prototype

1 Launch and log into Tinkercad® (pages 15-16).

2 Start a new Circuits project.

3 Drag an UNO board to the workspace.

4 Drag a short breadboard to the workspace.

5 Drag two (2) 220 micro servos to the workspace.

6 Place two (2) 10k potentiometers on the breadboard in the same column.

7 Connect with jumper wires:
- a ground pin on the UNO board to the cathode(-) column
- the 5v pin on the UNO board to the anode(+) column
- the anode(+) column to the same row as the anode(+) lead for both potentiometers. Two jumper wires
- the cathode(-) column to the same row as the cathode(-) lead for both potentiometers. Two jumper wires
- analog pin zero (A0) to the same row as the input lead (center) of one potentiometer
- analog pin one (A1) to the same row as the input lead (center) of the other potentiometer
- the anode(+) column to the same row as the anode(+) lead for both servos. Two jumper wires

Project 6 Servos with 10K potentiometers

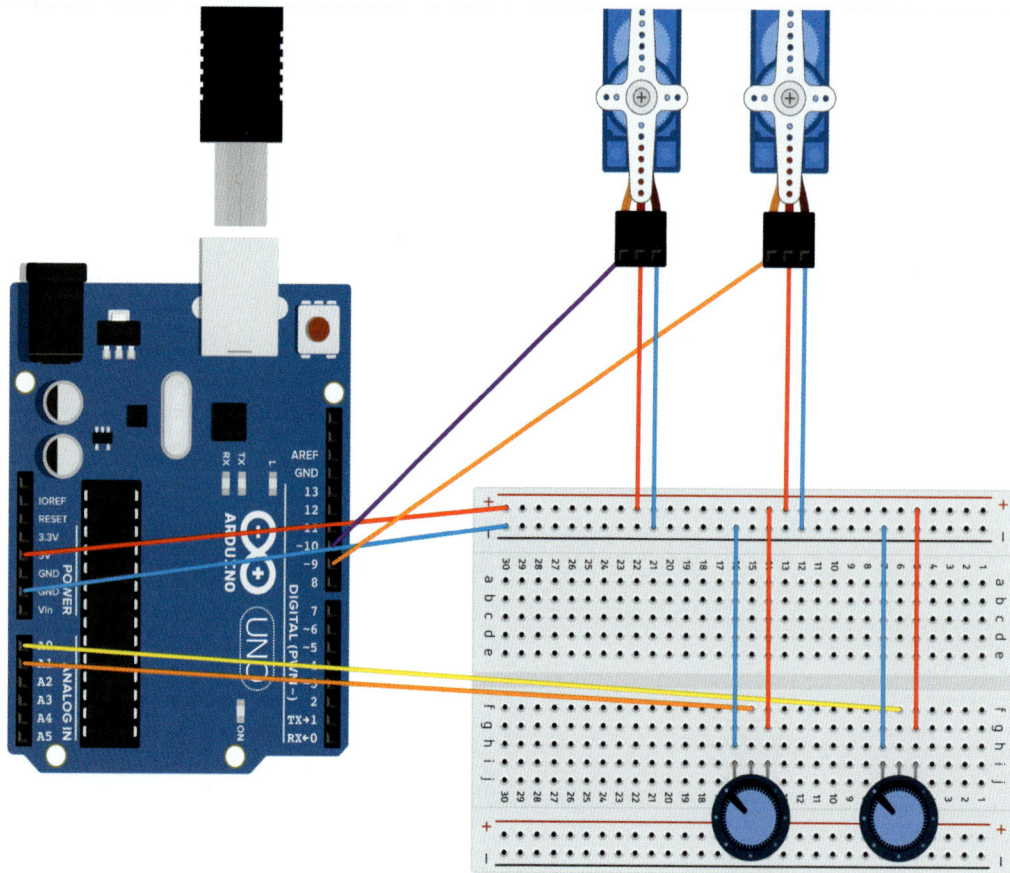

```
1  //Project 6
2  //Controlling Servos with 10k potentiometers
3  //thearduinoclassroom.com
4  //Copyright 2019, Isabel Mendiola and Peter Haydock
5
6  #include <Servo.h>  //Library for servos
7
8  Servo myservo1;
9  Servo myservo2;
10
11 int potpin1 = A0; //Analog connection to 10K potentiometer
12 int potpin2 = A1;
13 int val1;
14 int val2;
15 void setup()
16 {
17 myservo1.attach(9); //Digital connection to servos
18 myservo2.attach(10);
19 }
20 void loop()
21 {
22 val1 = analogRead(potpin1);
23 val2 = analogRead(potpin2);
24
25 val1 = map(val1, 0, 1023, 0, 179); //Add values for servo motion
26 val2 = map(val2, 0, 1023, 0, 179);
27
28 myservo1.write(val1);
29 myservo2.write(val2);
30
31 delay(15);
32 }
```

- the cathode(-) column to the cathode(-) connection for both servos. Two jumper wires
- digital pin nine (9) to the data connection for one servo
- digital pin 10 to the data connection for the other servo

Coding - Digital Prototype

8 Enter the code in Tinkercad® to match the code on page 71.

9 Start the simulation. Rotate the potentiometers and observe the digital prototype.

Project 6
Spin It Up

```
#include <Servo.h>  //Library for servos
Servo myservo1;
Servo myservo2;
int potpin1 = A0; //Analog connection to 10k
potentiometer
int potpin2 = A1;
int val1;
int val2;
void setup()
{
    myservo1.attach(9); //Digital connection to
servos
    myservo2.attach(10);
}
void loop()
{
   val1 = analogRead(potpin1);
   val2 = analogRead(potpin2);
   val1 = map(val1, 0, 1023, 0, 179); //Add
values for servo motion
   val2 = map(val2, 0, 1023, 0, 179);
   myservo1.write(val1);
   myservo2.write(val2);
   delay(15);
}
```

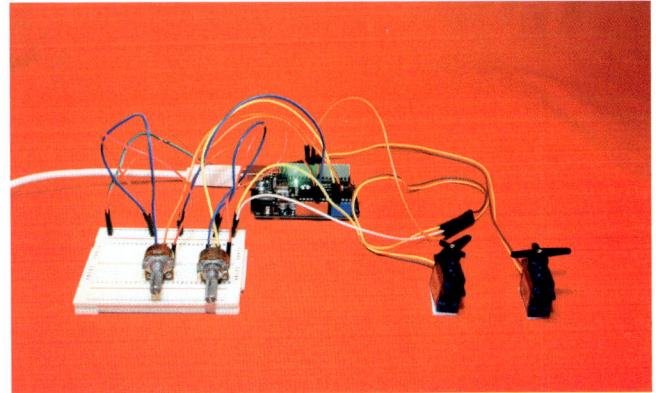

Connection - This 3-D printed prosthetic hand can be coupled with an UNO board, servos, and sensors to replace a hand lost due to an injury, accident, or other trauma.

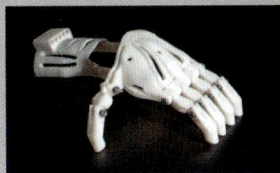

Scientists and engineers place the UNO microcontroller in the prosthetics's arm section to receive input from electronics attached to the living part of the arm. The signals from the electronics are used to control servos in the robotic hand simulating a real hand's function. Sensors in the finger tips control the amount of pressure applied to a grasped object. This allows the robotic hand to differentiate between objects like a rock and an egg.

Engineering Design - Project Build

10 Assemble the servos. Use a Phillips head screwdriver to place the armature on the motor.

11 Place two (2) 10k potentiometers on the breadboard in the same column.

12 Connect with jumper wires:
- a ground pin on the UNO board to the cathode(-) column
- the 5v pin on the UNO board to the anode(+) column
- the anode(+) column to the same row as the anode(+) lead for both potentiometers. Two jumper wires
- the cathode(-) column to the same row as the cathode(-) lead for both potentiometers. Two jumper wires
- analog pin zero (A0) to the same row as the input lead (center) of one potentiometer
- analog pin one (A1) to the same row as the input lead (center) of the other potentiometer
- the anode(+) column to the anode(+) connection for both servos. Two jumper wires
- the cathode(-) column to the cathode(-) connection for both servos. Two jumper wires
- digital pin nine (9) to the data connection for one servo
- digital pin 10 to the data connection for the other servo

Connection - The servos in this robotic arm and hand are used to hold and place parts in the manufacturing of different machines.

This arm and hand combination is at the National Institute of Standards and Technology (NIST) as a part of research on advanced material handling and manufacturing. NIST is at the forefront of researching different attachments to the arm for industrial use.

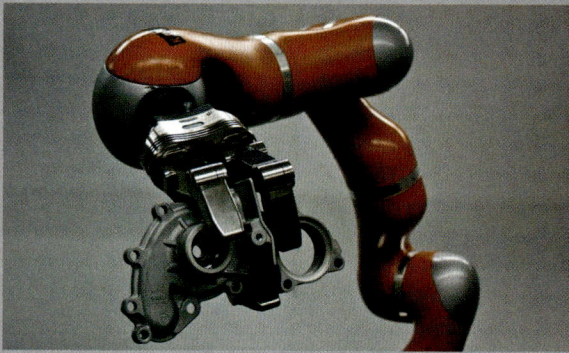

```
//Project 6
//Controlling Servos with Potentiometers
//thearduinoclassroom.com
//Copyright 2019, Isabel Mendiola and Peter Haydock

#include <Servo.h>  //Library for servos
Servo myservo1;
Servo myservo2;
int potpin1 = A0; //Analog connection to 10K potentiometer
int potpin2 = A1;
int val1;
int val2;
void setup()
{
    myservo1.attach(9); //Digital connection to servos
    myservo2.attach(10);
}
void loop()
{
    val1 = analogRead(potpin1);
    val2 = analogRead(potpin2);
    val1 = map(val1, 0, 1023, 0, 175); //Add values for servo motion
    val2 = map(val2, 0, 1023, 0, 175);
    myservo1.write(val1);
    myservo2.write(val2);
    delay(15);
}
```

13 Launch the Arduino® IDE software (pages 11-13) to make sure the board is communicating with the computer.

Go to the Tools menu and verify that the correct board is selected from the Board Manager menu. If not, select the correct board from the options listed.

Then confirm that the right Port is selected. If not, select the port that lists the UNO board from the options listed.

Coding - Project Build

14 Either copy and paste the code from the Tinkercad® prototype or type the code on page 71 into the Arduino® IDE software (pages 11-13). The library for this project needs to be installed. See the book web page for the link to the servo file and page 62 on how to install the library.

15 Save the sketch (rename as needed)

16 Verify the sketch.

17 Upload the sketch to the UNO board.

18 Rotate the potentiometers and observe the motion of each.

19 Document this project, discuss the project, and complete the extensions assigned from the next page. ✓

Connection - Canadarm2 also know as the Mobile Servicing System is on the International Space Station (ISS). It provides support for repairs to the ISS, helps with space walks by the astronauts (shown at bottom), and is also used to capture and launch satellites or resupply ships. The Canadarm2 has several servo joints to provide the maximum reach, flexibility and rotation in movement.

In this photo, the Canadian-built Dextre is attached to the end of Canadarm2. Dextre is also known as the Special Purpose Dextrous Manipulator (SPDM) and is used for holding satellites while they are repaired and launched.

Photo Credits: NASA

Project 6
Spin It Up

Project Reflection and Summative Activities, Discussion Starters, Extensions, and Problem

A = Essential

Reflection/Summative Activities:
- Have each team member document in their project/classroom journal who was on their team, what went well, what they could improve upon, what they would do differently if they were to do the project again, and verify that every person in the group can do the project.

- Collaboratively create signage for the classroom that documents the preparatory steps needed for each project.

For Discussion:
- What purpose does the line of code with the "map" statement serve?

- Compare and contrast a servo with a hydraulic piston?

Extensions:
- Adjust the map statement to change the range of motion of each servo.

- Attach a drinking straw to one end of each servo and edit the code to wave the straw quickly back and forth.

B = Recommended

Professional Connection:
- Research and write a short description of a profession that uses servos in your area. Be sure to include the education the profession requires, any laws or regulations that guide their use, and what activities are performed.

Extensions:
- Edit the sketch to create a device using one of the output devices from Projects Two through Five to indicate a condition of the servo (ex. light a red LED when the servo rotates to its maximum clockwise position).

- Submit your project to the book page on the website. Register and Login to submit at: thearduinoclassroom.com

From the homepage:
Click
    Books
Click
    UNO Edition Vol. 1

C = Optional

Extension:
- Using the servos, design and implement a series of movements to rotate through at least 5 positions both going clockwise and counter clockwise.

Problem:
- Identify a problem that requires a servo and create the solution. Include in your report the following:

Description of the problem.

Document how the problem is solved by this project.

What other resources are needed to solve the problem.

Design and produce the light display.

A project rubric can be found on page 236.

Anticipatory Sets
(from page 68)

Anticipatory Sets:
- Where might you see servos used? List at least three different places or situations that use this technology, describe the device that contains the servos, and explain the purpose of the servos. *Responses may include toys, automatic door openers, robots*

- What are the advantages and disadvantages of designing a machine with a servo? *Servos may wear out over time. Servos can turn rotational motion into linear motion with forward and backward motion.*

A

Reflection/Summative Activity: *Responses Will Vary*

For Discussion:
- What purpose does the line of code with the "map" statement serve? *The line of code maps the input value of the potentiometer to the output value of the servo. It turns the number from a value of 0 to 1023 to a value of 0 to 179.*

- Compare and contrast a servo with a hydraulic piston? *Servos use electrical energy to create rotational motion in a motor to drive the axle which can be converted to a linear or partial rotational motion. The hydraulic piston uses Pascal's Principle to move liquid in one volume which in turn moves liquid in another volume. The movement of the liquid can be translated into linear motion.*

Extensions:
- Adjust the map statement to change the range of motion of each servo. *Responses Will Vary*

- Attach a drinking straw to one end of each servo and edit the code to wave the straw quickly back and forth. *Responses Will Vary*

B

Professional Connection: *Responses Will Vary*

Extensions:
- Edit the sketch to create a device using one of the output devices from Projects Two through Five to indicate a condition of the servo (ex. light a red LED when the servo rotates to its maximum clockwise position). *Responses Will Vary*

- Submit your project to the book page on the website. *Responses Will Vary*

C

Extension:
- Using the servos, design and implement a series of movements to rotate through at least 5 positions both going clockwise and counter clockwise. *Responses Will Vary*

Problem: *Responses Will Vary*

Project 7
Count on Me

Lesson Integration:	Groupings:	Level:	Time to Complete:
Physics - Ohm's Law, Circuits, Time Mathematics - Counting, Cycles, Periodicity	1 - 2 Change Groupings	Starter	30 min. for the project and 45 min. for extensions.

Objectives:
- Build a device that can count from 0 to 9999 on a seven-segment display.
- Build a digital design and an Arduino® Prototype.

Prerequisite Skills:
- Time measurement (milliseconds)
- Arduino® IDE software (pages 11-13)
- Fritzing (pages 64 -66)
- Libraries (pages 62)

Purpose and Skills:
- Control a four digit seven segment LED

STEAM Connections:
Science - Metric time, Circuits
Technology - Code, Program Settings, Simulators, Digital Design, Electronic Components
Engineering - Designing, Building, and Using a Machine, Prototyping, Applied Physics
Allied Arts - Patterns
Math - Counting

Key Vocabulary:
N/A

Project Introduction:
- Introduce the groups to the purpose of the project, the skills developed, the standards met, and the goal of the project.

Anticipatory Sets:

- Where can you see the use of number displays? List at least three different places or situations where this technology is used, describe the devices, and explain the purpose of the devices.

- What kinds of controls and wiring might be involved to work a seven segment display?

Project 7
Count on Me

Educational Standards

ISTE Standards for Students
- Empowered Learner 1a, 1b, 1d
- Knowledge Constructor 3a, 3b, 3d
- Innovative Designer 4a, 4b, 4c, 4d
- Computational Thinker 5a, 5b, 5c, 5d
- Creative Communicator 6b, 6c, 6d
- Global Collaborator 7c

US Computer Science Standards
- Project correlations can be found on the book's web page.

US NGSS - Middle School
DCIs
- MS-PS4-1 PS4.A
- MS-PS4-2 PS4.A PS4.B
- MS-PS4-3 PS4.C

Cross Cutting Concepts
- Cause and Effect
- Scale, Proportion and Quantity
- Structure and Function
- Patterns

Science and Engineering Practices
- Planning and Carrying Out Investigations
- Obtaining, Evaluating, and Communicating Information
- Analyzing and Interpreting Data
- Scientific Knowledge is Based on Empirical Evidence
- Using Mathematics and Computational Thinking

US NGSS - High School
- Correlations can be found on the book's web page.

US Common Core Language Arts and Mathematics
- Correlations can be found on the book's web page.

Step-by-Step 1-20

Materials List:
- Computer with IDE software
- Connection to the Internet
- USB Cable
- UNO or UNO Compatible Microcontroller
- Short Breadboard
- Four (4) Digit Seven (7) Segment Display (14 pin)
- 12 Jumper Wires (Male to Male)

Engineering Design - Digital Prototype

1. Launch Fritzing (pages 64 - 66).

2. Click on "Breadboard."

3. Drag an UNO board to the workspace.

4. Drag a breadboard to the workspace.

5. Drag one four digit seven segment LED to the breadboard such that the leads cross over halves of the breadboard. Set the LED to cathode(-). Note that the bottom of the LED has the "dot."

6. Connect the following with jumper wires across the top of the LED:
   - digital pin 13 to the same row as pin 14 on the LED
   - digital pin 12 to the same row as pin 13 on the LED
   - digital pin 11 to the same row as pin 12 on the LED
   - digital pin 10 to the same row as pin 11 on the LED
   - digital pin nine (9) to the same row as pin 10 on the LED
   - digital pin two (2) to the same row as pin 9 on the LED

Pins on the LED display are numbered counter-clockwise from the pin on the front (with the part number) left around to the back left. For the digital prototype and project build, the LED has 14 pins. Pins seven (7) and eight (8) on the right end are not used.

7 Connect the following with jumper wires across the bottom of the LED:

- digital pin eight (8) to the same row as pin one (1) on the LED
- digital pin seven (7) to the same row as pin two (2) on the LED
- digital pin six (6) to the same row as pin three (3) on the LED
- digital pin five (5) to the same row as pin four (4) on the LED
- digital pin four (4) to the same row as pin five (5) on the LED
- digital pin three (3) to the same row as pin six (6) on the LED

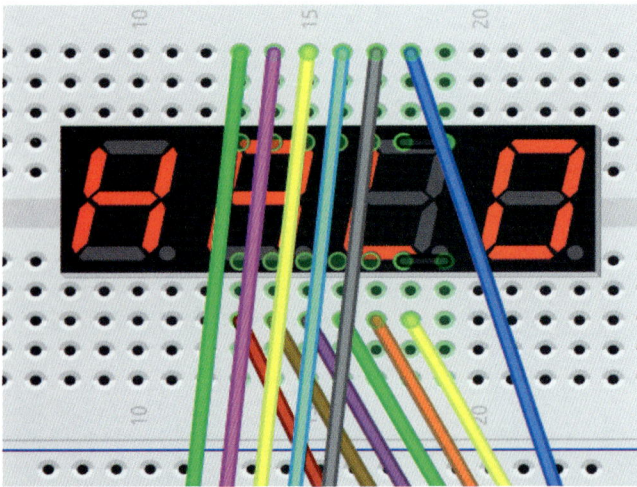

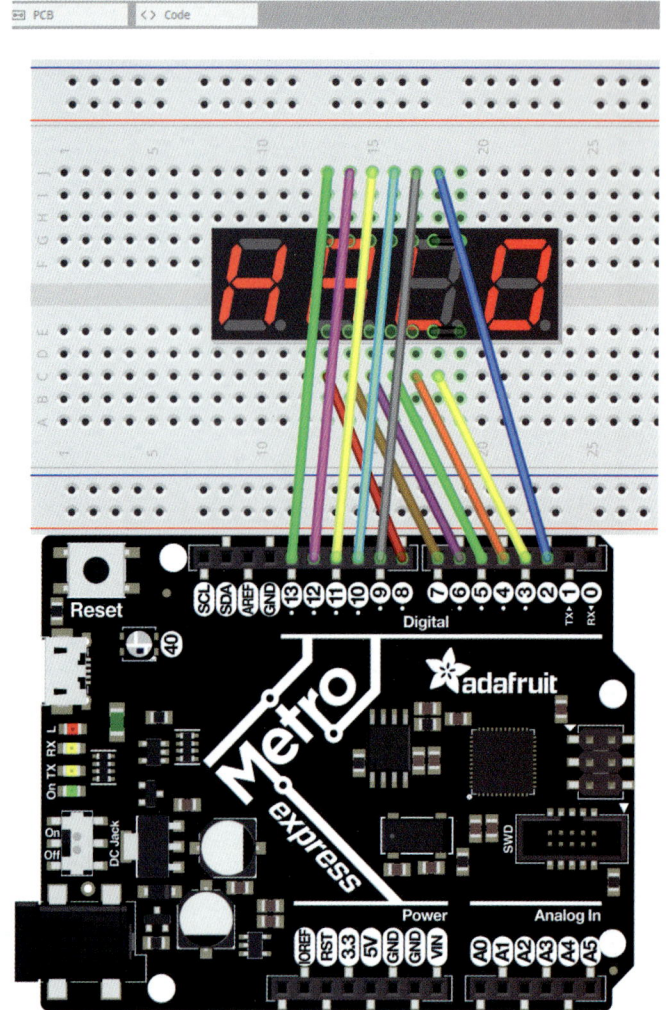

Connection - NASA uses a large clock to show the mission time at Cape Canaveral.

Photo credit: NASA/Jim Grossmann

Coding - Digital Prototype

8 Optional. Enter the code in Tinkercad® to match the code on this page. The library is already installed in Tinkercad®.

9 Verify the code in Tinkercad.

Engineering Design - Project Build

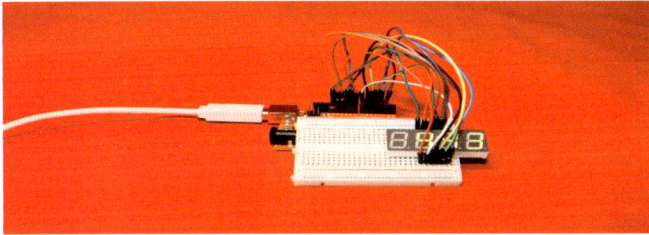

10 Place the seven segment LED on the breadboard such that the leads cross over the two halves of the breadboard.

11 Connect the following with jumper wires across the top of the LED:
- digital pin 13 to the same row as pin 14 on the LED
- digital pin 12 to the same row as pin 13 on the LED
- digital pin 11 to the same row as pin 12 on the LED
- digital pin 10 to the same row as pin 11 on the LED
- digital pin nine (9) to the same row as pin 10 on the LED
- digital pin two (2) to the same row as pin 9 on the LED

```
#include <SevSeg.h>
SevSeg sevseg; //Library for the 4 digit segment display
void setup()
{
  byte numDigits = 4;
  byte digitPins[] = {13,10,9,3}; // The four digits of the segment display
  byte segmentPins[] = {12,2,5,7,8,11,4,0}; // Segment display lines in alphabetical order.
  byte hardwareConfig = COMMON_CATHODE;
  sevseg.begin(hardwareConfig, numDigits, digitPins, segmentPins);
  sevseg.setBrightness(30);
}
//Number displayed
int num=0;
void loop()
{
  static int loop = 0; //Delay Loop Start
  loop++;
  if ( loop > 1000  ) {
   num++; //Add 1 to LED counter
   loop = 0; //Reset Delay
  }
  if ( num > 9999 ) // Reset count at 10,000
    num = 0;
  sevseg.setNumber(num,0);
  sevseg.refreshDisplay();
}
```

12 Connect the following with jumper wires across the bottom of the LED:
- digital pin eight (8) to the same row as pin one (1) on the LED
- digital pin seven (7) to the same row as pin two (2) on the LED
- digital pin six (6) to the same row as pin three (3) on the LED
- digital pin five (5) to the same row as pin four (4) on the LED
- digital pin four (4) to the same row as pin five (5) on the LED
- digital pin three (3) to the same row as pin six (6) on the LED

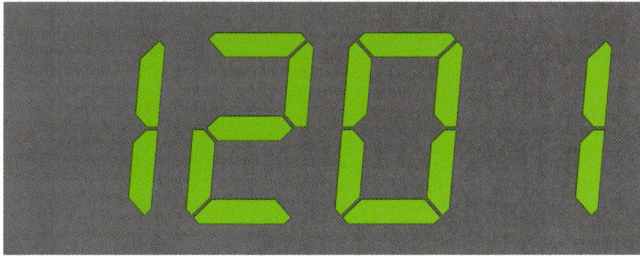

13 Connect the UNO board and computer using the USB cable.

14 Launch the Arduino® IDE software (pages 11-13) to make sure the board is communicating with the computer.

Go to the Tools menu and verify that the correct board is selected from the Board Manager menu. If not then select the correct board from the options listed.

Then confirm that the right Port is selected. If not then select the port that lists the UNO board from the options listed.

Coding - Project Build

15 Either copy and paste the code from the Tinkercad® prototype or type the code from page 79 into the Arduino® IDE software (pages 11-13). The library for this project needs to be installed. See the book web page for the link to the SevenSeg file and page 62 on how to install the library.

16 Save the sketch (rename as needed).

17 Verify the sketch.

18 Upload the sketch to the UNO board.

19 Observe the seven segment display on the breadboard.

20 Document this project, discuss the project, and complete the extensions assigned from page 82.

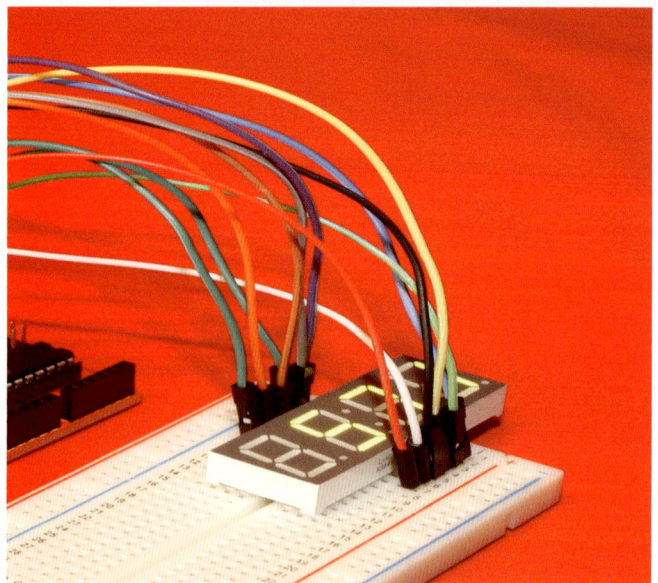

Connection - Measuring 2.9 x 12.19 x 2.44 meters, this is "Meantime" by artist Darren Almond. The artist had the clock shipped across the Atlantic Ocean in 2000 on a cargo ship. The artwork was photographed in Venice, Italy and New York City.

Mike Smith Studio in London England fabricated this 24 hour analog seven segment display clock which included a speaker to amplify the sounds the machinery made inside the shipping container.

The photos here show (clockwise from the top f the page) the finished work on display, the fabrication of the seven segment elements and the loading of the work on the first voyage it made by cargo ship.

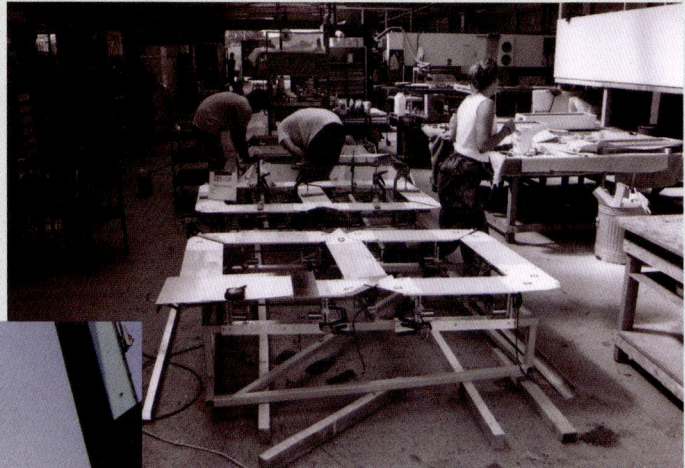

Photo Credits: Mike Smith Studio

Project 7
Count on Me

Project Reflection and Summative Activities, Discussion Starters, Extensions, and Problem

A = Essential

Reflection/Summative Activities:

- Have each team member document in their project/classroom journal who was on their team, what went well, what they could improve upon, what they would do differently if they were to do the project again, and verify that every person in the group can do the project.

- As a team create a graphic organizer summarizing the skills and knowledge obtained in doing this project.

For Discussion:

- Why is it called a four digit seven segment LED?

- List the limitations of using an seven segment LED for displaying characters.

Extension:

- Using the four digit seven segment LED, design, code and build an accurate timer (in seconds).

B = Recommended

Extensions:

- Reverse the counting so that the numbers appear in order from 9999 to zero (0).

- Edit the code to only display a number entered in the code from an equation, variable, constant, or other input. Implement the extension.

C = Optional

Extensions:

- Add or edit lines of code to perform a countdown and then count up. Implement the extension.

- Edit the code to display the position of a potentiometer or servo. Implement the extension.

- Submit your project to the book page on the website. Register and Login to submit at: thearduinoclassroom.com

From the homepage:
Click
Books
Click
UNO Edition Vol. 1

A project rubric can be found on page 236.

Anticipatory Sets
(from page 76)

Anticipatory Sets:
- Where can you see the use of number displays? List at least three different places or situations where this technology is used, describe the devices, and explain the purpose of the devices.
Places like digital clocks, timers, crosswalks, raceways,

- What kinds of controls and wiring might be involved to work a seven segment display?
Each segment has to be controlled as well as input and output current for the circuit.

A

Reflection/Summative Activities:
Responses Will Vary

For Discussion:
- Why is it called a seven segment LED?
The LED has seven dashed lines that can individually turned on to make characters.

- List the limitations of using an seven segment LED for displaying characters.
Only four characters at a time can be made. There are no diagonal segments to make letters like "X"s or "Z"s

Extension:
- Using the four digit seven segment LED, design, code and build an accurate timer (in seconds).
Replace the delay value to be "1000."

B

Extensions:
- Reverse the counting so that the numbers appear in order from 9999 to zero (0).
Responses Will Vary

- Edit the code to only display a number entered in the code from an equation, variable, constant, or other input. Implement the extension.
Responses Will Vary

C

Extension:
- Add or edit lines of code to perform a countdown and then count up. Implement the extension.
Responses Will Vary

- Edit the code to display the position of a potentiometer or servo. Implement the extension.
Responses Will Vary

- Submit your project to the book page on the website.
Responses Will Vary

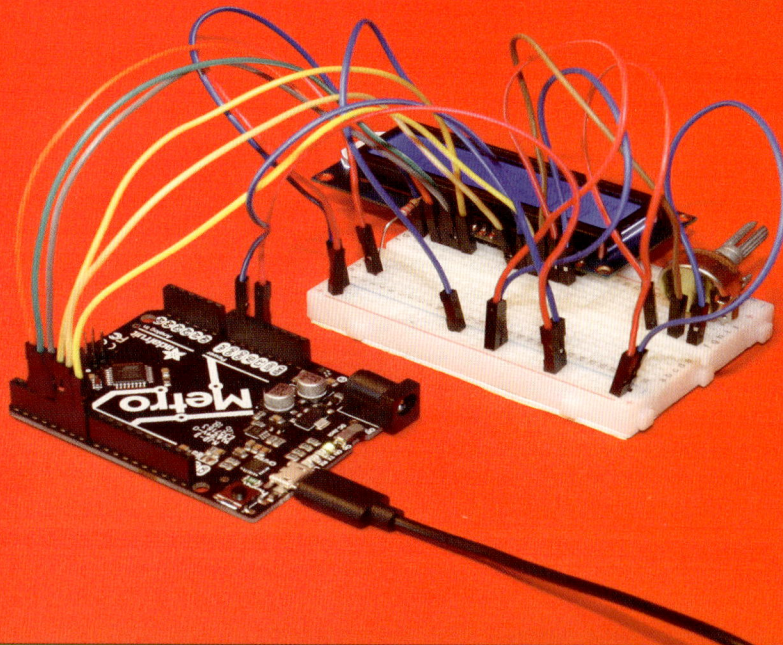

Getting Started

Lesson Integration:	Groupings:	Level:	Time to Complete:
Physics - Ohm's Law, Circuits, Voltage, Resistance Mathematics - Ratios	1 - 2	Intermediate	45 min. for the project and 45 min. for extensions.

Objectives:
- Investigate how circuits and electronic components interact with electrical energy.
- Control an LCD in a prototype and machine with a potentiometer.

Prerequisite Skills:
- Time measurement (milliseconds)
- Arduino® IDE software (pages 11-13)
- Tinkercad® (pages 15-16)
- Libraries (Page 62)

STEAM Connections:
Science - Circuits, Resistance, Ohm's law
Technology - Code, Program Settings, Simulators, Digital Design, Electronic Components, LCD, Potentiometer, Analog Data, Communications
Engineering - Designing, Building, and Using a Machine, Prototyping, Applied Physics
Allied Arts - Communication, Fonts
Math - Position Coordinates

Purpose and Skills:
- Control an LCD display
- Change displayed text to create a sign

Key Vocabulary:
LCD - Liquid Crystal Display. A device that applies a small current to a small area of liquid crystal to make the liquid opaque.

Project Introduction:
- Introduce the groups to the purpose of the project, the skills developed, the standards met, and the goal of the project.

Anticipatory Sets:
- Where would you see an LCD used? List at least three different places or situations where his technology is used, describe the devices, and explain the purpose of the devices.

Project 8
LCD Billboard

Educational Standards

ISTE Standards for Students
- Empowered Learner 1a, 1b, 1d
- Knowledge Constructor 3a, 3b, 3c, 3d
- Innovative Designer 4a, 4b, 4c, 4d
- Computational Thinker 5a, 5b, 5c, 5d
- Creative Communicator 6a, 6b, 6c, 6d
- Global Collaborator 7c, 7d

US Computer Science Standards
- Project correlations can be found on the book's web page.

US NGSS - Middle School
DCIs
- MS-PS4-3 PS4.C

Cross Cutting Concepts
- Cause and Effect
- Scale, Proportion and Quantity
- Structure and Function

Science and Engineering Practices
- Planning and Carrying Out Investigations
- Obtaining, Evaluating, and Communicating Information
- Analyzing and Interpreting Data
- Scientific Knowledge is Based on Empirical Evidence
- Using Mathematics and Computational Thinking

US NGSS - High School
- Correlations can be found on the book's web page.

US Common Core Language Arts and Mathematics
- Correlations can be found on the book's web page.

Materials List:
- Computer with IDE software
- Connection to the Internet
- USB Cable
- UNO or UNO Compatible Microcontroller
- Short Breadboard
- 10k Potentiometer
- 16x2 LCD
- 16 Jumper Wires (Male to Male)

The LCD may require some soldering. Use caution when soldering as the tip of the iron is hot and can burn. The soldering should be done by or under the supervision of a responsible adult.

Engineering Design - Digital Prototype

1 Launch and log into Tinkercad® (pages 15-16).

2 Start a new Circuits project.

3 Drag an UNO board to the workspace.

4 Drag a breadboard to the workspace.

5 Drag one 10k potentiometer to one end of the breadboard such that the leads span three rows in a column.

6 Place a 16x2 LCD on the other end of the breadboard (same half) as the potentiometer leaving two columns exposed for jumper wires.

7 Place a 220 Ohm resistor on the breadboard such that it connects across the halves of the breadboard from the row that has the second LED pin from the LCD to the same row in the other half.

9 Connect with jumper wires the following:
Use the technical diagrams on the following page to assist in jumper wire placement.
- the five volt (5v) pin on the UNO board to the anode(+) column on the opposite half of the breadboard from the LCD

The Arduino Classroom § 85

- a ground (GND) pin on the UNO board to the cathode(-) column on the opposite half of the breadboard from the LCD and potentiometer
- from the anode(+) column to a pin in the same row as the anode(+) lead on the potentiometer
- from the cathode(-) column to a pin in the same row as the cathode(-) lead of the potentiometer
- from the same row as the center pin of the potentiometer to a pin in the same row as the VO lead of the LCD
- from the anode(+) column to a pin in the same row as the VCC lead of the LCD
- from the cathode(-) column to a pin in the same row as the GND lead of the LCD
- from digital pin seven (7) on the UNO board to a pin in the same row as the RS lead of the LCD
- from the cathode(-) column to a pin in the

- same row as the RW lead of the LCD
- from digital pin six (6) on the UNO board to a pin in the same row as the E lead of the LCD
- from digital pin five (5) on the UNO board to a pin in the same row as the DB4 lead of the LCD
- from digital pin four (4) on the UNO board to a pin in the same row as the DB5 lead of the LCD
- from digital pin three (3) on the UNO board to a pin in the same row as the DB6 lead of the LCD
- from digital pin two (2) on the UNO board to a pin in the same row as the DB7 lead of the LCD
- from the anode(+) column to a pin in the same row as the resistor
- from the cathode(-) column to a pin in the same row as the first LED lead of the LCD ✓

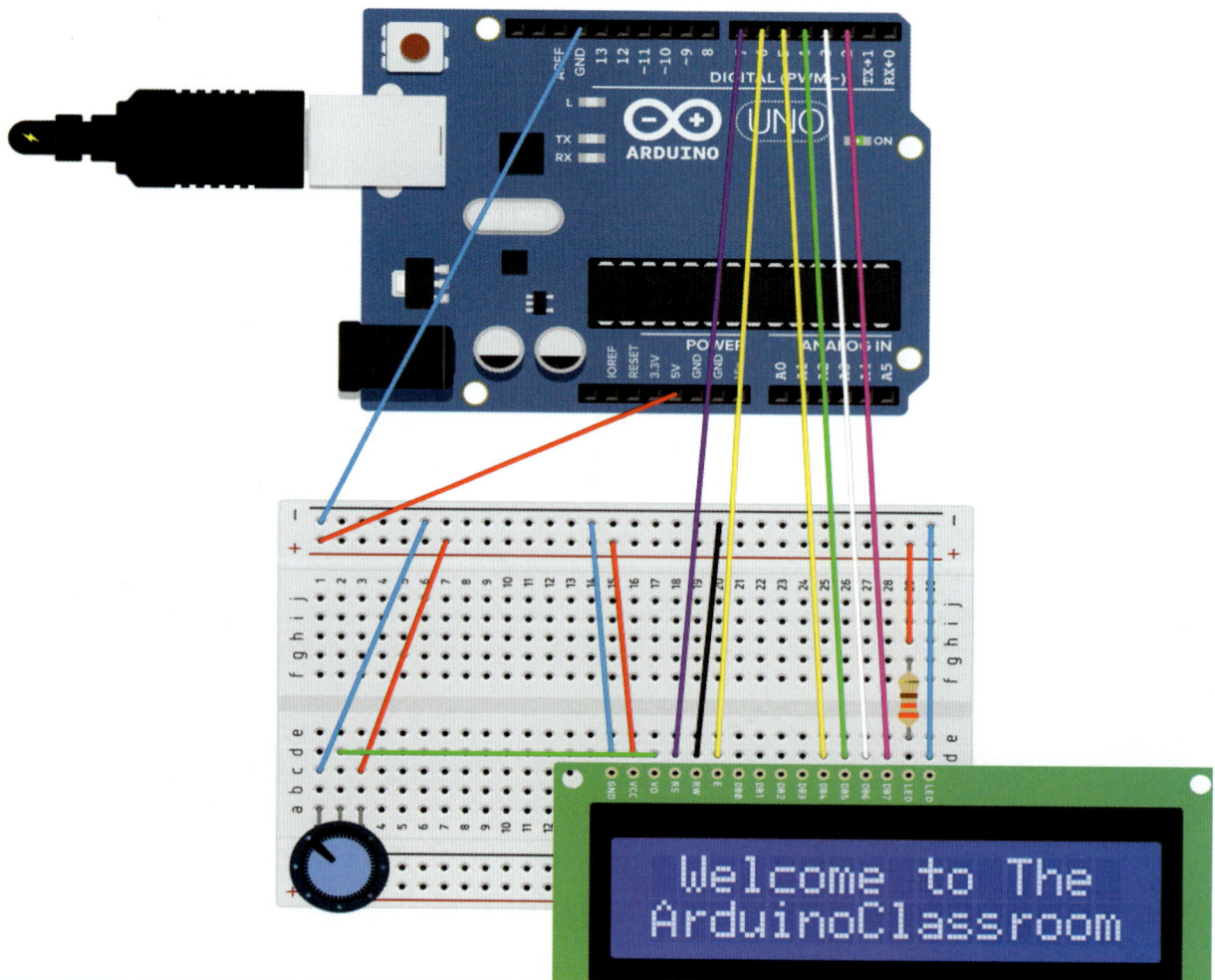

Coding - Digital Prototype

10 Enter the code in Tinkercad® to match the code on this page. Note that this code uses a library.

11 Start the simulation and observe the digital prototype. ✓

Engineering Design - Project Build

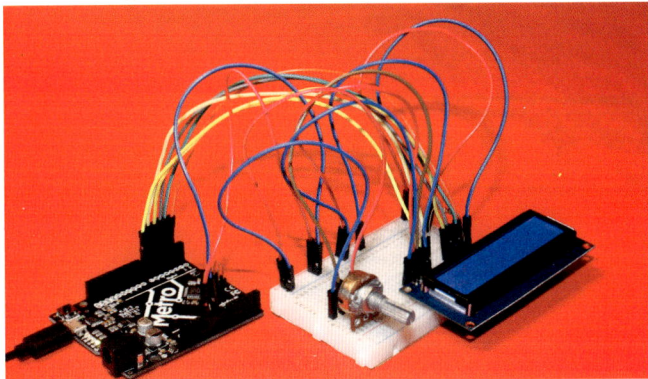

12 Place the potentiometer on one end of the breadboard spanning three rows in a column.

13 Place a 16x2 LCD on the other end but on the same half of the breadboard leaving two columns exposed for jumper wires.

14 Place a 220 Ohm resistor on the breadboard such that it connects across the halves of the breadboard from the row that has the second LED pin from the LCD to the same row in the other half.

15 Use the tables on this page to place the jumper wires. For additional help, use step-by-step number 9 and the diagram on the previous pages.

```
#include <LiquidCrystal.h> //add library for reading LCD
LiquidCrystal lcd(7, 6, 5, 4, 3, 2);// set microcontroller pins

void setup() {
  lcd.begin(16, 2);
}
void loop() {
  lcd.setCursor(1, 0);
  lcd.print("Welcome to The");// add the text to the first line of the LCD
  delay (1000); // time in milliseconds for displaying the text in the second line.
  lcd.setCursor(0, 1);
  lcd.print("ArduinoClassroom");  // add the text to the second line of the LCD}
}
```

From	To (LCD L to R)
Breadboard Cathode - Column	LCD GND (Pin 1) Same Row
Breadboard Anode + Column	LCD VCC (Pin 2) Same Row
Potentiometer Center Pin Same Row	LCD VO (Pin 3) Same Row
UNO Digital Pin D7	LCD RS (Pin 4) Same Row
Breadboard Cathode - Column	LCD RW (Pin 5) Same Row
UNO Digital Pin D6	LCD E (Pin 6) Same Row
UNO Digital Pin D5	LCD DB4 (Pin 11) Same Row
UNO Digital Pin D4	LCD DB5 (Pin 12) Same Row
UNO Digital Pin D3	LCD DB6 (Pin 13) Same Row
UNO Digital Pin D2	LCD DB7 (Pin 14) Same Row
Breadboard Anode + Column	Breadboard (Pin 15) Same Row with Resistor
Breadboard Cathode - Column	LCD LED (Pin 16) Same Row

From	To
UNO 5v	Breadboard Anode + Column
UNO GND	Breadboard Cathode- Column
Breadboard Anode + Column	Potentiometer Anode + Same Row
Breadboard Cathode - Column	Potentiometer Cathode - Same Row

16 Review the simplified circuit diagram for this project.

The circuit diagram here focuses on the 5v and ground connections. The potentiometer controls the brightness of the display through the VO input. The other 11 jumper wires are data connections for the display. These connections only carry electricity when data is sent.

Also note that the end connections on the LCD are ground connections while the second from each end is a 5v connection.

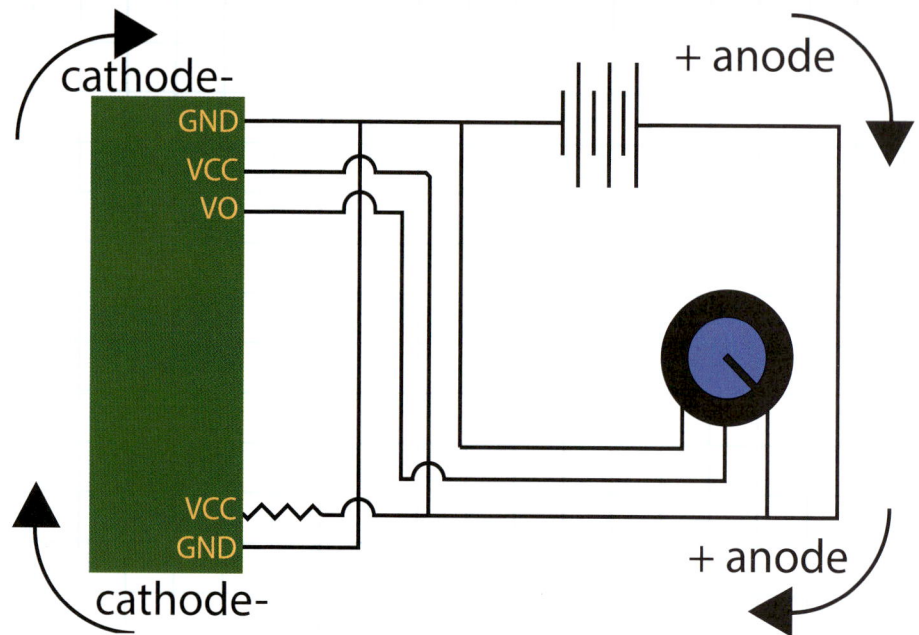

cathode-

GND
VCC
VO

+ anode

VCC
GND

cathode-

+ anode

17 Connect the UNO board and computer using the USB cable.

18 Launch the Arduino® IDE software (pages 11-13) to make sure the board is communicating with the computer.

Coding - Project Build

19 Either copy and paste the code from the Tinkercad® prototype or type the code from page 87 into the Arduino® IDE software (pages 11-13). The LCD library is part of the base IDE software so the sketch should work without any extra downloads.

20 Save the sketch (rename as needed).

21 Verify the sketch.

22 Upload the sketch to the UNO board.

23 Observe the LCD on the breadboard. Make sure to adjust the potentiometer to find the best setting to see the text on the LCD. Document these settings.

24 Document this project, discuss the project, and complete the extensions assigned from the next page.

```
TAC_UE_V1_P8_LCD_Billboard | Arduino 1.8.9 (Windows Store 1.8.21.0)
File Edit Sketch Tools Help

  TAC_UE_V1_P8_LCD_Billboard
//Project 8
//LCD Billboard
//thearduinoclassroom.com
//Copyright 2019, Isabel Mendiola and Peter Haydock

#include <LiquidCrystal.h> //add library for reading LCD
LiquidCrystal lcd(7, 6, 5, 4, 3, 2);// set microcontoller pins

void setup() {
  lcd.begin(16, 2);
}
void loop() {
  lcd.setCursor(1, 0);
  lcd.print("Welcome to The");;// add the text to the first line of the LCD
  delay (1000); // time in milliseconds for displaying the text in the second line.
  lcd.setCursor(0, 1);
  lcd.print("ArduinoClassroom");  // add the text to the second line of the LCD)
}

Done Saving.
```

Connection - LCD and LED signs are easy to install, maintain, program, and most importantly can be seen in a wide variety of lighting conditions. Public transportation systems like the Washington D.C. Metro, the San Francisco BART and the Italian national train system all use LCD and LED signs to communicate with passengers. As seen in this photo, trains in the Firenze (Florence), Italy Santa Maria Novella station are waiting to depart. LED screens showing train information like train number, departure time, and track number guide the 3 million visitors to Firenze through the station each year.

Project 8
LCD Billboard

Project Reflection and Summative Activities, Discussion Starters, Extensions, and Problem

A = Essential	B = Recommended	C = Optional

A = Essential

Reflection/Summative Activities:

- Have each team member document in their project/ classroom journal who was on their team, what went well, what they could improve upon, what they would do differently if they were to do the project again, and verify that every person in the group can do the project.

- As a class, create a PowerPoint presentation summarizing the class' experience with the project. Include the observations of the rotation of the potentiometer.

For Discussion:

- What is the function of the potentiometer in this project?

- What are the advantages, disadvantages in using the LCD as compared to the seven segment LED?

- What limitations are there for the 16x2 LCD?

- This project can be made energy independent from the computer using a 9v battery and the barrel connector. When would this configuration be used and why?

Extensions:

- Change the text that appears on the LCD.

- Adjust where the new text starts on the LCD.

B = Recommended

Technology Connections:

- Create a list of places you would see an LCD display used like the ones shown in this project.

- Research to compare and contrast how an LCD and an LED function.

Extension:

- Use the following code to move the text across the LCD screen. Hint - place at the start of the void loop().

```
{
  for (int positionCounter
= 0; positionCounter < 13;
positionCounter++) {
  lcd.scrollDisplayLeft();
  delay (100);
  }
  for (int positionCounter
= 0; positionCounter < 29;
positionCounter++) {
  lcd.scrollDisplayRight();
  delay (100);
  }
  for (int positionCounter
= 0; positionCounter < 16;
positionCounter++) {
  lcd.scrollDisplayLeft();
  delay (100);
  }
```

- Submit your project to the book page on the website. Register and Login to submit at: thearduinoclassroom.com

From the homepage:
Click
   Books
Click
   UNO Edition Vol. 1

C = Optional

Extensions:

- Add a blinking LED or piezo buzzer to the project. Loop the sound or blinking with repeating words on the screen.

- Using a 9v battery and the barrel connector, make the LCD portable. HINT: Upload the code to the UNO board first using the USB cable. Detach the cable. Attach the barrel connector with the battery connected. The code will run immediately.

Problem:

- Identify a problem that would be solved by the LCD when powered by the 9v battery like the one you have been working with in this project. How could this project be a solution? Include in your report the following:

Description of the problem.

Description of how the problem is solved by this project.

What other resources you would need to solve the problem.

Design and produce the LCD solution.

A project rubric can be found on page 236.

Anticipatory Sets
(from page 84)

Anticipatory Sets:
- Where would you see a LCD used? List at least three different places or situations where this technology is used, describe the devices, and explain the purpose of the devices.
LCDs are used in places that need to convey information such as signs that change often as with sensor outputs or schedules.

A

Reflection/Summative Activities:
Responses Will Vary

For Discussion:
- What is the function of the potentiometer in this project?
To control the brightness of the LCD display.

- What are the advantages and disadvantages to using and LCD compared to the seven segment LCD?
The LCD display can show many more characters to include full words and short sentences. The display can be cleared to show even longer text. A disadvantage is the number of jumper wires to control the LCD.

- What limitations are there for this LCD?

This LCD does not have color.

Extensions:
- Change the text that appears on the LCD.
Responses Will Vary

Change where the new text starts on the LCD.
Responses Will Vary

B

Technology Connections:
- Create a list of places you would see an LCD display used like the one here.
Responses Will Vary

- Compare and contrast an LCD and an LED.
Responses Will Vary. Answers should that the LCD has a liquid crystal that when a voltage is applied it changes color where the LED turns on when the voltage is applied. LCDs do not emit light, where a LED can emit light.

Extension:
- Using the following code create the appearance of text moving across the LCD. Hint - place at the start of the void loop().
Responses Will Vary

- Submit your project to the book page on the website.
Responses Will Vary

C

Extensions:
- Add a blinking LED or piezo buzzer to the project. Loop the sound or blinking with repeating words on the screen.
Responses Will Vary

Problem:
Responses Will Vary

Project 9
Lights On / Lights Off

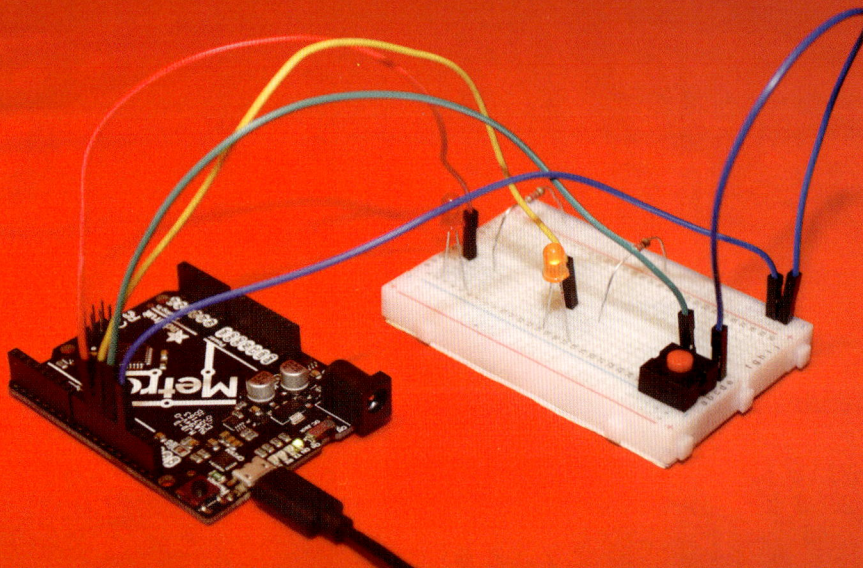

Getting Started			
Lesson Integration:	**Groupings:**	**Level:**	**Time to Complete:**
Physics - Ohm's Law, Circuits, Voltage, Resistance, Switches Mathematics - Ratios	1 - 2	Intermediate	30 min. for the project and 45 min. for extensions.

Objectives:
- Investigate how circuits and electronic components interact with electrical energy.
- Control lights in a prototype and machine with a switch.

Prerequisite Skills:
- Time measurement (milliseconds)
- Understanding of light spectrum
- Arduino® IDE software (pages 11-13)
- Tinkercad® (pages 15-16)

Purpose and Skills:
- Introduce the use of a pushbutton switch
- Use conditional programming (IF / ELSE)

STEAM Connections:
Science - Circuits
Technology - Code, Program Settings, Simulators, Digital Design, Electronic Components, LCD, Potentiometer, Analog Data, Communications, Safety Controls, Switches, Conditional Statements
Engineering - Designing, Building, and Using a Machine, Prototyping, Applied Physics

Allied Arts - Communication, Safety
Math - Ratios, Conversions

Key Vocabulary:
Pushbutton Switch- A device that when pushed created a closed circuit. When not pushed the circuit is open.
Switch - A part in an electrical circuit that through an action opens (disconnects) or closes (connects) a circuit.

Project Introduction:
- Introduce the groups to the purpose of the project, the skills developed, the standards met, and the goal of the project.

Anticipatory Sets:
- Where would you see a switch or pushbutton used? List at least three different places or situations where this technology is used, describe the devices, and explain the purpose of the devices.

- Describe what you think is happening to the electricity in a pushbutton?

Project 9
Lights On / Lights Off

Educational Standards

ISTE Standards for Students
- Empowered Learner 1a, 1b, 1d
- Knowledge Constructor 3a, 3b, 3c, 3d
- Innovative Designer 4a, 4b, 4c, 4d
- Computational Thinker 5a, 5b, 5c, 5d
- Creative Communicator 6b, 6c, 6d
- Global Collaborator 7c, 7d

US Computer Science Standards
- Project correlations can be found on the book's web page.

US NGSS - Middle School
DCIs
- MS-PS4-3 PS4.C
- MS-PS4-1 PS4.A

Cross Cutting Concepts
- Cause and Effect
- Scale, Proportion and Quantity
- Structure and Function

Science and Engineering Practices
- Planning and Carrying Out Investigations
- Obtaining, Evaluating, and Communicating Information
- Analyzing and Interpreting Data
- Scientific Knowledge is Based on Empirical Evidence
- Using Mathematics and Computational Thinking

US NGSS - High School
- Correlations can be found on the book's web page.

US Common Core Language Arts and Mathematics
- Correlations can be found on the book's web page.

Step-by-Step 1-23

Materials List:
- Computer with IDE software
- Connection to the Internet
- USB Cable
- UNO or UNO Compatible Microcontroller
- Short Breadboard
- Two (2) LEDs
- Two (2) 220 Ohm Resistors
- Five (5) Jumper Wires (Male to Male)
- Small Pushbutton

Engineering Design - Digital Prototype

1. Launch and log into Tinkercad® (pages 15-16).

2. Start a new Circuits project.

3. Drag an UNO board to the workspace.

4. Drag a breadboard to the workspace.

5. Place one small pushbutton to the breadboard such that the leads span the center of the breadboard. Look at the prototype technical diagram on this page for details on the placement.

6. Place two (2) LEDs on the breadboard so that each lead is in the same column

7. Place two (2) 220 Ohm resistors on the breadboard so that one lead is in the same row as each LED's cathode(-) lead and the other lead is in the cathode(-) column.

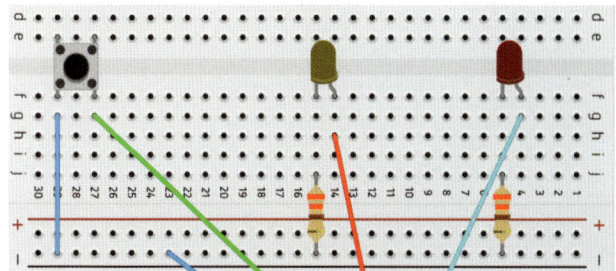

8 Connect with a jumper wires the following:
- Digital pin 10 to the same row as the anode(+) lead of one of the LEDs
- Digital pin 11 to the same row as the anode(+) lead of one of the LEDs
- Digital pin 12 to the same row as one of the pushbutton leads
- from the same row as the other pushbutton lead on the same half of the board to the cathode(-) column
- from the cathode(-) column to a ground (GND) pin on the UNO board

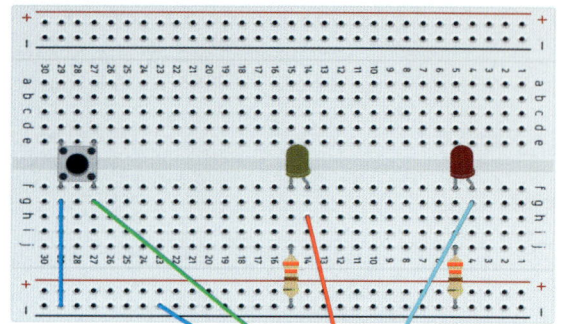

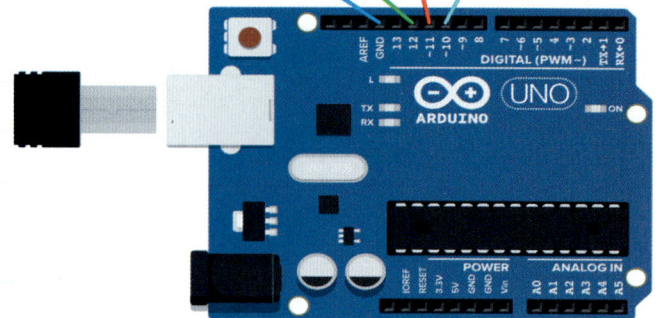

Coding - Project Build

9 Enter the code in Tinkercad® to match the code on this page. Note the use of "if" and "else."

10 Start the simulation. Observe the digital prototype. Use the pushbutton switch.

```
void setup()
{

  pinMode(12, INPUT_PULLUP);// PUSHBUTTON
  pinMode(11, OUTPUT);// LED 1
  pinMode(10, OUTPUT);// LED 2
}

void loop()
  {
int pushed = digitalRead(12);// READS PUSHBUTTON INFO
  if(pushed == LOW){
  digitalWrite(11, LOW);// PUSHED CHANGES LIGHT TO LED 2
  digitalWrite(10, HIGH);//
  }
else{
  digitalWrite(11, HIGH);
  digitalWrite(10, LOW);


  }
}
```

To use this pushbutton switch, two things have to be implemented in a project. First the hardware has to be set up and second the code has to be instructed what to do. The switch has two positions, open and closed. When the switch is not pressed it is open and the code cannot complete the red LED circuit through the switch so it leaves the yellow LED on. By pressing the button, the code can complete the circuit and turn on the red LED.

```
Text                              ⬇  📖  ✳  1 (Arduino Uno R3)  ▼

1  //Project 9
2  //Pushbutton´s LEDS
3  //thearduinoclassroom.com
4  //Copyright 2019, Isabel Mendiola and Peter Haydock
5
6  void setup()
7  {
8
9    pinMode(12, INPUT_PULLUP);// PUSHBUTTON
10   pinMode(11, OUTPUT);// LED 1
11   pinMode(10, OUTPUT);// LED 2
12 }
13
14 void loop()
15   {
16   int pushed = digitalRead(12);// READS PUSHBUTTON INFO
17     if(pushed == LOW){
18     digitalWrite(11, LOW);// PUSHED CHANGES LIGHT TO LED 2
19     digitalWrite(10, HIGH);//
20   }
21
22   else{
23     digitalWrite(11, HIGH);
24     digitalWrite(10, LOW);
25
26   }
27 }
```

Engineering Design - Project Build

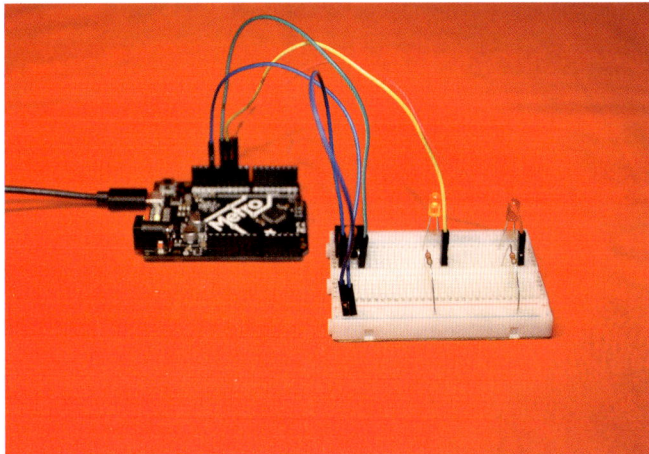

Connection - Pushbutton switches are found in many places to control devices and machines. In this case a pedestrian walk light is controlled by a pushbutton to allow people to cross the street. Often the activation of the pushbutton sets in motion a longer red light for cross traffic to give time to people to cross. When the pushbutton is not pressed the timing of the stoplight is optimized for cars and trucks to pass through the intersection.

11) Place one small pushbutton on the breadboard such that the leads span the same column on the breadboard. Note this pushbutton has only two leads.

12) Place two (2) LEDs on the breadboard so that each lead is in the same column

13) Place two (2) 220 Ohm resistors on the breadboard so that one lead is in the same row as each LED's cathode(-) lead and the other lead is in the cathode(-) column

14) Connect with a jumper wires the following:
- Digital pin 10 to the same row as the anode(+) lead of one of the LEDs
- Digital pin 11 to the same row as the anode(+) lead of one of the LEDs.
- Digital pin 12 to the same row as one of the pushbutton leads
- from the same row as the other pushbutton lead on the same half of the board to the cathode(-) column
- from the cathode(-) column to a ground (GND) pin on the UNO board

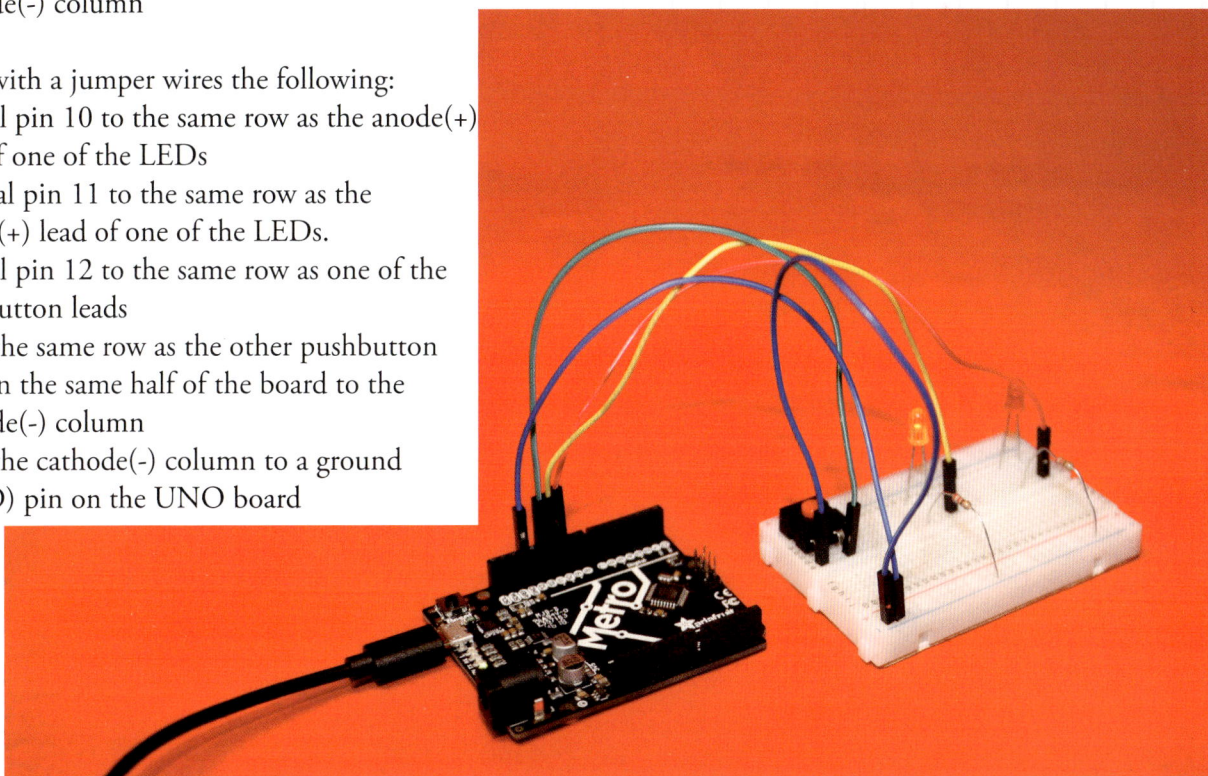

15 Review the circuit diagram for this project.

16 Connect the UNO board and computer using the USB cable.

17 Launch the Arduino® IDE software (pages 11-13) to make sure the board is communicating with the computer.

Go to the Tools menu and verify that the correct board is selected from the Board Manager menu. If not, select the correct board from the options listed.

Then confirm that the right Port is selected. If not, select the port that lists the UNO board from the options listed.

Coding - Project Build

18 Either copy and paste the code from the Tinkercad® prototype or type the code from page 94 into the Arduino® IDE software (pages 11-13).

19 Save the sketch (rename as needed).

20 Verify the sketch.

21 Upload the sketch to the UNO board.

cathode-   + anode

cathode-   + anode

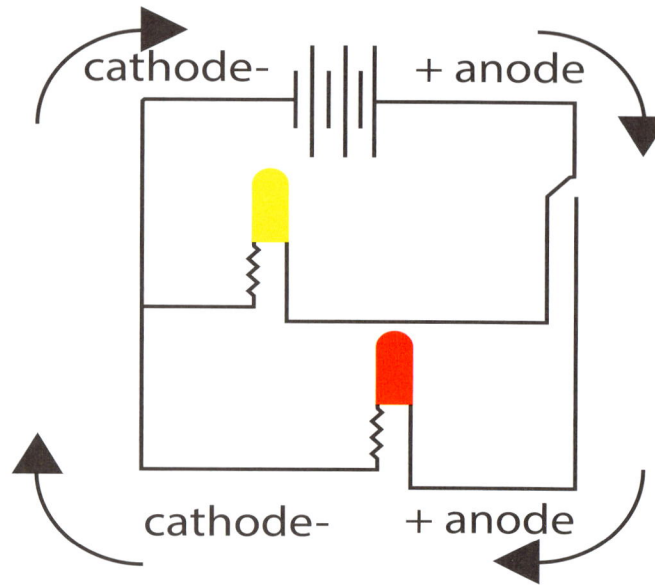

```
TAC_UE_V1_P9_Light_on_Light_off | Arduino 1.8.9 (Windows Store 1.8.21.0)    —    □    ×
File Edit Sketch Tools Help

TAC_UE_V1_P9_Light_on_Light_off

//Project 9
//Pushbutton's LEDS
//thearduinoclassroom.com
//Copyright 2019, Isabel Mendiola and Peter Haydock

void setup() {

  pinMode(12, INPUT_PULLUP);// PUSHBUTTON
  pinMode(11, OUTPUT);// LED 1
  pinMode(10, OUTPUT);// LED 2
}

void loop() {

  int pushed = digitalRead(12);// READS PUSHBUTTON INFO
  if(pushed == LOW){
    digitalWrite(11, LOW);// PUSHED CHANGES LIGHT TO LED 2
    digitalWrite(10, HIGH);//

  }else{
    digitalWrite(11, HIGH);
    digitalWrite(10, LOW);

  }
}
```

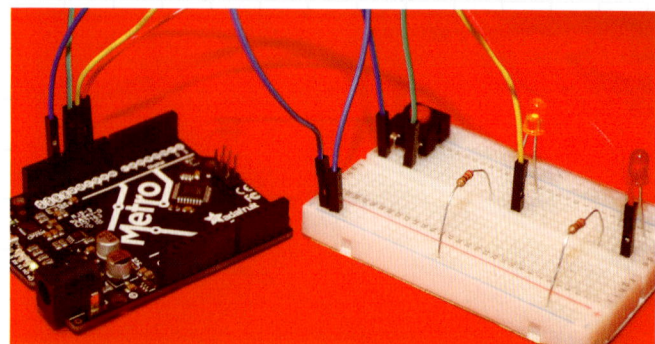

Arduino/Genuino Uno on COM4

22 Observe the LEDs on the breadboard. The pushbutton switch in this project behave differently than the pushbuttons in the digital prototype and must be pushed twice to cycle from off to on to off again

23 Document this project, discuss the project, and complete the extensions assigned from the next page. ✓

Connection - Changing the light color and lighting pattern to indicate a change in the conditions is an important safety tool. Designers and engineers will change the patterns of blinking, colors, and even location of the lights to convey more information to people and help them understand what to do.

The sign on this page is now common at busy intersections to make people think twice about crossing a street. Because some people are unaccustomed to interpreting the light patterns, cities have found that written instructions reinforce what to do as seen in this sign.

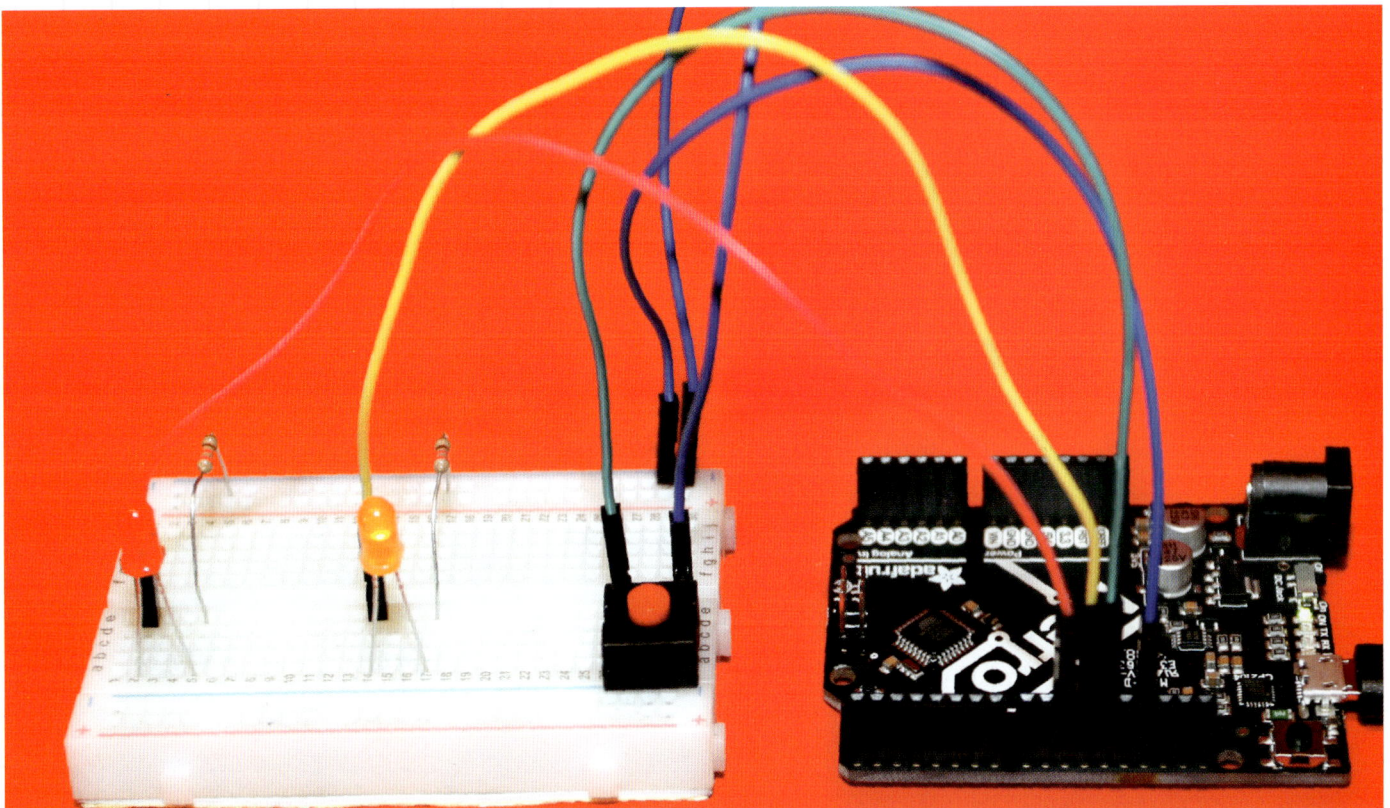

START CROSSING Watch For Vehicles

DON'T START Finish Crossing If Started — FLASHING

STEADY DON'T CROSS

TO CROSS → PUSH BUTTON

Project 9
Lights On / Lights Off

Project Reflection and Summative Activities, Discussion Starters, Extensions, and Problem

A = Essential

Reflection/Summative Activities:

- Evaluate your expertise and your team member(s) on a continuum from novice to expert (0 to 4) for:

Working as team member
Defining tasks
Using new resources
Communicating
Overcoming Obstacles
Collaborating

Support your evaluation with evidence.

- Compare your self-evaluation to those your team member(s).

- The classroom guide will provide their evaluation to you for additional feedback.

For Discussion:

- List three advantages and disadvantages to using a light with a switch.

- What coding is used to enable the pushbutton switch to control an action?

Extensions:

- Change the code to turn on and turn off both lights at the same time.

- Simplify the project to only turn on one light when the pushbutton is on.

- Is there a way to do this project without code? If so set it up and implement it.

B = Recommended

Project Connection:

- Survey the placement and use of light switches in your home. Create a light switch useage chart for members of your family to record their use of the light controlled by that switch. Determine the length of the study. Include the time and date of turning on and off the light. Describe how the lights are used including how often.

Extensions:

- Analyze the data collected in the Project Connection.

As a class compare the data.

Make recommendations to your household about reducing electricity use (including to whom and what actions to be taken).

Review the study and recommend changes for a second trial.

- Change the design of the project. Ex. change the colors, the pattern, how the are lights controlled.

- Submit your project to the book page on the website. Register and Login to submit at: thearduinoclassroom.com

From the homepage:
Click
    Books
Click
    UNO Edition Vol. 1

C = Optional

Extensions:

- Change the code to have the lights blink in two different patterns. One when the button is pushed and when not.

- Investigate other types of switches and explain how they differ from the switch in this project.

Problem:

- Identify a problem that would be solved by lights that can be switched manually. How could this project (or variation of the project you design) be a solution? Include in your report the following:

Description of the problem.

Description of how the problem is solved by this project.

What other resources you would need to solve the problem.

Design and produce the solution.

A project rubric can be found on page 236.

Anticipatory Sets
(from page 92)

Anticipatory Sets:

- Where would you see a switch or pushbutton used? List at least three different places or situations where this technology is used, describe the devices, and explain the purpose of the devices. *Pushbuttons are used in calculators, telephones, kitchen appliances, magnetic locks and various other mechanical and electronic devices used either in homes or industries.*

- When you think of an electrical switch, describe what you think is happening to the electricity? *When the switch is open, the electricity is not allowed to complete the circuit. When it is closed, the electricity can complete the circuit and power the devices in the circuit.*

A

Reflection/Summative Activities:
Responses Will Vary

For Discussion:
- What are the advantages and disadvantages to using a switch to turn off or on a light? *Consumption of less power. Controlling when a light is on and which light is on. Control a light at a distance. Accept other correct/plausible replies.*

- What coding is used to enable the pushbutton switch to control an action? *The "if" and "else" lines of code in the loop.*

Extensions:
- Change the code to turn on and turn off both lights at the same time. *Responses Will Vary*

- Simplify the project to only turn on one light when the pushbutton is on. *Responses Will Vary*

- Is there a way to do this project without code? Is so set it up and implement it. *Yes this is a possibility. The setup would use the other unused leads of the switch while remaining connected to the 5v and ground. The lights will not be controllable beyond on and off though.*

B

Project Connections:
Responses Will Vary

Extensions:
- Analyze the data collected in the Project Connection. *Responses Will Vary*

- Change the design of the project. *Responses Will Vary*

Submit your project to the book page on the website.

Responses Will Vary

C

Extensions:
- Change the code to have the lights blink in two different patterns. One when the button is pushed and when not. *Responses Will Vary*

- Investigate other types of switches and explain how they differ from this project. *Responses Will Vary*

Problem:
Responses Will Vary

Project 10
Pushbutton Counting

Lesson Integration:	Groupings:	Level:	Time to Complete:
Physics - Ohm's Law, Circuits, Voltage, Resistance Biology - Population Census Mathematics - Ratios, Counting, Integers	1 - 2 Change Groupings	Intermediate	45 min. for the project and 45 min. for extensions.

Objectives:
- Investigate how circuits and electronic components interact with electrical energy.
- Control a counting device

Prerequisite Skills:
- Time measurement (milliseconds)
- Arduino® IDE software (pages 11-13)
- Tinkercad® (pages 15-16)

Purpose and Skills:
- Build a pushbutton counter
- Introduce addition and subtraction with variables

STEAM Connections:
Science - Circuits, Resistance, Ohm's Law, Ecology, Populations, Census
Technology - Code, Program Settings, Simulators, Digital Design, Electronic Components, LCD, Analog Data, Communications, Switches,

Conditional Statements
Engineering - Designing, Building, and Using a Machine, Prototyping, Applied Physics
Allied Arts - Sociology, Behavior, Demographics
Math - Addition, Subtraction, Counting

Key Vocabulary:
Census - Counting the occurrence of a thing or action

Project Introduction:
- Introduce the groups to the purpose of the project, the skills developed, the standards met, and the goal of the project.

Anticipatory Sets:
- Where would you see a pushbutton counter used? List at least three different places or situations where pushbutton technologies are used, describe the devices, and explain the purpose of the devices.

Project 10
Pushbutton Counting

Educational Standards

ISTE Standards for Students
- Empowered Learner 1a, 1b, 1d
- Knowledge Constructor 3a, 3b, 3c, 3d
- Innovative Designer 4a, 4b, 4c, 4d
- Computational Thinker 5a, 5b, 5c, 5d
- Creative Communicator 6a, 6b, 6c, 6d
- Global Collaborator 7c, 7d

US Computer Science Standards
- Project correlations can be found on the book's web page.

US NGSS - Middle School
DCIs
- MS-PS4-3 PS4.C
- MS-ESS3-4 ESS3.C

Cross Cutting Concepts
- Cause and Effect
- Scale, Proportion and Quantity
- Structure and Function

Science and Engineering Practices
- Planning and Carrying Out Investigations
- Obtaining, Evaluating, and Communicating Information
- Analyzing and Interpreting Data
- Scientific Knowledge is Based on Empirical Evidence
- Using Mathematics and Computational Thinking

US NGSS - High School
- Correlations can be found on the book's web page.

US Common Core Language Arts and Mathematics
- Correlations can be found on the book's web page.

Step-by-Step 1-23

Materials List:
- Computer with IDE software
- Connection to the Internet
- USB Cable
- UNO or UNO Compatible Microcontroller
- Short Breadboard
- Two (2) 220 Ohm resistors
- 16x2 LCD Display
- 18 Jumper Wires (Male to Male)

Engineering Design - Digital Prototype

1. Launch and log into Tinkercad® (pages 15-16).

2. Start a new Circuits project.

3. Drag an UNO board to the workspace.

4. Drag a breadboard to the workspace.

5. Drag two small pushbutton switches to one end of the breadboard such that the leads span the center of the breadboard, but with the buttons occupying separate rows.

6. Place a 16x2 LCD on the other end but on the same half of the breadboard leaving two columns exposed for jumper wires.

7. Place two 220 Ohm resistors on the breadboard such that each connects across the halves of the breadboard from the second pin in from each end of the LCD to the same row in the other half.

8. Connect with jumper wires the following:
   Use the technical diagrams on the following page to assist in jumper wire placement.
   - the five volt (5v) pin on the UNO board to the anode(+) column on the opposite half of the breadboard from the LCD and potentiometer

- a ground (GND) pin on the UNO board to the cathode(-) column on the opposite half of the breadboard from the LCD
- from digital pin nine (9) to a pin in the same row as the lead of the first pushbutton switch
- from the cathode(-) column to a pin in the same row as the other lead of the first pushbutton switch
- from digital pin nine (8) to a pin in the same row as the lead of the second pushbutton switch
- from the cathode(-) column to a pin in the same row as the other lead of the second pushbutton switch
- from the cathode(-) column to a pin in the same row as the GND lead of the LCD
- from the anode(+) column to a pin in the same row as the resistor that connects to the same row as the VCC lead of the LCD
- from the cathode(-) column to a pin in the same row as the VO lead of the LCD
- from digital pin seven (7) on the UNO board to a pin in the same row as the RS lead of the LCD

- from the cathode(-) column to a pin in the same row as the RW lead of the LCD
- from digital pin six (6) on the UNO board to a pin in the same row as the E lead of the LCD
- from digital pin five (5) on the UNO board to a pin in the same row as the DB4 lead of the LCD
- from digital pin four (4) on the UNO board to a pin in the same row as the DB5 lead of the LCD
- from digital pin three (3) on the UNO board to a pin in the same row as the DB6 lead of the LCD
- from digital pin two (2) on the UNO board to a pin in the same row as the DB7 lead of the LCD
- from the anode(+) column to a pin in the same row as the second resistor
- from the cathode(-) column to a pin in the same row as the first LED lead of the LCD

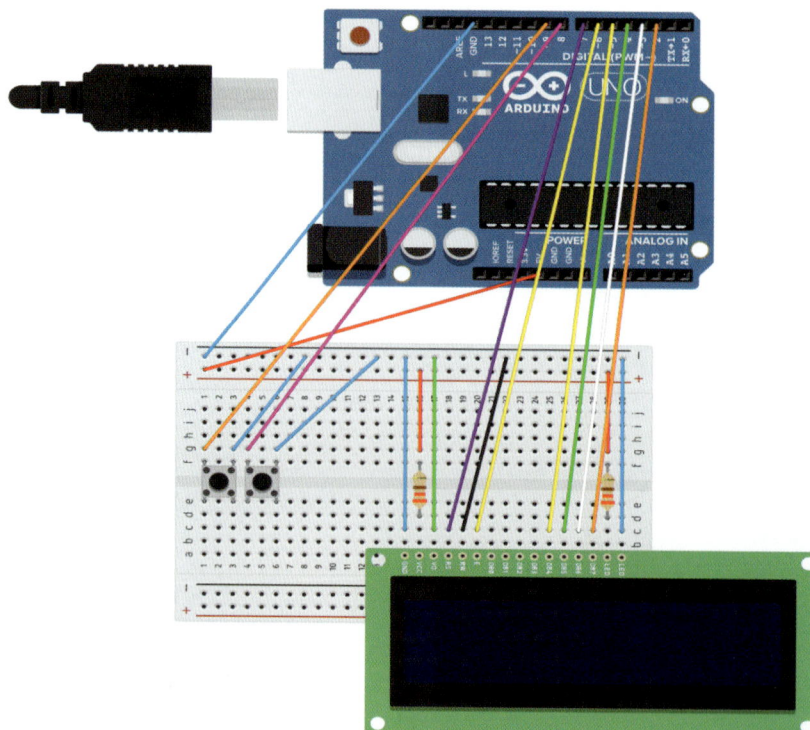

Coding - Digital Prototype

9 Enter the code in Tinkercad® to match the code on this page.

10 Start the simulation. Click on each button and observe the digital prototype. ✓

Connection - The Mayan numbering system is base 20 with subdivisions of ones (1) and fives (5). Modern counting uses base 10 with subdivisions of one (1).

Each digit of a Mayan number ranges from zero to 19 with the places being 20 times the previous.

One bar is equal to five (5).

One dot is equal to one (1).

Three (3) bars are 15.

Three (3) bars with four (4) dots under is 19.

Zero (0) was based on the cacao bean.

A larger number like 101 is one (1) bar over one (1) dot.

```
#include <LiquidCrystal.h>      // Add the library for reading the LCD
LiquidCrystal lcd(7,6,5,4,3,2);  //  Add digital microcontroller pins
const int  Up_buttonPin   = 8;   //  Pushbutton 2
const int  Down_buttonPin = 9;   //  Pushbutton 1
int count = 0;
void setup()
{
  lcd.begin(16, 2);  //Columns and rows of the LCD
  lcd.print("- and + counting");//Type the message for the display
  pinMode(8,INPUT);
  pinMode(9,INPUT);
  digitalWrite(8,HIGH);
  digitalWrite(9,HIGH);
}
void loop()
{
  if(digitalRead(8) == LOW)      // Pushbutton +
  {
  count++;                 // Increase Count by 1
  lcd.begin(16, 2);          //Columns and rows of the LCD
  lcd.print("- and + counting");//Type the message for the display
  lcd.setCursor(0, 1 );
  lcd.print(count);
  delay(500);
  }
  if(digitalRead(9) == LOW)      // Pushbutton -
  {
  count--;                 // Decreasing Count by 1
  lcd.begin(16, 2);             //Columns and rows of the LCD
  lcd.print("- and + counting");//Type the message for the display
  lcd.setCursor(0, 1 );
  lcd.print(count);
  delay(500);
  }
}
```

Engineering Design - Project Build

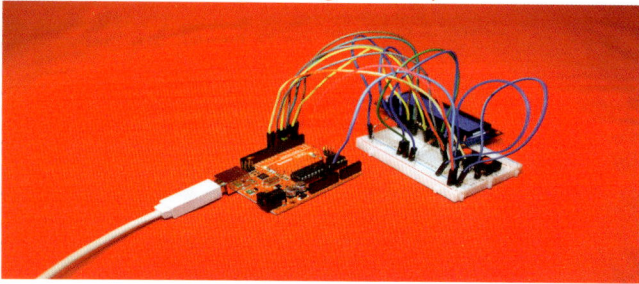

11 Place two small pushbutton switches on one end of the breadboard such that the leads span the center of the breadboard, but with the buttons occupying separate rows.

12 Place a 16x2 LCD on the other end of the breadboard but on the same half of the breadboard leaving two columns exposed for jumper wires.

13 Place two 220 Ohm resistors on the breadboard such that each connects across the halves of the breadboard from the second pin in from each end of the LCD to the same row in the other half.

14 Use the table on this page to place the jumper wires. For additional help, use instruction numbered 8 and diagram on the previous pages.

From	To (LCD L to R)
Breadboard Cathode- Column	LCD GND (pin 1) Same Row
Breadboard Anode + Column	LCD VCC (pin 2) Resistor 1 Same Row
Breadboard Cathode- Column	LCD VO (pin 3) Same Row
UNO Digital Pin D7	LCD RS (pin 4) Same Row
Breadboard Cathode- Column	LCD RW (pin 5) Same Row
UNO Digital Pin D6	LCD E (pin 6) Same Row
UNO Digital Pin D5	LCD DB4 (pin 11) Same Row
UNO Digital Pin D4	LCD DB5 (pin 12) Same Row
UNO Digital Pin D3	LCD DB6 (pin 13) Same Row
UNO Digital Pin D2	LCD DB7 (pin 14) Same Row
Breadboard Anode + Column	LCD VCC (pin 15) Resistor 2 Same Row
Breadboard Cathode- Column	LCD LED (pin16) Same Row

From	To
UNO 5v	Breadboard Anode + Column
UNO GND	Breadboard Cathode- Column
UNO Digital Pin D9	Pushbutton 1 Anode + Lead Same Row
Breadboard Cathode - Column	Pushbutton 1 Cathode - Lead Same Row
UNO Digital Pin D8	Pushbutton 2 Anode + Lead Same Row
Breadboard Cathode - Column	Pushbutton 2 Cathode - Lead Same Row

Connection - Mechanical counters activated by pushing a button or "clicker" are popular census taking machines. This photo shows the interior of a counter. One drawback to these counters is they can only count up.

15 Review the simplified circuit diagram for this project.

16 Connect the UNO board and computer using the USB cable.

17 Launch the Arduino® IDE software (pages 11-13) to make sure the board is communicating with the computer.

Coding - Project Build

18 Either copy and paste the code from the Tinkercad® prototype or type the code from page 103 into the Arduino® IDE software (pages 11-13). Note: The LCD library should already be installed.

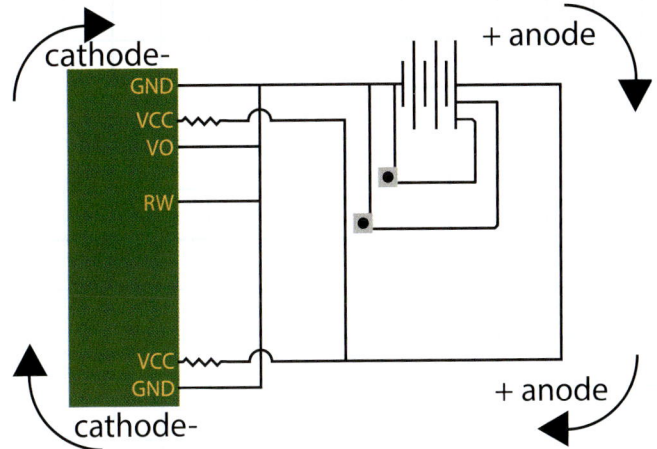

```
TAC_UE_V1_P10

//Project 10
//LCD Counter using two Pushbuttons
//thearduinoclassroom.com
//Copyright 2019, Isabel Mendiola and Peter Haydock

#include <LiquidCrystal.h>          // Add the library for reading the LCD
LiquidCrystal lcd(7,6,5,4,3,2);     //  Add digital microntroller pins
const int  Up_buttonPin  = 8;       //  Pushbutton 2
const int  Down_buttonPin = 9;      //  Pushbutton 1
int count = 0;

void setup()
{
  lcd.begin(16, 2);  //Columns and rows of the LCD
  lcd.print("- and + counting");//Type the message for the display
  pinMode(8,INPUT);
  pinMode(9,INPUT);
  digitalWrite(8,HIGH);
  digitalWrite(9,HIGH);
}
void loop()
{
  if(digitalRead(8) == LOW)      // Pushbutton up
  {
  count++;                             // Increase Count by 1
    lcd.begin(16, 2);                //Columns and rows of the LCD
  lcd.print("- and + counting"); //Type the message for the display
  lcd.setCursor(0, 1 );
  lcd.print(count);
  delay(500);
  }
  if(digitalRead(9) == LOW)      // Pushbutton down
  {
  count--;                             // Decreasing Count by 1
  lcd.begin(16, 2);                //Columns and rows of the LCD
  lcd.print("- and + counting");   //Type the message for the display
  lcd.setCursor(0, 1 );
  lcd.print(count);
  delay(500);
  }
}
Done compiling.
```

19 Save the sketch (rename as needed).

20 Verify the sketch.

21 Upload the sketch to the UNO board.

22 Test the project and count up and down using the pushbuttons. Note: The pushbutton has to be pushed twice to do only one count otherwise the count up or down will continue.

23 Document this project, discuss the project, and complete the extensions assigned from the next page.

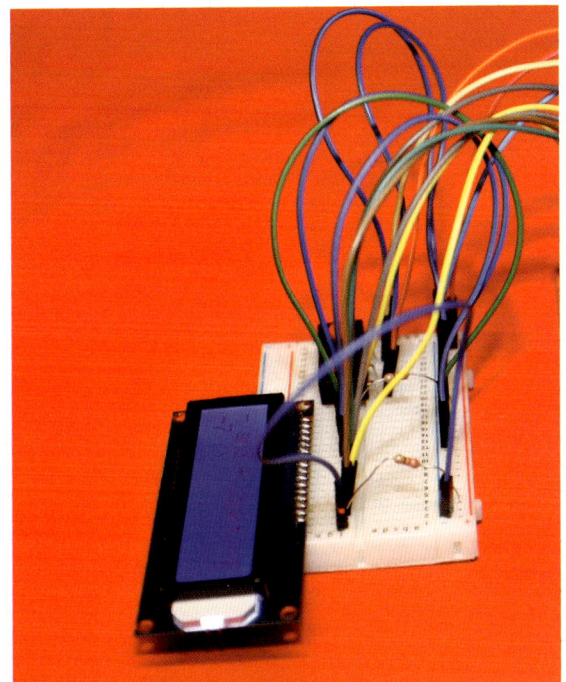

A = Essential

Reflection/Summative Activities:

- Have each team member document in their project/classroom journal who was on their team, what went well, what they could improve upon, what they would do differently if they were to do the project again, and verify that every person in the group can do the project.

- Exchange journals within the team and write responses to their reflection.

For Discussion:

- Several lines of code are repeated for writing on the LCD. Why would the text displayed be rewritten for each button push?

Extensions:

- Change the lines of code to add or subtract by twos, threes, or other intervals.

- Change the code to allow each button to only add or only subtract but by different intervals. ex. button 1 adds by ones and button 2 adds by tens

- Change the code to start with a number other than zero (0).

B = Recommended

STEM Connection:

- Investigate a technique in STEM that takes a census. Present your findings as poster or display. Explain the reason for the census, the technique used, what the information tells those taking the census, and what might be done with the information.

- Submit a photo of your poster or display on the website. Register and Login to submit at: thearduinoclassroom.com

From the homepage:
Click
    Books
Click
    UNO Edition Vol. 1

C = Optional

Extension:

- Change the code to perform a multiplication or division calculation when each button is pressed.

Problem:

- Using a local newspaper or local news website to identify a problem or question that would be solved by conducting a census. How could this project be a solution? Include in your report the following:

Description of the problem or question.

Description of how the problem is solved by this project.

What other resources you would need to solve the problem.

Carry out the census.

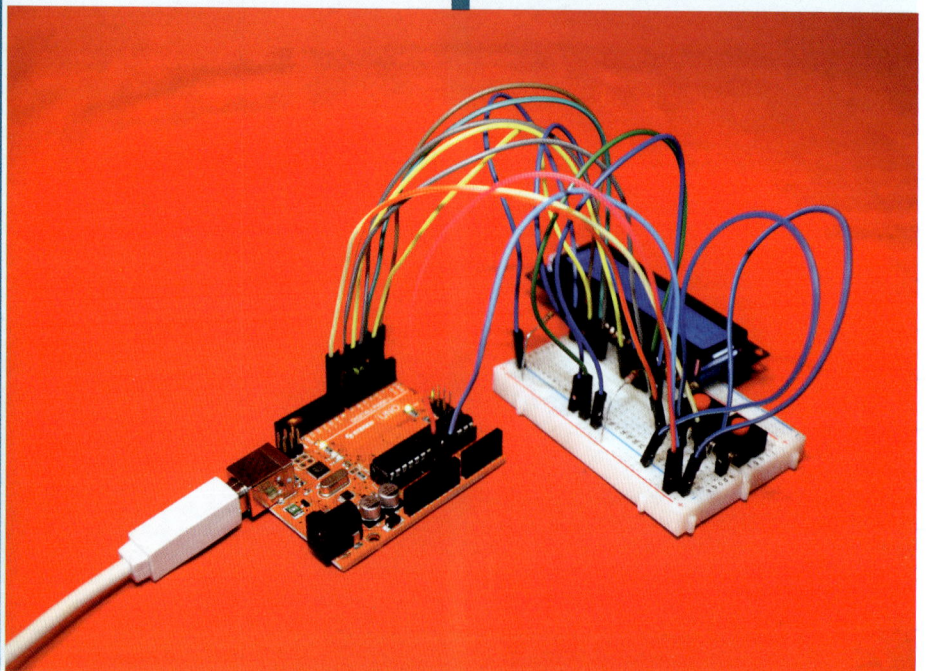

A project rubric can be found on page 236.

Anticipatory Sets
(from page 100)

Anticipatory Sets:
- Where would you see a pushbutton counter used? List at least three different places or situations where pushbutton counters are used, describe the devices, and explain the purpose of the devices.
Analog and digital pushbutton counters are used to count the number of people, animals or other things in less controlled environments. One place commonly seen is in museums where gallery observers note the number of people that enter a room to record foot traffic in a part of the exhibit. Another place would be in creating a census of animals in the wild in a location or as they pass a certain place. Traffic engineers will place counters that use pneumatic switches on roads to count the number of cars and trucks that pass on that part of the road.

A

Reflection/Summative Activities:
Responses Will Vary

For Discussion:
- Several lines of code are repeated for writing on the LCD. Why would the text displayed be rewritten for each button push? *In order to display the complete new number the previous number should be cleared. Have the team delete the code and operate the pushbuttons without the extra code to see.*

Extensions:
- Change the lines of code to add or subtract by twos, threes, or other interval.
Responses Will Vary

- Change the code to only add or only subtract.
Responses Will Vary

- Change the code to start with a number other than zero (0)
Responses Will Vary

B

STEM Connection:
Responses Will Vary

C

Extension:
- Change the code to perform a multiplication or division.
Responses Will Vary

Problem:
Responses Will Vary

Project 11
Music Maker

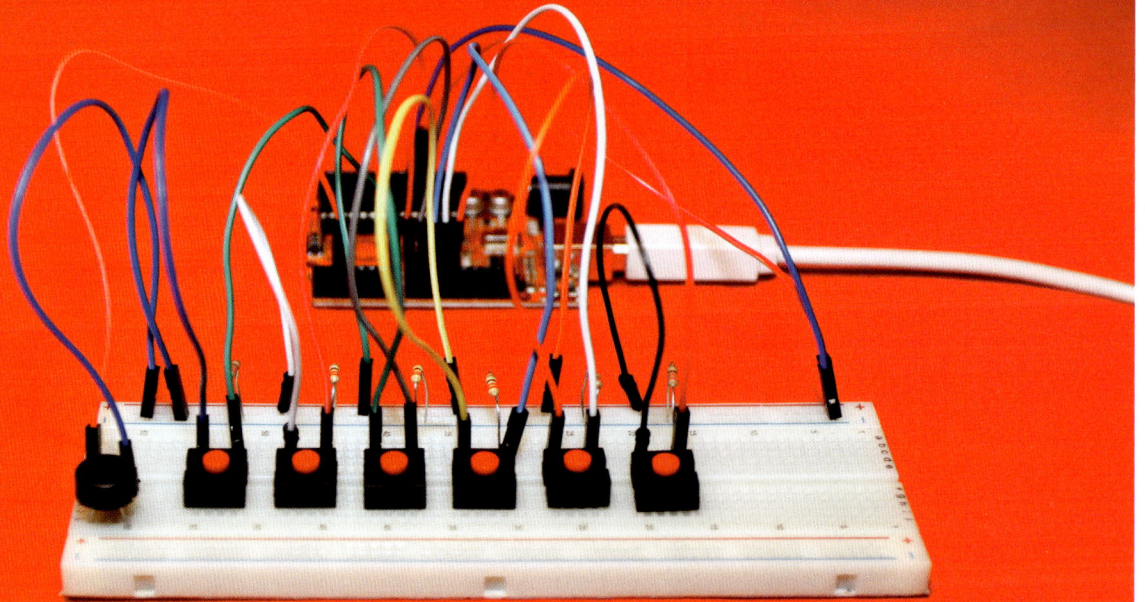

Getting Started

Lesson Integration:	Groupings:	Level:	Time to Complete:
Physics - Ohm's Law, Circuits, Voltage, Resistance, Sound, Frequency Mathematics - Time, Ratios	1 - 2	Intermediate	45 min. for the project and 45 min. for extensions.

Objectives:
- Investigate how circuits and electronic components interact with electrical energy.
- Build a music making device

Prerequisite Skills:
- Time measurement (milliseconds)
- Understanding of sound and frequency
- Arduino® IDE software (pages 11-13)
- Tinkercad® (pages 15-16)

Purpose and Skills:
- Controlling a piezo
- Connecting and controlling different combinations of inputs and outputs

STEAM Connections:
Science - Sound, Circuits, Resistance, Ohm's Law, Sound, Frequency
Technology - Code, Program Settings, Simulators, Digital Design, Electronic Components, Analog Data, Communications, Switches, Conditional Statements
Engineering - Designing, Building, and Using a Machine, Prototyping, Applied Physics
Allied Arts - Music, Harmonics
Math - Ratios, Conversions, Scales

Key Vocabulary:
Harmonic - A musical note related to a primary tone by some fractional part (1/2, 1/3, 1/4, etc..).

Project Introduction:
- Introduce the groups to the purpose of the project, the skills developed, the standards met, and the goal of the project.

Anticipatory Sets:
- Where would you see music making devices used? List at least three different places or situations where music making technologies are used, describe the devices, and explain the purpose of the devices.

Project 11
Music Maker

Step-by-Step 1-22

Materials List:
- Computer with IDE software
- Connection to the Internet
- USB Cable
- UNO or UNO Compatible Microcontroller
- Long Breadboard
- Piezoelectric Buzzer (anode +)
- Six (6) 220 Ohm Resistors
- Six (6) Pushbutton Switches
- 16 Jumper Wires (Male to Male)

Engineering Design - Digital Prototype

1. Launch and log into Tinkercad® (pages 15-16).

2. Start a new Circuits project.

3. Drag an UNO board to the workspace.

4. Drag a breadboard to the workspace.

5. Drag one piezo buzzer to the breadboard such that the leads span two rows in a column. Review the diagrams on this page and the next page for assistance.

6. Place six (6) pushbutton switches evenly spaced and spanning the center of the breadboard.

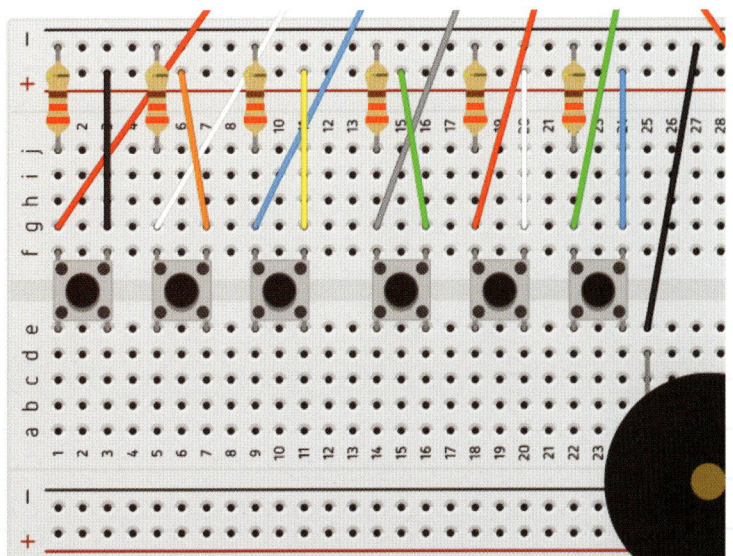

7 Place six (6) 220 Ohm resistors on the breadboard such that each connects the cathode(-) column and the row with one lead from the pushbutton.

8 Connect with jumper wires the following:
- the five volt (5v) pin on the UNO board to the anode(+) column
- a ground (GND) pin on the UNO board to the cathode(-) column on the breadboard with the connected resistors
- the same row with the other leg of each pushbutton switch on the same side of the breadboard as the resistors with the anode(+) column which is also connected to the 5v connection. This is for all 6 pushbuttons
- digital pins seven (7) through 12 to the same row as each row connected to a resistor.
- the cathode(-) column to the row with the cathode(-) lead of the piezo buzzer
- the same row connected to the anode(+) lead of the piezo buzzer and pin three (3) on the UNO board

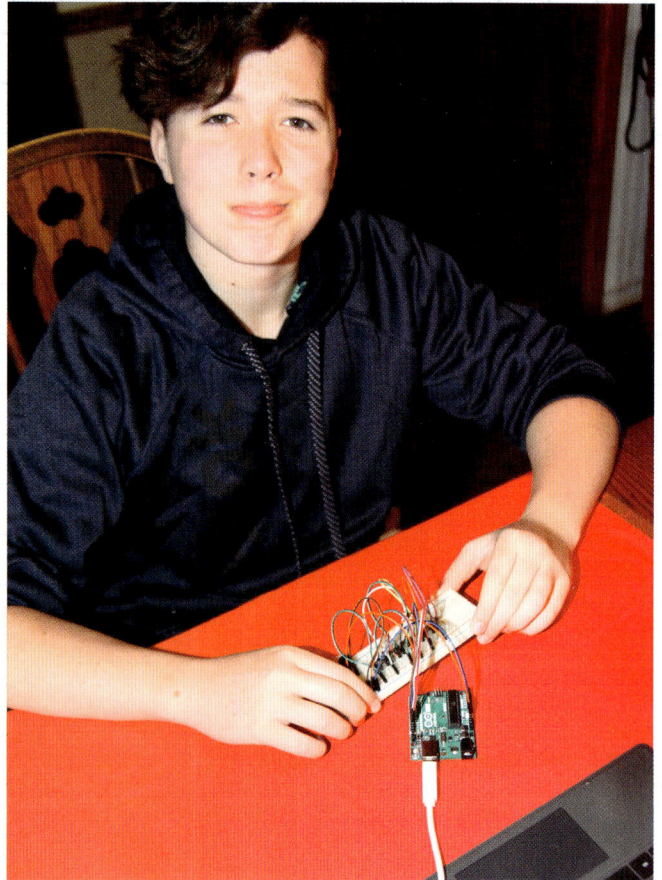

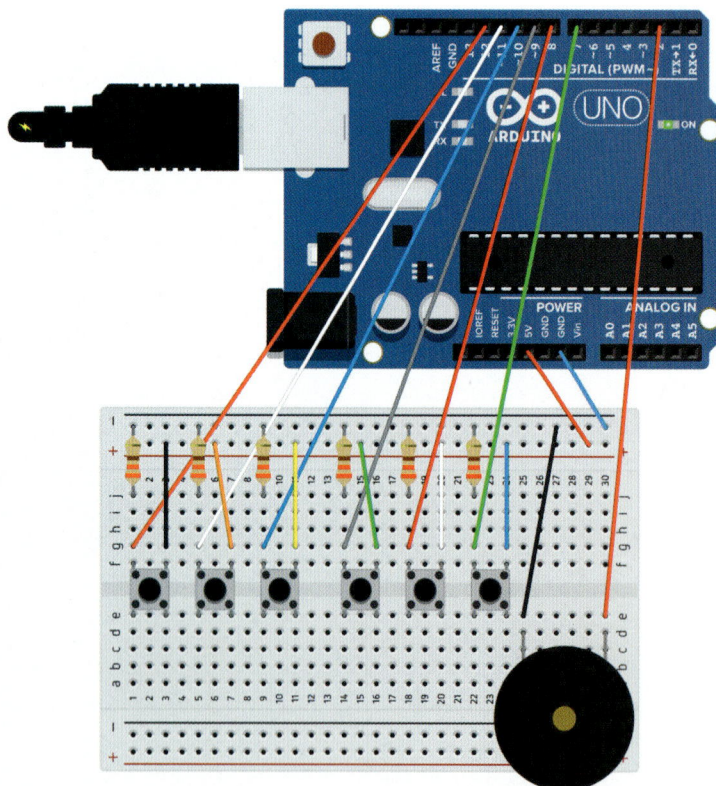

```
#define FrequencyD5 576.65 // Frequencies for
#define FrequencyE5 647.27 // equal-tempered scale,
#define FrequencyF5 685.76 // ex. A4 = 434 Hz
#define FrequencyG5 769.64
#define FrequencyA5 864.00
#define FrequencyB5 969.81

const float Frequency[] = {FrequencyD5, FrequencyE5, FrequencyF5,
          FrequencyG5, FrequencyA5, FrequencyB5};
void setup()           // sets the communication with Arduino IDE
{
}
void loop()  //Sound repeats every time the button is pushed.
{
 int button = 0, freq = 0;   for (int pin = 7; pin <= 12; pin++) {
  if (digitalRead(pin) == HIGH) {
    button = pin - 7;
    freq = float(Frequency[button]);
    tone(2, freq, 50);  // change the tone of the six pushbuttons
  }
 }
 delay(10);   // small number long tones.
}
```

Coding - Digital Prototype

9 Enter the code in Tinkercad® to match the code on this page. Notice the use of "float" to set the variable values to use decimals and not just integers.

10 Start the simulation. Click on the push buttons and observe the digital prototype.

Connection - Much western music is based on a seven-note, or heptatonic scale. The scale repeats each octave higher or lower after the seven notes. The most recognizable western seven-note scale is the diatonic scale where there are five whole tone intervals and two half tone interval. This scale can be played on a piano from a "C" key to the next "C" key using only the white keys. Other versions of the heptatonic scale exist and have different tone intervals within the octave.
There is also a pentatonic scale that covers the same octave, but with only five tones. Many cultures have used this scale for folk or traditional music including the eastern cultures of China, Korea, and Japan. Also some English, Celtic, and West African forms use this scale. Modern forms of Gospel, Bluegrass, Jazz, and Rock can also use the pentatonic scale. Jazz greats like John Coltrane and Duke Ellington (pictured here) used pentatonic scales in their music.

Engineering Design - Project Build

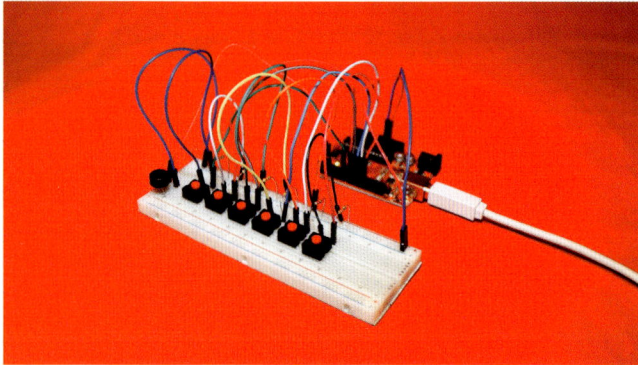

11 Drag one piezo buzzer to the breadboard such that the leads span to rows in a column.

12 Place six (6) pushbutton switches evenly spaced and spanning the same column on the breadbaord.

13
- Place six (6) 220 Ohm resistors on the breadboard such that each connects the cathode(-) column and the row with one lead from the pushbutton.

14 Connect with jumper wires the following:
- the five volt (5v) pin on the UNO board to the anode(+) column
- a ground (GND) pin on the UNO board to the cathode(-) column on the breadboard with the connected resistors
- the same row with the other leg of each pushbutton switch on the same side of the breadboard as the resistors with the anode(+) column which is also connected to the 5v connection. This is for all 6 pushbuttons.
- digital pins seven (7) through 12 to the same row as each row connected to a resistor
- the cathode(-) column to the row with the cathode(-) lead of the piezo buzzer
- the same row connected to the anode(+) lead of the piezo buzzer and pin three (3) on the UNO board

15 Connect the UNO board and computer using the USB cable.

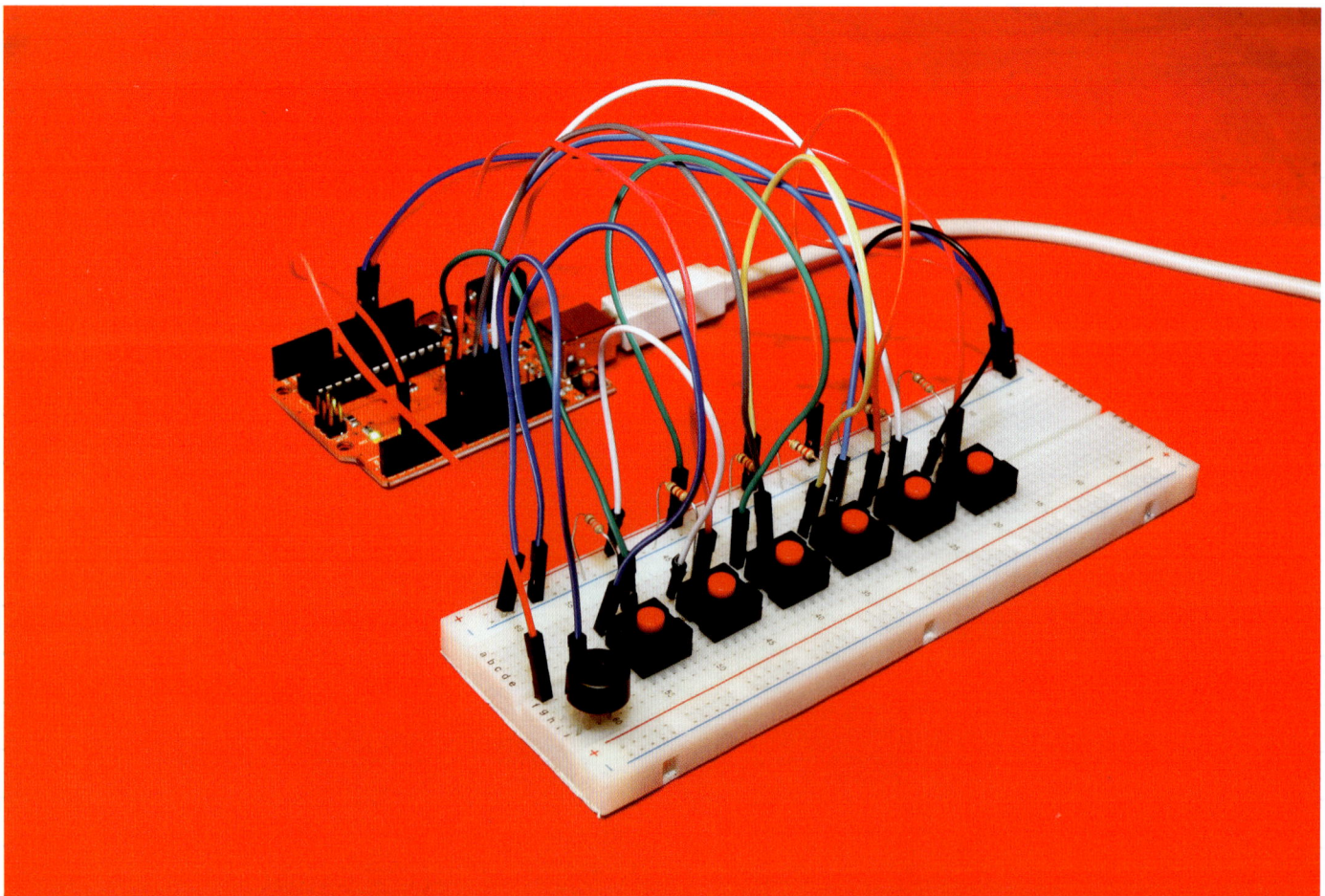

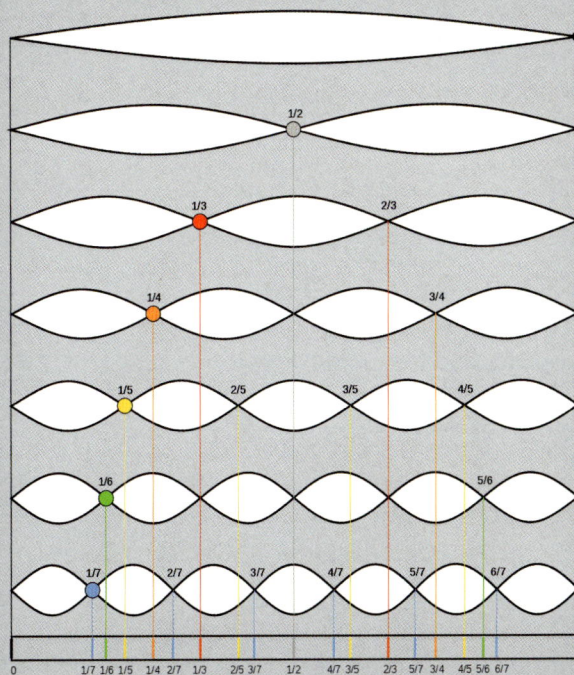

16 Launch the Arduino® IDE software (pages 11-13) to make sure the board is communicating with the computer.

Go to the Tools menu and verify that the correct board is selected from the Board Manager menu. If not, select the correct board from the options listed.

Then confirm that the right Port is selected. If not, select the port that lists the UNO board from the options listed.

Coding - Project Build

17 Either copy and paste the code from the Tinkercad® prototype or type the code from page 111 into the Arduino® IDE software (pages 11-13).

18 Save the sketch (rename as needed).

19 Verify the sketch.

20 Upload the sketch to the UNO board.

21 Press the pushbutton switches to test the tones. Recall the buttons need to be pushed twice "off" or "on." Make sure every button is in the off position to start.

22 Document this project, discuss the project, and complete the extensions assigned from the next page.

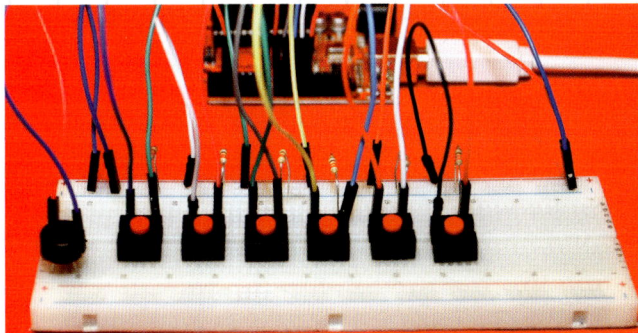

Connection - The octave is central to music in both heptatonic (seven note) and pentatonic (five note) scales. Each scale will produce different primary tones. Depending on the instrument different harmonic tones will also be produced. These harmonic tone are mathematically related through ratios to the primary tone.

When tones an octave apart are generated at the same time the harmonics are particularly evident. The tones resonate with each other and reinforce each other.

This graphic shows the waves generated by a sound and how ratios of the primary tone would be mathematically related to form harmonics.

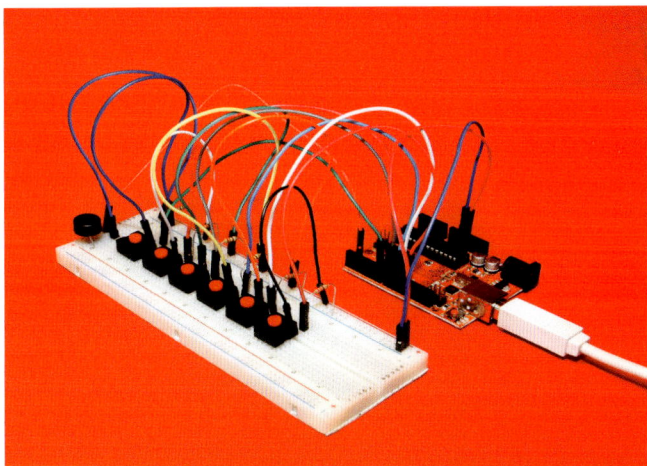

Project 11
Music Maker

A = Essential

Reflection/Summative Activity:

- Have each team member document in their project/ classroom journal who was on their team, what went well, what they could improve upon, what they would do differently if they were to do the project again, and verify that every person in the group can do the project.

On a scale of 1-5 have each team member rate their ability to listen and recall what they have heard.

For Discussion:

- What purpose is there to generating sounds that have controlled tones?

Classroom Challenge:

- Pick one team to generate a pattern of seven notes. CHALLENGE each team to replicate the pattern fastest. The team that does fastest creates the new classroom challenge.

Extensions:

- Change the tones played when each pushbutton switch is pressed.

- Change the length of time the tone plays when each pushbutton switch is pressed.

- Determine the upper and lower tone limits of the piezo.

B = Recommended

Professional Connection:

- Research and write a short description of an musician, singer or type of music and include the scale and scale type that they/it used. Do a classroom presentation about the music including samples.

Extensions:

- Compare and contrast the tones generated by the music maker to an instrument of your choice.

- As a class, play musical "telephone" with a 10 note pattern introduced by the classroom guide.

- Draw a circuit diagram for this project.

- Investigate the different types of scales (heptatonic and pentatonic) and replicate the scale in the code.

- Submit your project on the book page on the website. Register and Login to submit at: thearduinoclassroom.com

From the homepage:
Click
    Books
Click

    UNO Edition Vol. 1

C = Optional

Earth Science Connection:

- Research and write a short report on the different types of waves earthquakes produce.

Extension:

- Change the code to have each button play a different short set of tones.

Problem:

- Identify a project that would use the skills you have been working with in this project. How could this project be a solution? Include in your report the following:

Description of the problem

Description of how the problem is solved by this project

What other resources you would need to solve the problem.

Design and produce the solution.

A project rubric can be found on page 236.

Anticipatory Sets
(from page 132)

Anticipatory Sets:
- Where would you see music making devices used? List at least three different places or situations where music making devices are used, describe the devices, and explain the purpose of the devices.
Responses Will Vary

A

Reflection/Summative Activity:
Responses Will Vary

For Discussion:
- What purpose is there to generating sounds that have controlled tones ?
Responses Will Vary

Extensions:
- Change the tones played when each pushbutton switch is pressed.
Responses Will Vary

- Change the length of time the tone plays when each pushbutton switch is pressed.
Responses Will Vary

- Determine the upper and lower tone limits of the piezo.
Responses Will Vary

B

Professional Connection:
Responses Will Vary

Extensions:
- Draw a circuit diagram for this project.
Responses Will Vary

- Investigate the different types of scales (heptatonic and pentatonic) and replicate the scale in the code.
Responses Will Vary

- Submit your project to the book page on the website.

Responses Will Vary

C

Earth Science Connection:
- Research and write a short report on the different types of waves earthquakes produce.
Responses Will Vary

Extension:
- Change the code to have each button play a different short musical set of tones?
Responses Will Vary

Problem:
Responses Will Vary

Project 12
Calculator

Lesson Integration:		Groupings:	Level:	Time to Complete:
Physics - Ohm's Law, Circuits, Voltage, Resistance Mathematics - Arithmetic, Integers, Decimals		1 - 2	Advanced	60 min. for the project and 45 min. for extensions.

Objectives:
- Investigate how circuits and electronic components interact with electrical energy.
- Build a calculator

Prerequisite Skills:
- Time measurement (milliseconds)
- Arithmetic
- Arduino® IDE software (pages 11-13)
- Tinkercad® (pages 15-16)

Purpose and Skills:
- Control a keypad input to make a four function calculator

STEAM Connections:
Science - Circuits, Resistance, Ohm's law
Technology - Code, Program Settings, Simulators, Digital Design, Electronic Components, Analog Data, Conditional Statements, Calculator
Engineering - Designing, Building, and Using a Machine, Prototyping, Applied Physics

Allied Arts -
Math - Arithmetic, Ratios, Conversions, Decimals, Integers

Key Vocabulary:
Calculator - A device that performs mathematical operations

Project Introduction:
- Introduce the groups to the purpose of the project, the skills developed, the standards met, and the goal of the project.

Anticipatory Sets:
- Where would you see calculators used? List at least three different places or situations where this technology is used, describe the devices, and explain the purpose of the devices.

ISTE Standards for Students
- Empowered Learner 1d
- Knowledge Constructor 3b, 3c, 3d
- Innovative Designer 4a, 4b, 4c, 4d
- Computational Thinker 5a, 5b, 5c, 5d
- Creative Communicator 6a, 6b, 6c, 6d
- Global Collaborator 7c

US Computer Science Standards
- Project correlations can be found on the book's web page.

US NGSS - Middle School
DCIs
- MS-PS4-3 PS4.C

Cross Cutting Concepts
- Cause and Effect
- Scale, Proportion and Quantity
- Structure and Function

Science and Engineering Practices
- Planning and Carrying Out Investigations
- Obtaining, Evaluating, and Communicating Information
- Analyzing and Interpreting Data
- Scientific Knowledge is Based on Empirical Evidence
- Using Mathematics and Computational Thinking

US NGSS - High School
- Correlations can be found on the book's web page.

US Common Core Language Arts and Mathematics
- Correlations can be found on the book's web page.

Materials List:
- Computer with IDE software
- Connection to the Internet
- USB Cable
- UNO or UNO Compatible Microcontroller
- Long Breadboard
- 4x4 Keypad
- 10k Potentiometer
- 16x2 LCD
- 220 Ohm Resistor
- 24 Jumper Wires (Male to Male)

Engineering Design - Digital Prototype

1. Launch and log into Tinkercad® (pages 15-16).

2. Start a new Circuits project.

3. Drag an UNO board to the workspace.

4. Drag a breadboard to the workspace.

5. Drag one 10k potentiometer to the breadboard such that the leads span three rows in a column.

6. Drag a 4x4 keypad to the workspace.

7. Drag and 16x2 LCD to the breadboard such that the leads are across the other half of the breadboard from the potentiometer and across the same column.

8. Place a 220 Ohm resistor on the breadboard such that it connects the same row across the halves of the breadboard and the same row as the LED (VCC) pin of the LCD (second in from the right)

9. Connect with jumper wires the following:
   Use the technical diagram on the following page to assist in jumper wire placement.
   - the five volt (5v) pin on the UNO board to the anode(+) column on the same half of the breadboard as the potentiometer

Hover over the connection ends to identify the lead or pin connection designation.

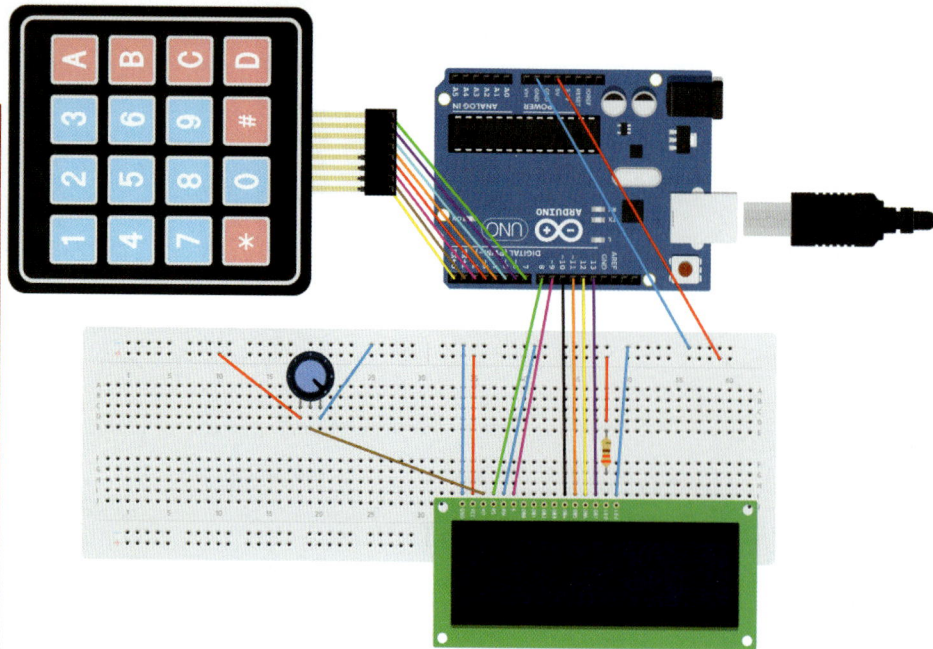

Column 4

- a ground (GND) pin on the UNO board to the cathode(-) column on the same half of the breadboard as the potentiometer
- from digital pin eight (8) on the UNO board to a pin in the same row as the RS lead of the LCD
- from the cathode(-) column to a pin in the same row as the RW lead of the LCD
- from digital pin nine (9) to a pin in the same row as the lead of the E lead of the LCD
- from digital pin 10 on the UNO board to a pin in the same row as the DB4 lead of the LCD
- from digital pin 11 on the UNO board to a pin in the same row as the DB5 lead of the LCD
- from digital pin 12 on the UNO board to a pin in the same row as the DB6 lead of the LCD
- from digital pin 13 on the UNO board to a pin in the same row as the DB7 lead of the LCD
- from the anode(+) column to a pin in the same row as the resistor that connects to the LED (VCC) lead of the LCD
- from the cathode(-) column to a pin in the same row as the GND lead of the LCD
- from the anode(+) column to a pin in the

same row as the VCC lead of the LCD
- from the cathode(-) column to a pin in the same row as the GND lead of the LCD
- from the same row as the center lead of the potentiometer to a pin in the same row as the VO lead of the LCD
- from the cathode(-) column to a pin in the same row as cathode(-) lead of the potentiometer
- from the anode(+) column to a pin in the same row as anode(+) lead of the potentiometer
- from digital pin four (4) to the column one (1) pin of the 4x4 calculator pad
- from digital pin five (5) to the column two (2) pin of the 4x4 calculator pad
- from digital pin six (6) to the column three (3) pin of the 4x4 calculator pad
- from digital pin seven (7) to the column four (4) pin of the 4x4 calculator pad
- from digital pin three (3) to the row four (4) pin of the 4x4 calculator pad
- from digital pin two (2) to the row three (3) pin of the 4x4 calculator pad
- from digital pin one (1) to the row two (2) pin of the 4x4 calculator pad
- from digital pin zero (0) to the row one (1) pin of the 4x4 calculator pad

Coding - Digital Prototype

11 Start the simulation. Try some calculations and observe the digital prototype. ✔

10 Enter the code in Tinkercad® to match the code on this page.

```cpp
#include <LiquidCrystal.h>
#include <Keypad.h>
LiquidCrystal lcd(8,9,10,11,12,13);
//connections to LCD
const byte Rows = 4; //first four
pins of the 4X4 keypad
const byte Columns = 4; //second
four pins of the 4X4 keypad
//keypad map
char keys [Rows] [Columns] = {
  {'1', '2', '3', '/'},
  {'4', '5', '6', '*'},
  {'7', '8', '9', '-'},
  {'C', '0', '=', '+'}
};
byte rowPins[Rows] = {0,1,2,3}; //
connection to the first set of four
rows of the keypad
byte colPins[Columns] = {4,5,6,7};
//connection to the second set of
four of the keypad
//keypad
Keypad myKeypad = Keypad(
makeKeymap(keys), rowPins,
colPins, Rows, Columns);
boolean valOnePresent = false;//
variables
boolean next = false;
boolean final = false;
String num1, num2;
float ans;
char operation;
void setup(){
  lcd.begin(16,2);
  lcd.setCursor(4,0); // LCD dis-
plays"My Arduino"
  lcd.print("My Arduino");
  lcd.setCursor(3,1); //LCD dis-
plays"Project"
  lcd.print("Project");
  delay(2300);
  lcd.clear(); //clears the display
(LCD)
}
void loop(){
  char key = myKeypad.getKey();
  if(key != NO_KEY && (key ==
'1'|| key == '2'|| key== '3'|| key ==
'4'|| key == '5'|| key == '6'||
    key == '7' || key == '8' || key ==
'9'|| key== '0'))
    {
      if (valOnePresent != true)
      {
        num1 = num1 + key;
        int numLength = num1.
length();
        lcd.setCursor(15 - num-
Length, 0); //to adjust one
whitespace for operator
        lcd.print(num1);
      }//end if()
      else
      {
        num2 = num2 + key;
        int numLength = num2.
length();
        lcd.setCursor(15 - num-
Length, 1);
        lcd.print(num2);
        final = true;
      }//end else()
    }//end if()
  else if (valOnePresent == false
&& key != NO_KEY && (key ==
'/' || key == '*' || key == '-' || key ==
'+'))
    {
      if (valOnePresent == false)
      {
        valOnePresent = true;
        operation = key;
        lcd.setCursor(15,0);
        lcd.print(operation);
      }//end if()
    }//end else if()

  else if (final == true && key !=
NO_KEY && key == '=')
    {
      if (operation == '+')
      {
        ans = num1.toInt() + num2.
toInt();
      }
      else if (operation == '-')
      {
        ans = num1.toInt() - num2.
toInt();
      }
      else if (operation == '*'){
        ans = num1.toInt() * num2.
toInt();
      }
      else if (operation == '/'){
        ans = float (num1.toInt()) /
float (num2.toInt());
      }
      lcd.clear();
      lcd.setCursor(15,0);
      lcd.autoscroll();
      lcd.print(ans);
      lcd.noAutoscroll();
    }
  else if (key != NO_KEY && key
== 'C')
    {
      lcd.clear();
      valOnePresent = false;
      final = false;
      num1 = "";
      num2 = "";
      ans = 0.0;
      operation = ' ';
    }
}
```

Engineering Design - Project Build

12 Place one 10k potentiometer to the breadboard such that the leads span three rows in a column.

13 Place a 16x2 LCD to the breadboard such that the leads are across the other half of the breadboard from the potentiometer and across the same column.

14 Place a 220 Ohm resistor on the breadboard such that it connects the same row across the halves of the breadboard and the same row as the LED (VCC) pin of the LCD (second from the right)

15 Use the three tables on this page to make the necessary connections with the jumper wires.

From	To (LCD L to R)
Breadboard Cathode - Column	LCD GND (Pin 1) Same Row
Breadboard Anode + Column	LCD VCC (Pin 2) Same Row
Potentiometer Center Pin Same Row	LCD VO (Pin 3) Same Row
UNO Digital Pin D8	LCD RS (Pin 4) Same Row
Breadboard Cathode - Column	LCD RW (Pin 5) Same Row
UNO Digital Pin D9	LCD E (Pin 6) Same Row
UNO Digital Pin D10	LCD DB4 (Pin 11) Same Row
UNO Digital Pin D11	LCD DB5 (Pin 12) Same Row
UNO Digital Pin D12	LCD DB6 (Pin 13) Same Row
UNO Digital Pin D13	LCD DB7 (Pin 14) Same Row
Breadboard Anode + Column	Breadboard (Pin 15) Same Row with Resistor
Breadboard Cathode - Column	LCD LED (Pin 16) Same Row

From	To
UNO 5v	Breadboard Anode + Column
UNO GND	Breadboard Cathode- Column
Breadboard Anode + Column	Potentiometer Anode + Same Row
Breadboard Cathode - Column	Potentiometer Cathode - Same Row

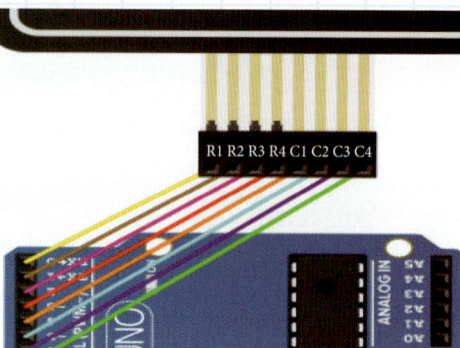

From UNO	To Keypad (left to right)
UNO Digital Pin D0	4 x 4 Calculator Pad Pin for Row 1
UNO Digital Pin D1	4 x 4 Calculator Pad Pin for Row 2
UNO Digital Pin D2	4 x 4 Calculator Pad Pin for Row 3
UNO Digital Pin D3	4 x 4 Calculator Pad Pin for Row 4
UNO Digital Pin D4	4 x 4 Calculator Pad Pin for Column 1
UNO Digital Pin D5	4 x 4 Calculator Pad Pin for Column 2
UNO Digital Pin D6	4 x 4 Calculator Pad Pin for Column 3
UNO Digital Pin D7	4 x 4 Calculator Pad Pin for Column 4

Project 12
Calculator

Connection - Mathematical and numerical skills have gone hand in hand with civilization and can be seen in all ancient civilizations. Record keeping, astronomical observations, calendar keeping, architecture, and taxation all rely on being able to perform mathematical calculations.

Among the earliest forms of writing found in the archaeological record are accounts of trades in the kingdom of Sumeria (modern day Iraq) preserved on clay tablets from more than 4500 years ago. This clay tablet is an accounting of silver for one of the governors in Sumeria from 2500 B.C.E.

Soon after the development of mathematical abilities, recording followed and then machines to make calculations faster emerged. Just prior to the invention of the electronic calculator, the slide rule (seen below) was the dominant device for making calculations in the mid 20th century.

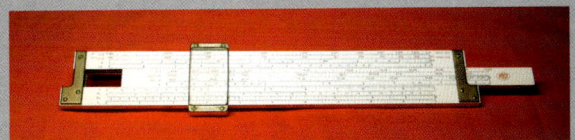

16 Connect the UNO board and computer using the USB cable.

17 Launch the Arduino® IDE software (pages 11-13) to make sure the board is communicating with the computer.

Go to the Tools menu and verify that the correct board is selected from the Board Manager menu. If not, select the correct board from the options listed.

Then confirm that the right Port is selected. If not, select the port that lists the UNO board from the options listed.

Coding - Project Build

18 Either copy and paste the code from the Tinkercad® prototype or type the from page 151 into the Arduino® IDE software (pages 11-13).

19 Install the keypad library (page 62) using the link on the "Links" page from the book web page.

20 Save the sketch (rename as needed).

21 Verify the sketch.

22 Upload the sketch to the UNO board.

23 Perform several calculations to see how the calculator works.

24 Document this project, discuss the project, and complete the extensions assigned from the next page.

A = Essential   **B** = Recommended   **C** = Optional

Reflection/Summative Activities:

- Have each team member document in their project/ classroom journal who was on their team, what went well, what they could improve upon, what they would do differently if they were to do the project again, and verify that every person in the group can do the project.

- How is your calculator like an abacus? Document your response.

Section Reflection/Summative Activities:

- Have the team create a graphic organizer summarizing the skills and knowledge they obtained in doing Projects Seven through Twelve.

- Document in your journal key vocabulary, inputs and controls used, and skills developed in this section.

For Discussion:

- What are the advantages and disadvantages to a calculator?

- What are the limitations to this design as compared to other calculators?

Extension:

- Program one of the keys to perform the squaring function.

Historical Connection:

- Research and write a short description of a calculating device used in the past. Be sure to include a photo or drawing of the device, explain how it counted or calculated, and for how long the device was used.

Extension:

- Program one of the keys to perform another mathematical function.

- Submit your project to the book page on the website. Register and Login to submit at: thearduinoclassroom.com

From the homepage:
Click
    Books
Click
    UNO Edition Vol. 1

Problem:

- Identify a problem that would be solved by the calculator like the one you have been working with in this project. How could this project be a solution? Include in your report the following:

Description of the problem.

Description of how the problem is solved by this project.

What other resources you would need to solve the problem.

Design and produce the solution.

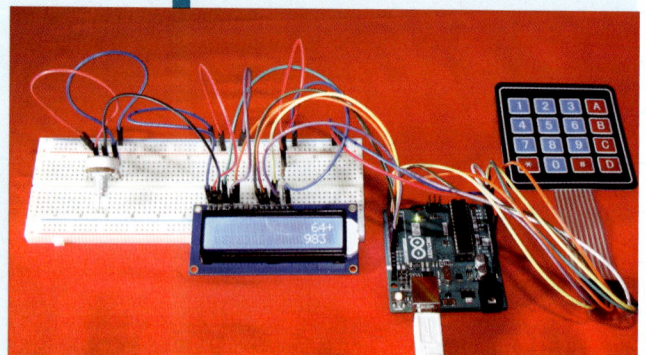

A project rubric can be found on page 236.

Anticipatory Sets
(from page 116)

Anticipatory Sets:
- Where would you see calculators used? List at least three different places or situations where calculators are used, describe the devices, and explain the purpose of the devices.
Calculators are used in math classes, businesses including accounting firms.

A

Reflection/Summative Activities:
Responses Will Vary

Section Reflection/Summative Activities:
Responses Will Vary

For Discussion:
- What are the advantages and disadvantages to a calculator?
Advantages include the speed of calculations, the accuracy of the calculations, the ability of handling larger numbers. Disadvantages include being subject to the errors of the user.

- What are the limitations to this design as compared to other calculators?
Wiring, coding, the board, the pad and breadboard. Other calculators are hardwired and self-contained.

Extension:
Program one of the keys to perform the squaring function.
Responses Will Vary

B

Professional Connection:
Responses Will Vary

Extension:
- Program one of the keys to perform another mathematical function.
Responses Will Vary

- Submit your project to the book page on the website.
Responses Will Vary

C

Problem:
Responses Will Vary

Section 3
Sensors

SKILLS

PROJECTS

The first 12 projects and 14 skills presented give an overview of the variety of capabilities one has with Arduino® projects. Built on what has been presented, the complexity of the projects will increase with the next 13 projects. But what comes next? At some point, you will want to go on the Internet and try projects not in this book. This section will address some things you should consider as you wade into the open source community.

Open Source Projects Overview

One of the best reasons to work with Arduino® projects is the fact that it is an open source platform with a large open source project community that constantly is competitive and cooperative at the same time. Every person in the ecosystem started just like you are starting, at the beginning with only interest driving them.

Plus because the Arduino® platform is based on C/C++ there are a multitude of coding experts that like to contribute to the Arduino® ecosystem. If you look online, there are dozens of sites with projects to pick from.

It has been our experience that most projects that are on open source sites fail the first time the project is implemented. Even some of the manufacturers or parts distributors that provide projects will fail in your first implementation.

As well meaning as the person or organization is that uploads the project, you will inevitably need to fix at least one issue. The project may even have a video showing the project working, but if you look closely, you might observe changes in the project in the video as compared to the parts lists, diagrams, code, or photos provided. **Do not give up.** Use your skills to troubleshoot the issue and you will be rewarded with a working project.

As a start, review page 67 of this book for troubleshooting ideas. This gives you a great place to start in trying to fix the problems you are encountering.

There are three technical reasons the projects fail:
1. Code
2. Design and Connections
3. Hardware

Page 67 goes into detail on things to look for in the code and some aspects of the design. We will add things to look for when building a project from an open source site. We will start with what we did not cover on page 67.

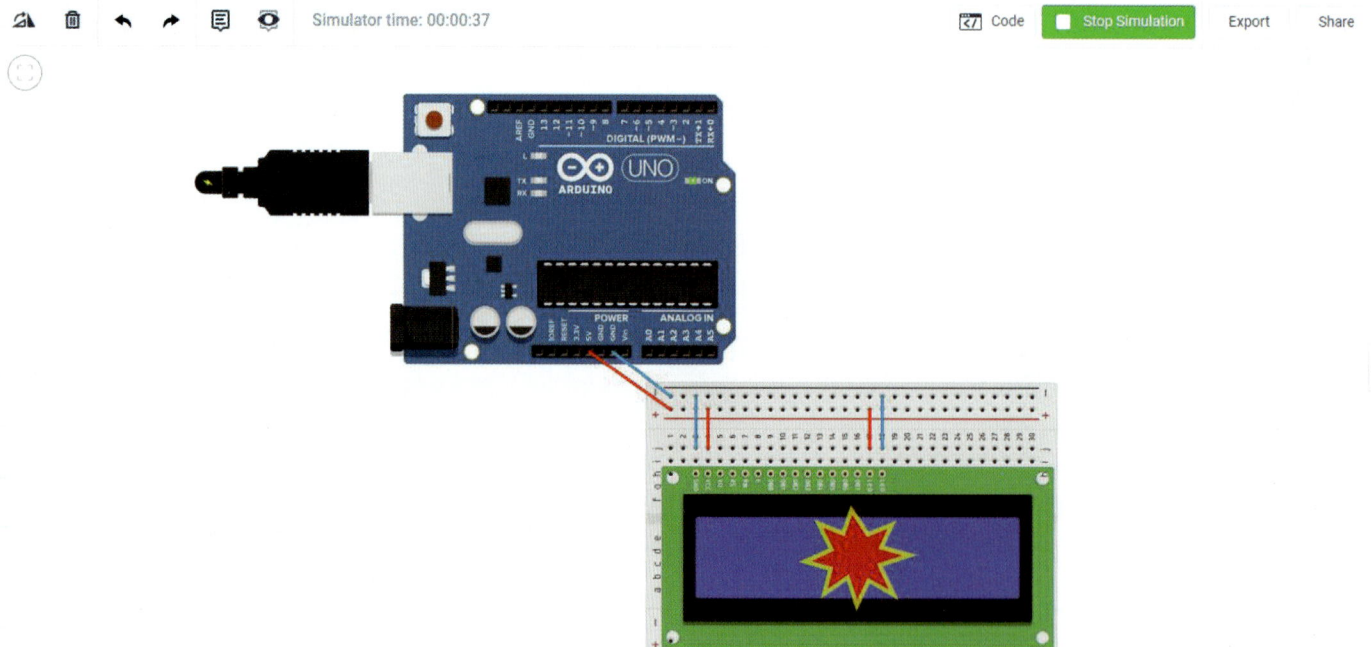

Hardware

Generally parts from different sources are the same. All LEDs are manufactured with an anode(+) lead and a cathode(-) lead. The anode(+) lead being longer. You can be safe in not having to worry about LEDs, resistors, or jumper wires. However once the parts get more specialized, differences will occur. For example Arduino® UNO boards are open source and manufactures may change pin locations, colors, and most often how the boards are labeled.

To start, familiarize yourself with the parts you have. Read their technical specifications or build simple projects with them to understand how they work. If you are purchasing a part for the first time for a project, review the documentation before ordering.

Understand the voltage requirements, the on-board connections, the input and output requirements, and capabilities.

The part may work best only with 3.3 volt inputs or may need to be assembled or soldered. The part may have anode(+) and cathode(-) connection requirements or not have any polarity to consider. It may work best with a specific resistor or its functioning may vary with each level of resistance.

Some manufactures have special connections and/or wiring that require an intermediary board to translate the signal date from the sensor.

One peculiarity we have found is hardware needing calibration. Some of the gas sensors often need to warm up for a length of time, then a reading determined using one sketch which then is entered into a second sketch that does the sensing.

Humidity, ambient light, temperature and sounds all may impact the operation of a sensor. Fortunately most code allows for calibration of the input and output.

Some distributors or online warehouses are so big, there might be several dozen versions of the "same" device which will require more homework on your part. We recommend finding a distributor / manufacturer that specializes in electronics and not an online warehouse. Look to see that they show the part in use, supply the Arduino® code, and show photos and diagrams.

Price should be a tertiary factor in your decision. The old adage of, "you get what you pay for," holds true. A cheap part off an auction site or discount site is still cheap. It is almost always worth it to pay more for a better part.

Once you know your hardware, you can look at the project's design and code to change these as needed to get things working.

Design and Connections

With your knowledge of the hardware now look at the design of the project and the connections.

Make sure anode(+) and cathode(-) sides of the circuit are connected correctly. Trace electricity from the UNO board through the hardware from anode(+) to cathode(-). Everything should complete a circuit.

If the hardware has three or more connections look to see if there are data inputs or outputs along with the electrical connections. Understand how the UNO board sends and receives data. Make note of the pin numbers the connections uses on the UNO board for referencing to the code.

Look at placement of hardware on the breadboard to ensure each circuit is complete. We have noticed many diagrams missing jumper wires!

Sometimes it is as simple as reseating the pins and wires to connect on a different part of the breadboard.

Code

Page 67 is a great resource for code troubleshooting. In addition to the information on that page, make sure that the correct library or libraries are downloaded, installed and referenced in the code.

And lastly, compare the code to known working code.

Project 13
Night Light

Lesson Integration:	Groupings:	Level:	Time to Complete:
Physics - Ohm's Law, Circuits, Voltage, Resistance, Light Earth Science - Length of Days, Seasons, Precession Mathematics - Ratios	1 - 2	Starter	30 min. for the project and 45 min. for extensions.

Objectives:
- Investigate how circuits and electronic components interact with electrical energy.
- Control light a light in a prototype and machine with a photoresistor

Prerequisite Skills:
- Arduino® IDE software (pages 11-13)
- Tinkercad® (pages 15-16)
- Serial Monitor and Plotter (pages 58-59)

Purpose and Skills:
- Control an LED light using a light sensor

STEAM Connections:
Science - Circuits, Resistance, Ohm's law, Light Intensity, Days, Seasons
Technology - Code, Program Settings, Simulators, Digital Design, Electronic Components, Analog Data, Conditional Statements, Resistor
Engineering - Designing, Building, and Using a

Machine, Prototyping, Applied Physics
Allied Arts - Lighting
Math - Ratios, Conversions, Comparisons, Logic

Key Vocabulary:
Photoresistor - A device that changes its electrical resistance based on the amount of light it senses.
Sensor - A device that measures a environment for an input (i.e. light, sound, temperature, etc..)

Project Introduction:
- Introduce the groups to the purpose of the project, the skills developed, the standards met, and the goal of the project.

Anticipatory Sets:
- Where would you see a night light or light sensitive switch used? List at least three different places or situations where light sensitive technologies are used, describe the devices, and explain the purpose of the devices.

Project 13
Night Light

Educational Standards

ISTE Standards for Students
- Empowered Learner 1a, 1b, 1d
- Knowledge Constructor 3a, 3b, 3c, 3d
- Innovative Designer 4a, 4b, 4c, 4d
- Computational Thinker 5a, 5b, 5c, 5d
- Creative Communicator 6a, 6b, 6c, 6d
- Global Collaborator 7c, 7d

US Computer Science Standards
- Project correlations can be found on the book's web page.

US NGSS - Middle School
DCIs
- MS-PS4-3   PS4.C
- MS-PS4-1   PS4.A
- MS-ESS1-1 ESS1.B

Cross Cutting Concepts
- Cause and Effect
- Scale, Proportion and Quantity
- Structure and Function

Science and Engineering Practices
- Planning and Carrying Out Investigations
- Obtaining, Evaluating, and Communicating Information
- Analyzing and Interpreting Data
- Scientific Knowledge is Based on Empirical Evidence
- Using Mathematics and Computational Thinking

US NGSS - High School
- Correlations can be found on the book's web page.

US Common Core Language Arts and Mathematics
- Correlations can be found on the book's web page.

Materials List:
- Computer with IDE software
- Connection to the Internet
- USB Cable
- UNO or UNO Compatible Microcontroller
- Short Breadboard
- LED
- Six (6) Jumper Wires (Male to Male)
- 220 Ohm Resistor
- 100 Ohm Resistor
- Photoresistor (LDR)

Engineering Design - Digital Prototype

1 Launch and log into Tinkercad® (pages 15-16).

2 Start a new Circuits project.

3 Drag an UNO board to the workspace.

4 Drag a breadboard to the workspace.

5 Place an LED across two rows of a column.

6 Place the photoresistor across two different rows of a column.

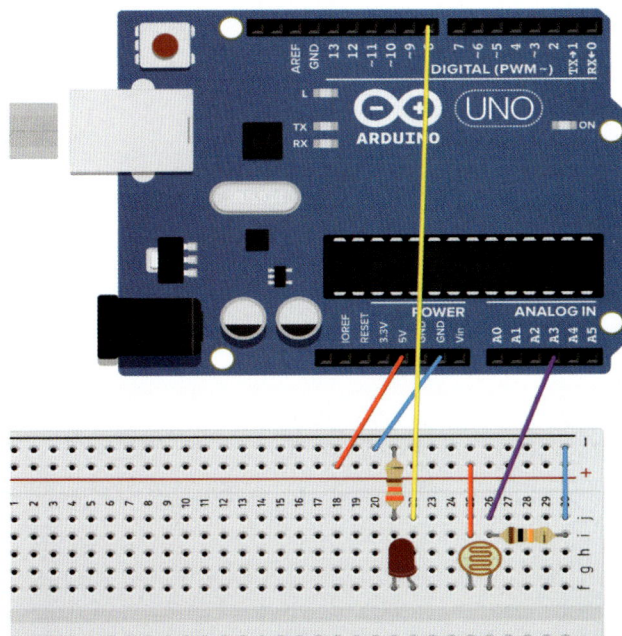

Project 13
Night Light

7 Place a 220 Ohm resistor on the breadboard to connect the cathode(-) column and the row that has the cathode(-) lead of the LED.

8 Place a 100 Ohm resistor on the breadboard to connect one lead of the photoresistor to an open row in the same column.

9 Connect with jumper wires the following:
- the five volt (5v) pin to the anode(+) column
- digital pin eight (8) to the same row as the anode(+) lead of the LED.
- a ground (GND) pin on the UNO board to the cathode(-) column on the breadboard.
- the anode(+) row to the row with the open lead of the photoresistor
- analog pin A3 to the same row shared by the 100 ohm resistor and the lead of the photoresistor.
- the cathode(-) row to the row with the other lead from the 100 Ohm resistor

Coding - Digital Prototype

10 Enter the code in Tinkercad® to match the code on this page.

11 Start the Simulation. Click on the photoresistor, change the light settings and observe the digital prototype.

Using any other resistor than the 100 Ohm resistor will require changing the code. Specifically the values in the conditional statements will need to be changed. A higher Ohm resister will require lower values while a lower Ohm resistor or no resistor will require higher values. In any of these cases the values will have to be calibrated to the conditions being measured.

```
int light = 0;        // Set light variable value
void setup() {
   Serial.begin(9600); //for reading serial monitor on tools section
   pinMode(8, OUTPUT); // Set digital pin 8 as OUTPUT for LED
}
void loop() {
   light = analogRead(A3); // reads the information provided by the photocell
   Serial.println(light);  // displays light values
   if(light > 670) {
      Serial.println("Light!");
      digitalWrite(8,LOW); // Led is off
   }
   else if(light > 330 && light < 670) {
      Serial.println("Dim light!");
      digitalWrite(8,LOW);  // Led is off
   }
   else {
      Serial.println("Dark!");
      digitalWrite(8,HIGH); // Led is on
   }
   delay(500); //Time in milliseconds
}
```

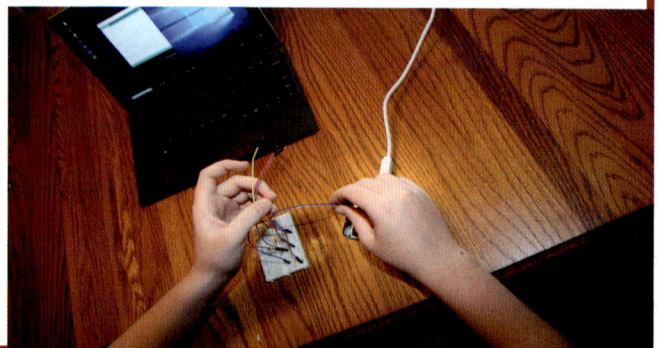

```
1   //Project 13
2   //Light detection with a LED and Photocell
3   //thearduinoclassroom.com
4   //Copyright 2019, Isabel Mendiola and Peter Haydock
5
6   int light = 0;              // Set light variable value
7   void setup() {
8       Serial.begin(9600); //for reading serial monitor on tools section
9       pinMode(8, OUTPUT); // Set digital pin 8 as OUTPUT for LED
10  }
11  void loop() {
12      light = analogRead(A3); // reads the information provided by the photocell
13      Serial.println(light);  // displays light values
14      if(light > 670) {
15          Serial.println("Light!");
16          digitalWrite(8,LOW); // Led is off
17      }
18      else if(light > 330 && light < 670) {
19          Serial.println("Dim light!");
20          digitalWrite(8,LOW);   // Led is off
21      }
22      else {
23          Serial.println("Dark!");
24          digitalWrite(8,HIGH); // Led is on
25      }
26      delay(500); //Time in milliseconds
27  }
```

Connection - A common night light with an LED light and a photoresistor on the front that activates the light when the room or hallway darkens.

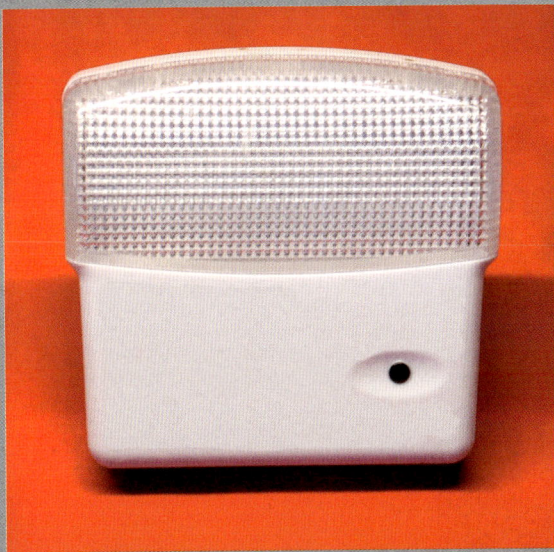

Engineering Design - Project Build

12 Place an LED across two rows of a column on the breadboard.

13 Place the photoresistor across two different rows of a column.

14 Place a 220 Ohm resistor on the breadboard to connect the cathode(-) column and the row that has the cathode(-) lead of the LED.

15 Place a 100 Ohm resistor on the breadboard to connect one lead of the photoresistor to an open row in the same column.

16 Connect with jumper wires the following:
- the five volt (5v) pin to the anode(+) column
- digital pin eight (8) to the same row as the anode(+) lead of the LED.
- a ground (GND) pin on the UNO board to the cathode(-) column on the breadboard.
- the anode(+) row to the row with the open lead of the photoresistor
- analog pin A3 to the same row shared by the 100 ohm resistor and the lead of the photoresistor.
- the cathode(-) row to the row with the other lead from the 100 Ohm resistor

17 Review the circuit diagram for this project.

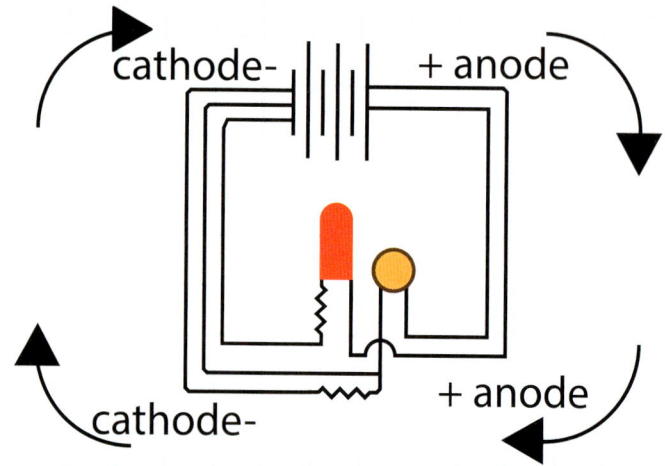

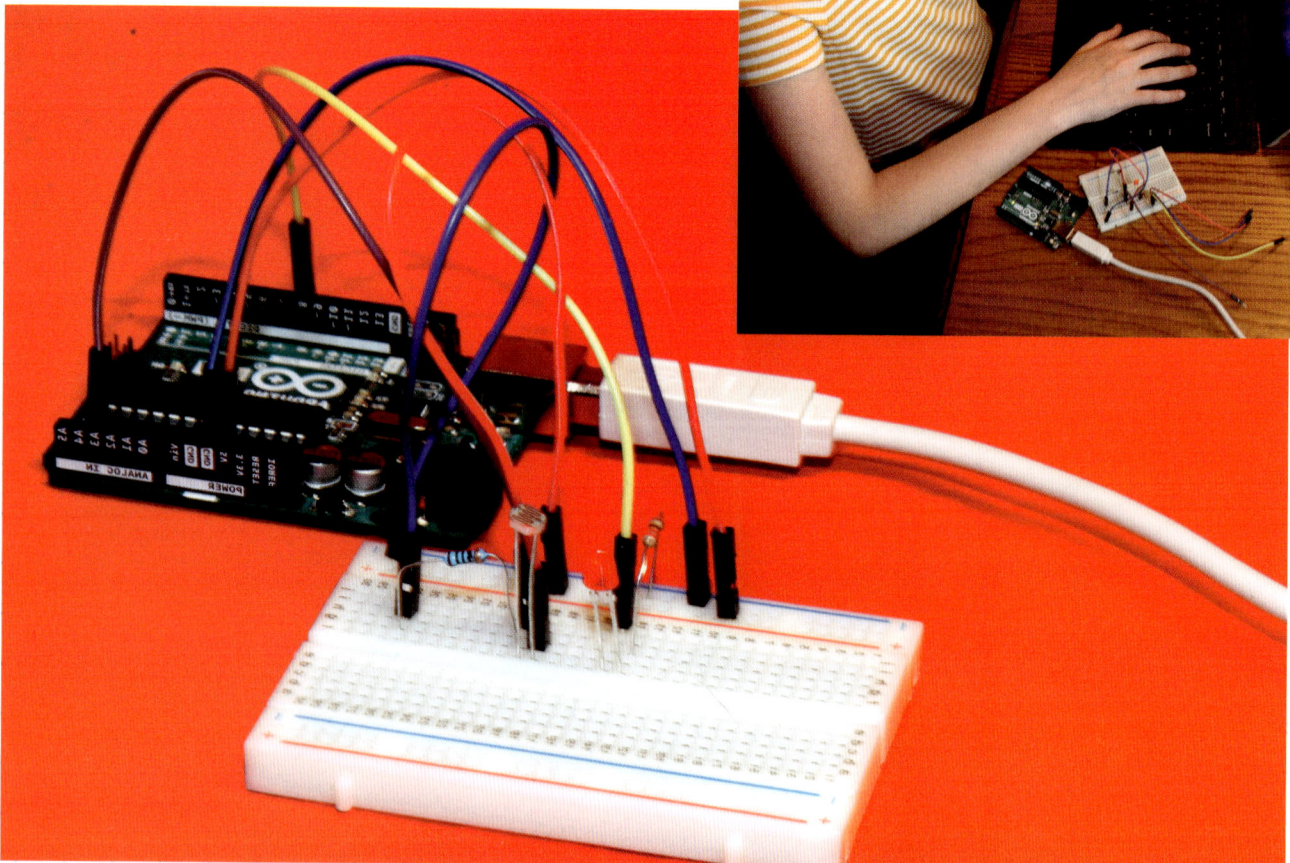

18 Connect the UNO board and computer using the USB cable.

19 Launch the Arduino® IDE software (pages 11-13) to make sure the board is communicating with the computer.

Go to the Tools menu and verify that the correct board is selected from the Board Manager menu. If not then select the correct board from the options listed.

Then confirm that the right Port is selected. If not then select the port that lists the UNO board from the options listed.

Coding - Project Build

20 Either copy and paste the code from the Tinkercad® prototype or type the code from page 130 into the Arduino® IDE software (pages 11-13).

21 Save the sketch (rename as needed).

22 Verify the sketch.

23 Upload the sketch to the UNO board.

24 Test the project by varying the amount of light the photoresistor receives.

25 Document this project, discuss the project, and complete the extensions assigned from the next page.

```
TAC_UE_V1_P13_Night_Light | Arduino 1.8.9 (Windows Store 1.8.21.0)     -  □  ×
File Edit Sketch Tools Help

  TAC_UE_V1_P13_Night_Light

//Project 13
//Light detection with a LED and Photocell
//thearduinoclassroom.com
//Copyright 2019, Isabel Mendiola and Peter Haydock

int light = 0;          // Set light variable value
void setup() {
    Serial.begin(9600); //for reading serial monitor on tools section
    pinMode(8, OUTPUT); // Set digital pin 8 as OUTPUT for LED
}
void loop() {
    light = analogRead(A3); // reads the information provided by the photocell
    Serial.println(light);  // displays light values
    if(light > 670) {
        Serial.println("Light!");
        digitalWrite(8,LOW); // Led is off
    }
    else if(light > 330 && light < 670) {
        Serial.println("Dim light!");
        digitalWrite(8,LOW);  // Led is off
    }
    else {
        Serial.println("Dark!");
        digitalWrite(8,HIGH); // Led is on
    }
    delay(500); //Time in milliseconds
}

Done Saving.
```

Connection - On the top of this streetlight is a light detector. It turns on and off the light depending on the amount of sunlight available.

Also, it can be adjusted to turn on with differing amounts of light. Generally it will turn on just as the sun is setting, but will also turn on if cloud cover blocks enough sunlight. A summer thunderstorm will generally be dark enough to turn on the light. Workers periodically clean the sensor on the light to ensure that the light is not turning on too early before or staying on too late after sunrise.

Light sensors combined with an ultra efficient LED or as in this case a sodium based light source reduce the amount of electricity used.

These sensors are especially useful in the higher northern and southern latitudes where daylight time can vary by several hours over the course of a year.

A = Essential

Reflection/Summative Activities:

- Have each team member document in their project/classroom journal who was on their team, what went well, what they could improve upon, what they would do differently if they were to do the project again, and verify that every person in the group can do the project.

- As an extension, create a proposal that would detail producing 100 or more night lights for a community initiative (ex. for a nursing home). Include in your proposal, the project costs, project specifications, and the needs and wants of the purchaser.

For Discussion:

- Why is there a need for photoresistor?

- Look at the structure of the code and show how the conditional statements (if, if else, else) could be changed to change the output of the LED ?

Extensions:

- Change and test the code by adjusting the sensitivity settings to change the response of the LED (blinking in dim light, full lighting in brighter conditions etc..).

- Plan and implement the use of multiple LEDs to respond to the level of light.

B = Recommended

Connection:

- If you were designing an electronic or bionic eye, what part could a photoresistor play? What other kinds of devices would you need to make an artificial eye?

Extensions:

- Change the code to record the brightest or darkest light level over an hour/day/year.

- Design and implement changes to the project to light an RGB LED different colors depending on the input from the photoresistor.

- Design and implement changes to the project to operate at lest one servo depending on the input from the photoresistor.

- Submit your project to the book page on the website. Register and Login to submit at: thearduinoclassroom.com

From the homepage:
Click
 Books
Click
 UNO Edition Vol. 1

C = Optional

Extensions:

- Add or edit lines of code to the sketch to light an RGB LED different colors depending on the input from the photoresistor.

- Change the code to use the Serial Plotter and record light levels over the course of a set time frame (hour(s) or day(s)).

- Change the design and code to use the LCD Display instead of the Serial Monitor.

- Calibrate the sensitivity levels to specific lumen levels.

Problem:

- Review a product catalog (online or in print) and identify an opportunity to include a photoresistor controlled light. How could this project be a solution? Include in your report the following:

Description of the problem.

Description of how the problem is solved by this project.

What other resources you would need to solve the problem.

Design and produce the solution.

A project rubric can be found on page 236.

Anticipatory Sets
(from page 128)

Anticipatory Sets:
- Where would you see a night light or light sensitive switch used? List at least three different places or situations where pushbutton counters are used, describe the devices, and explain the purpose of the devices.
Automated street lights, night lights that turn on after dark, garden lights.

A

Reflection/Summative Activities:
Responses Will Vary

For Discussion:
- What are the advantages and disadvantages in using a photoresistor?
The photoresistor has many levels of sensitivity that can provide data to the sketch. It is a separate device that has to be programmed or could fail.

- Look at the structure of the code and show how the conditional statements (if, if else, else) could be changed to change the output of the LED ?
One could reverse the logic, change the data sensitivity points for the LED or eliminate one or two of the points.

Extensions:
- Change and test the code by adjusting the sensitivity settings to change the response of the LED (blinking in dim light, full lighting in brighter conditions etc...).
Responses Will Vary

- Plan and implement the use of multiple LEDs to respond to the level of light.
Responses Will Vary

B

Connection:
Responses Will Vary

Extensions:
- Change the code to record the brightest or darkest light level over an hour/day/year.
Responses Will Vary

- Design and implement changes to the project to light an RGB LED different colors depending on the input from the photoresistor.
Responses Will Vary

- Design and implement changes to the project to operate at lest one servo depending on the input from the photoresistor.
Responses Will Vary

- Submit your project to the book page on the website. Register and Login to submit at: thearduinoclassroom.com
Responses Will Vary

C

Extensions:
- Change the code to use the Serial Plotter and record light levels over the course of a set time frame (hour(s) or day(s)).
Responses Will Vary

- Change the design and code to use the LCD Display instead of the Serial Monitor.
Responses Will Vary

- Calibrate the sensitivity levels to specific lumen levels.
Responses Will Vary

Problem:
Responses Will Vary

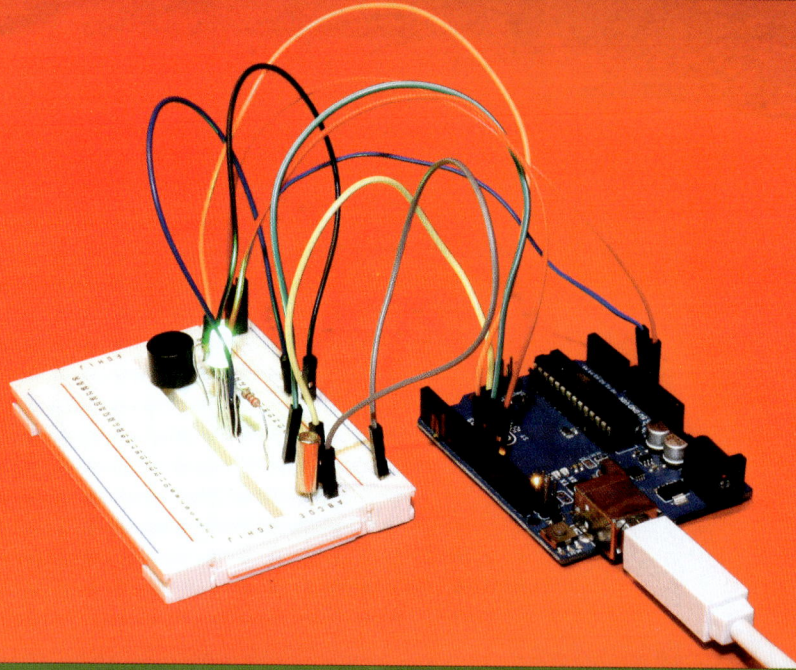

Getting Started

Lesson Integration:	Groupings:	Level:	Time to Complete:
Physics - Ohm's Law, Circuits, Voltage, Resistance, Light Earth Science - Earthquakes, Seismology Mathematics - Logic	1 - 2	Starter	45 min. for the project and 45 min. for extensions.

Objectives:
- Investigate how circuits and electronic components interact with electrical energy.

Prerequisite Skills:
- Time measurement (milliseconds)
- Sound Frequency
- Arduino® IDE software (pages 11-13)
- Fritzing (pages 64-66)

STEAM Connections:
Science - Circuits, Resistance, Ohm's Law, Seismology
Technology - Code, Program Settings, Simulators, Digital Design, Electronic Components, Analog Data, Conditional Statements
Engineering - Designing, Building, and Using a Machine, Prototyping, Applied Physics
Allied Arts - Colors, Sound, Frequency
Math - Logic, Boolean Operators

Purpose and Skills:
- Use input from a sensor to control outputs
- Use conditional code to change LED and sound states

Key Vocabulary:
Seismology - the study of earthquakes and other movements of Earth from plate tectonics, landslides, and explosions.

Project Introduction:
- Introduce the groups to the purpose of the project, the skills developed, the standards met, and the goal of the project.

Anticipatory Sets:
- Where would you see a bump detector used? List at least three different places or situations where bump detector technologies are used, describe the devices, and explain the purpose of the devices.

Note: The sensor for this project is not in Tinkercad® nor is it in Fritzing. The step-by-step instructions explain how to substitute an equivalent part in Fritzing.

Project 14
Sensing Jolts, Bumps, and Rattles

Step-by-Step 1-25

Materials List:
- Computer with IDE software and Fritzing
- Connection to the Internet
- USB Cable
- UNO or UNO Compatible Microcontroller
- Short or Long Breadboard
- RGB LED (Anode +)
- 9 Jumper Wires (Male to Male)
- Piezoelectric Buzzer (Anode +) Piezo
- Tilt Ball Sensor

Engineering Design - Digital Prototype

1 Download, install, and launch (as needed) Fritzing (pages 64-66) to build a digital prototype.

2 Start a new sketch in Fritzing.

3 Drag an UNO board to the workspace.

4 Place an RGB LED across four rows of the same column of the breadboard. Recall that RGB LEDs may be either anode(+) or cathode(-) and may have different lead arrangements for the color diodes. For the prototype use the Parts Inspector to examine the properties of the RGB LED.

5 Place a 220 Ohm resistor with both leads in the same column to connect the same row as the red lead of the RGB LED and an empty row

6 Place a 220 Ohm resistor with both leads in the same column to connect the same row as the green lead of the RGB LED and an empty row

7 Place a tilt switch (as a substitute) on the breadboard across two rows of the same column as the piezo and RGB LED.

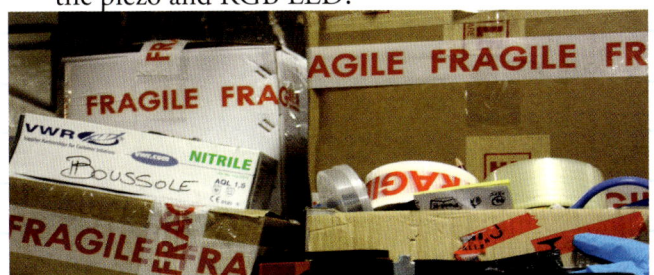

8 Connect with jumper wires the following:
- a ground (GND) pin on the UNO board to the cathode(-) column on the breadboard.
- the 5v pin to the anode(+) column on the breadboard.
- digital pin nine (9) to the same row as the anode(+) lead of the piezo
- the cathode(-) row to the same row as the cathode(-) lead of piezo
- digital pin 12 to the same row open lead of the resistor that connects to the row with the red lead of the RGB LED
- digital pin 11 to the same row open lead of the resistor that connects to the row with the green lead of the RGB LED
- the anode(+) row of the breadboard to the same row as the anode(+) lead of the RGB LED. Unlike Tinkercad®, Fritzing has an anode(+) RGB LED and can be used.
- the cathode(-) row to the row as one lead of the tilt ball sensor.

- digital pin 10 to the same row as the other lead of the tilt ball sensor.

Coding - Digital Prototype

9 Optional. Enter the code in Tinkercad® to match the code on this page.

10 Start the simulation. If the code has been entered correctly. The code will run without an error, but nothing will happen on the stage.

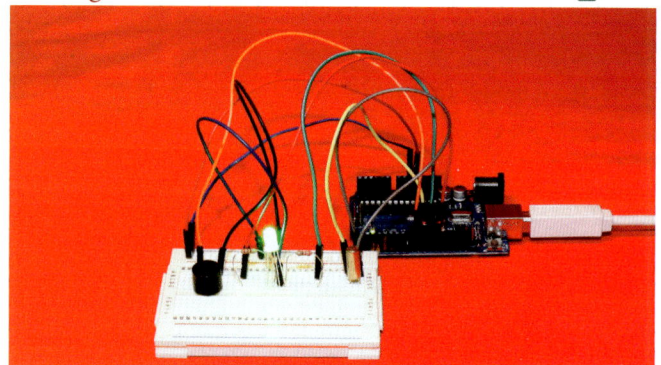

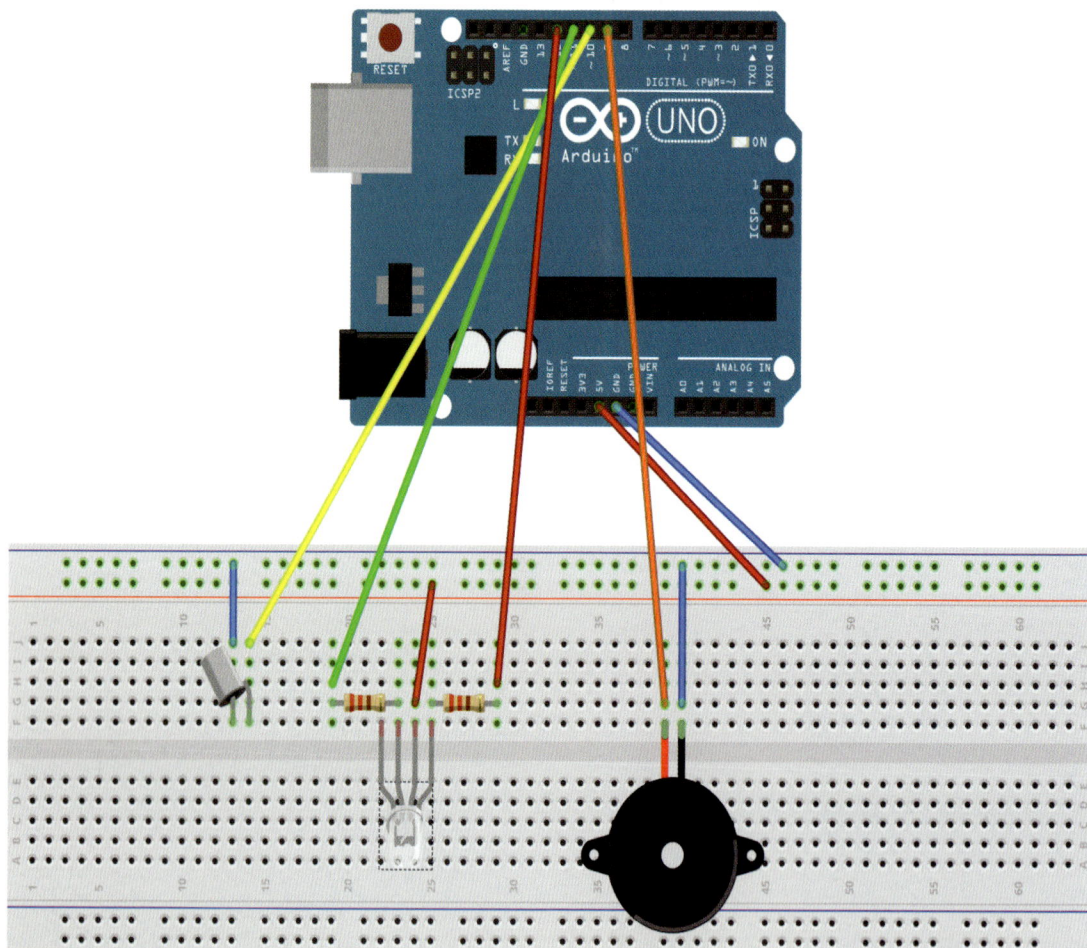

```
int redPin = 12;  // Declare variables
int greenPin = 11;
int sensor = 10;
int buzzer = 9;
int val =0;

void setup ()
{
 pinMode (redPin, OUTPUT) ; // red
 pinMode (greenPin, OUTPUT) ; // green
 pinMode (sensor, INPUT) ;  // sensor
 pinMode (buzzer, OUTPUT) ;  // buzzer
 Serial.begin(9600);
}
void loop ()
{
 val = digitalRead (sensor) ; // Reads sensor
values
Serial.println(val); // Print Sensor Value in Seri-
al Monitor
 if (val > 0) // LED changes color when motion
detected
  {
  digitalWrite (redPin, HIGH);
  digitalWrite (greenPin, LOW);
  tone (buzzer, 552);    // Start Buzzer
  delay (500);        //Buzzer Time
  noTone (buzzer); //Turns off any residual
  signal to the buzzer
  delay (100);
  tone (buzzer, 552);    // Start Buzzer
  delay (100);        //Buzzer Time
  noTone (buzzer); //Turns off any residual
  signal to the buzzer
  }
  else
  {
  digitalWrite (redPin, LOW);
  digitalWrite (greenPin, HIGH);
  noTone (buzzer);
  }
}
```

Engineering Design - Project Build

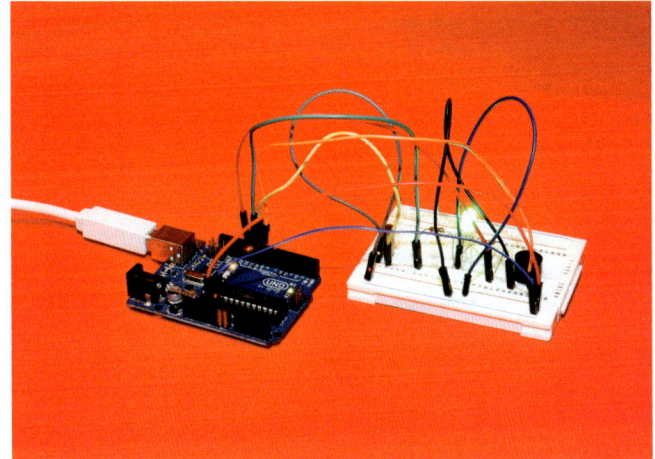

11   Place one piezo on the breadboard across two rows of a column

12   Place the cathode(-) RGB LED across four rows of the same column.

13   Place a 220 Ohm resistor with both leads in the same column to connect the same row as the red lead of the RGB LED and an empty row

14   Place a 220 Ohm resistor with both leads in the same column to connect the same row as the green lead of the RGB LED and an empty row

15   Place the tiltball sensor in the same column as the piezo and RGB LED.

Connection - An Amouri seismograph from the early 1800s run by the US Weather Service to measure earthquakes.

16 Connect with jumper wires the following:

- a ground (GND) pin on the UNO board to the cathode(-) column on the breadboard.
- the 5v pin to the anode(+) column on the breadboard.
- digital pin nine (9) to the same row as the anode(+) lead of the piezo
- the cathode(-) row to the same row as the cathode(-) lead of piezo
- digital pin 12 to the same row open lead of the resistor that connects to the row with the red lead of the RGB LED
- digital pin 11 to the same row open lead of the resistor that connects to the row with the green lead of the RGB LED
- the cathode(-) row of the breadboard to the same row as the cathode(-) lead of the RGB LED
- the cathode(-) row to the row as one lead of the tiltball sensor.
- digital pin 10 to the same row as the other lead of the tiltball sensor.

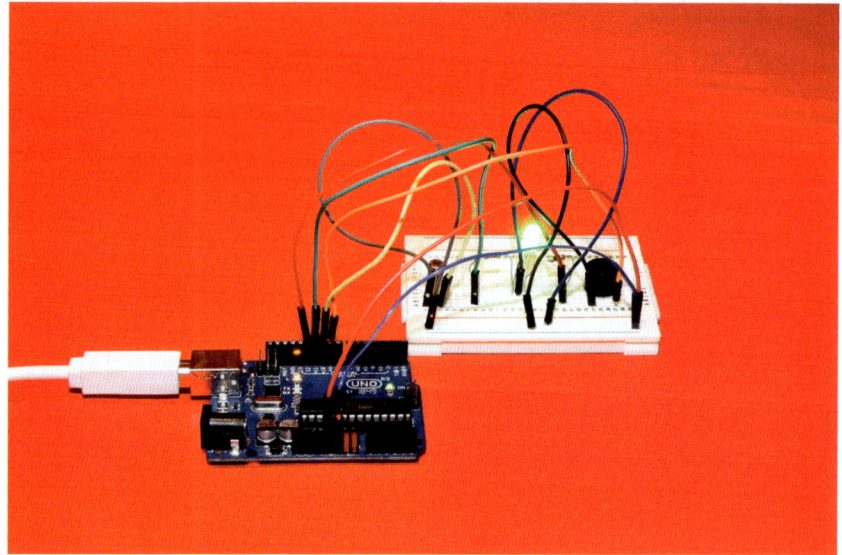

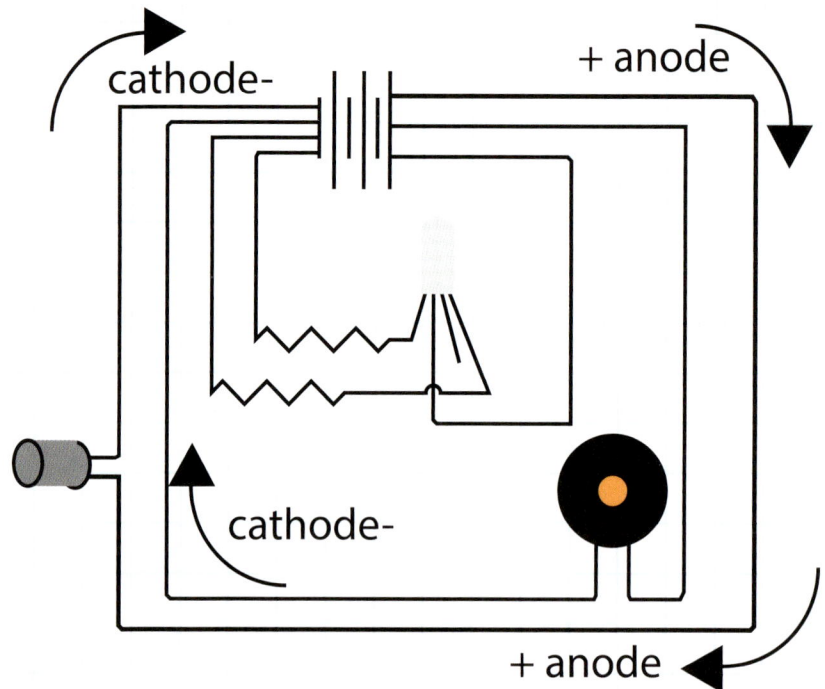

17 Study the circuit diagram for this project.

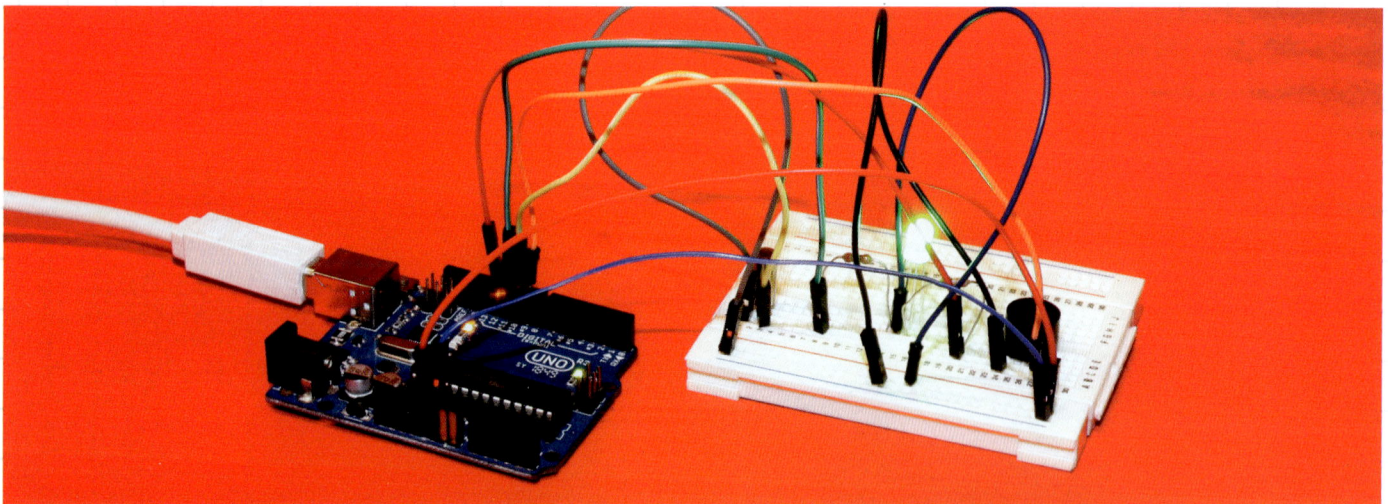

cathode-   + anode

cathode-

+ anode

18 Connect the UNO board and computer using the USB cable.

19 Launch the Arduino® IDE software (pages 11-13) to make sure the board is communicating with the computer.

Go to the Tools menu and verify that the correct board is selected from the Board Manager menu. If not, select the correct board from the options listed.

Then confirm that the right Port is selected. If not, select the port that lists the UNO board from the options listed.

Coding - Project Build

20 Either copy and paste the code from the Tinkercad® prototype or type the code from page 139 into the Arduino® IDE software (pages 11-13).

21 Save the sketch (rename as needed).

22 Verify the sketch.

```
TAC_UE_V1_P14_Adafruit_Tiltball_Sensor
//Project 14
//Tilt Ball Sensor
//thearduinoclassroom.com
//Copyright 2019, Isabel Mendiola and Peter Haydock

int redPin = 12;  // Declare variables
int greenPin = 11;
int sensor = 10;
int buzzer = 9;
int val =0;

void setup ()
{
  pinMode (redPin, OUTPUT) ; // red
  pinMode (greenPin, OUTPUT) ; // green
  pinMode (sensor, INPUT) ;  // sensor
  pinMode (buzzer, OUTPUT) ;  // buzzer
  Serial.begin(9600);
}
void loop ()
{
  val = digitalRead (sensor) ; // Reads sensor values
  Serial.println(val); // Print Sensor Value in Serial Monitor

  if (val > 0)
  {
  digitalWrite (redPin, HIGH);
  digitalWrite (greenPin, LOW);
  tone(buzzer, 552);    // Alarm tones
  delay (500);
  noTone(buzzer);
  delay (100);           //Time between each beep
  tone(buzzer, 552);
  delay (500);
  noTone(buzzer);
  }
  else
  {digitalWrite (redPin, LOW);
  digitalWrite (greenPin, HIGH);
  noTone(buzzer);
  }
}
```

23 Upload the sketch to the UNO board.

24 Test the project with bumps and other movements of the breadboard.

25 Document this project, discuss the project, and complete the extensions assigned from the next page.

Connection - The ShockLog® 298 from SpotSee™ monitors shipments for vibrations, bumps, and shocks. It records the time and intensity of the events and communicates them to a recipient for evaluation.
Companies use devices like this to monitor shipments of equipment and products that are sensitive to excessive movement or temperatures. Power generation and transmission equipment, electronics, and packaged goods are excellent examples of products that benefit from a technology like this.
SpotSee™ builds add-ons for the ShockLog® 298 that can track the location of impacts as well as temperature, air pressure, and humidity conditions of the package. There is even an add-on that can send the data in real-time via cellular phone call. Monitoring technologies like this identify where and when a package underwent a stress as it shipped, helping to avoid costly delays to projects and protect a company's investment.

Photo Courtesy of SpotSee™

Project 14
Sensing Jolts, Bumps, and Rattles
Project Reflection and Summative Activities, Discussion Starters, Extensions, and Problem

A = Essential

Reflection/Summative Activities:
- Think of a medical facility like a hospital, clinic or doctor's office and in pairs discuss where a bump detector would be useful.

- Record a summary in your journal. Share your observations and considerations with the class and amend your journal with additional notes.

For Discussion:
- What are the advantages and disadvantages to using both light and sound to indicate a sensor has been triggered?

- Why would one want to record and announce a bump to an object?

- Investigate what other kinds of sensors would work instead of the shock sensor?

Extensions:
- Change the pattern of the alarm (ex. light pattern, tone pattern, tone frequency, pause length).

- Change the code to make the green light blink in a regular pattern when no shock is detected.

- Change the code to make the bump detector more or less sensitive to movements.

B = Recommended

Technology Connection:
- Research and write short description of a device that uses tone and lights in combination with a sensor to indicate a change of state.

Professional Connection:
- Interview a professional that works with electrical or plumbing systems, health or environmental controls, or safety systems. In the interview, ask about devices or controls that they would install for their clients that uses, tones or light to indicate or warn about a change in vibration state.

Extension:
- Adjust the pattern of the alarm (light pattern, tone pattern, tone frequency, pause length). Change the light color permanently to another color to indicate a that a bump/shock as occurred.

- Submit your project to the book page on the website. Register and Login to submit at: thearduinoclassroom.com

From the homepage:
Click
    Books
Click

    UNO Edition Vol. 1

C = Optional

Extensions:
- Document the alarm and ring tones on a cell phone. Compare and contrast and the tones and as a class record preferences for alarms and ringing tones.

- Change the code in the project to have the alarm play a musical tune instead of a single tone.

- Design and implement a change in the project to activate a servo when a bump is detected.

Problem:
- Identify a problem that would be solved by a bump detector like the one you have been working with in this project. How could this project be a solution? Include in your report the following:

Description of the problem.

Description of how the problem is solved by this project.

What other resources you would need to solve the problem.

Design and produce the solution.

A project rubric can be found on page 236.

Anticipatory Sets
(from page 136)

Anticipatory Sets:
- Where would you see a bump detector used? List at least three different places or situations where bump detectors are used, describe the devices, and explain the purpose of the devices.
A bump detector is integrated into air bag deployment systems in car safety systems, package shipment tracking and quality, some alarms in museums and businesses on display cases.

Reflection/Summative Activities:
Responses Will Vary

For Discussion:
- What are the advantages and disadvantages to using both light and sound to indicate a sensor has been triggered?
Using both sound and light will call attention to the alarm if someone is not looking at the sensor or if they are making too much noise to hear the alarm.

- Why would one want to record and announce a bump to an object?
Some small bumps may damage fragile or sensitive objects and not be discernible without a sensor. Also some materials that are manufactured must be made without any bumps or shocks to form strong internal bonds or crystals. The sensor could note the shocks.

- Investigate what other kinds of sensors would work instead of the shock sensor?
Responses Will Vary

A

Extensions:
- Change the pattern of the alarm (ex. light pattern, tone pattern, tone frequency, pause length).
Responses Will Vary

- Change the code to make the green light blink in a regular pattern when no shock is detected.
Responses Will Vary

- Change the code to make the bump detector more or less sensitive to movements.
Responses Will Vary

B

Technology Connection:
Responses Will Vary

Professional Connection:
Responses Will Vary

Extension:
- Change the pattern of the alarm (light pattern, tone pattern, tone frequency, pause length) AND
Change the light color permanently to another color to indicate a that a bump/shock as occurred.
Responses Will Vary

- Submit your project to the book page on the website.
Responses Will Vary

C

Extensions:
- Document the alarm and ring tones on a cell phone. Compare and contrast and the tones and as a class record preferences for alarms and ringing tones.
Responses Will Vary

- Change the code in the project to have the alarm play a musical tune instead of a single tone.
Responses Will Vary

- Design and implement a change in the project to activate a servo when a bump is detected.
Responses Will Vary

Problem:
Responses Will Vary

Project 15
Polychrome

Lesson Integration:	Groupings:	Level:	Time to Complete:
Physics - Ohm's Law, Circuits, Voltage, Resistance, Light, Color, Reflectivity, Spectrum Biology - Plants, Photosynthesis Chemistry - Properties of Matter, Organic Chemistry, Pigments Mathematics - Ratios	1 - 2	Intermediate	45 min. for the project and 45 min. for extensions.

Objectives:
-Investigate how circuits and electronic components interact with electrical energy.

Prerequisite Skills:
- Understanding of light spectrum
- Arduino® IDE software (pages 11-13)
- Fritzing (pages 64-66)
- Serial Monitor and Plotter (pages 58-59)
- Libraries (Page 62)

Purpose and Skills:
- Use a sensor to analyze RGB colors

STEAM Connections:
Science - Light, Circuits, Resistance, Ohm's law, Photosynthesis, Spectrum
Technology - Code, Program Settings, Simulators, Digital Design, Electronic Components, Analog Data
Engineering - Designing, Building, and Using a Machine, Prototyping, Applied Physics
Allied Arts - Colors, Pigments, RGB Colors

Math - Ratios, Conversions, Data Graphing

Key Vocabulary:
Photosynthesis - The process that plants use to absorb energy from light and convert it to stored chemical energy.
Absorption - Light strikes a surface and some or all of its energy is transferred to the surface and stored as heat, chemical, or electrical energy.
Reflection - Light energy is returned to the environment as light energy.

Project Introduction:
- Introduce the groups to the purpose of the project, the skills developed, the standards met, and the goal of the project.

Anticipatory Sets:
- Where would you see a color detector used? List at least three different places or situations where color detector technologies are used, describe the devices, and explain the purpose of the devices.

Project 15
Polychrome

Educational Standards

ISTE Standards for Students
- Empowered Learner 1a, 1b, 1d
- Knowledge Constructor 3a, 3b, 3c, 3d
- Innovative Designer 4a, 4b, 4c, 4d
- Computational Thinker 5a, 5b, 5c, 5d
- Creative Communicator 6a, 6b, 6c, 6d
- Global Collaborator 7c, 7d

US Computer Science Standards
- Project correlations can be found on the book's web page.

US NGSS - Middle School
DCIs
- MS-PS4-1 PS4.A
- MS-LS1-3 LS1.A
- MS-LS1-4 LS1.B
- MS-LS1-5 LS1.B
- MS-LS1-6 LS1.C, PS3.D
- MS-LS1-8 LS1.D
- MS-LS2-3 LS2.B

Cross Cutting Concepts
- Cause and Effect
- Scale, Proportion and Quantity
- Structure and Function
- Patterns

Science and Engineering Practices
- Planning and Carrying Out Investigations
- Engaging in Argument from Evidence
- Obtaining, Evaluating, and Communicating Information
- Analyzing and Interpreting Data
- Scientific Knowledge is Based on Empirical Evidence
- Constructing Explanations and Designing Solutions
- Using Mathematics and Computational Thinking

US NGSS - High School
- Correlations can be found on the book's web page.

Step-by-Step 1-26

Materials List:
- Computer with IDE software and Fritzing
- Connection to the Internet
- USB Cable
- UNO or UNO Compatible Microcontroller
- TCS3200 Color Sensor
- Seven (7) Jumper Wires (Male to Female)

Engineering Design - Digital Prototype

1. Download, install, and launch (as needed) Fritzing (pages 64-66) to build a digital prototype.

2. Visit The Arduino® Classroom website and register for the website.

3. Navigate to the "UNO Edition Vol. 1" page by hovering over "Books" on the main menu and clicking on the "UNO Edition Vol. 1" link.

4. On the right of the page find and then click on the "Volume 1 Links" link.

5. Click on the "Fritzing Part Download" link for Project 13 which will take you to the page to download the file.

6. Extract the file.

7. Install the file (page 66).

8. Start a new sketch in Fritzing.

9. Drag an UNO board to the workspace.

10. Delete the breadboard.

11. Drag the TCS3200 Sensor on to the stage.

US Common Core Language Arts and Mathematics
- Correlations can be found on the book's web page.

12 Connect with jumper wires the following:

- the 5v pin to the Vcc /Vdd pin on the color detector
- a ground (GND) pin on the UNO board with the ground (GND) pin on the color detector
- pin eight (8) on the UNO board to the "S3" pin
- pin nine (9) on the UNO board to the "S2" pin
- pin 10 on the UNO board to the "OUT" pin
- pin 11 on the UNO board to the "S1" pin
- pin 12 on the UNO board to the "S0" pin

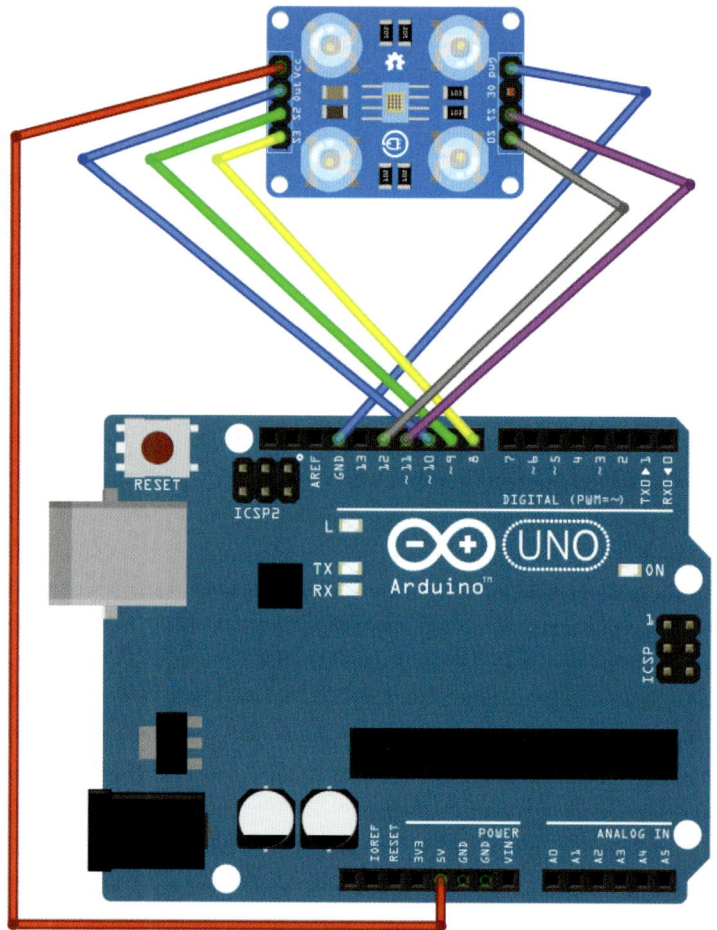

Connection - When you observe an object and its color, you are actually seeing the reflected color or colors. Any color not seen is being absorbed. For example, when you see a red ball, the ball is absorbing all of the other colors (blue, green, yellow and orange) except for red, which is reflected to your eyes.

When you look at a typical plant, you mostly see the color green. All of the other visible colors absorbed with green reflected. The plant absorbs the energy from reds, oranges, blues, and violets to power the photosynthesis process which make sugars for the plant.

Two of the most important chemicals in plants for photosynthesis are chlorophyll a and b. The graphic here shows the light colors that chlorophyll a and b absorb. See how the absorption is lowest at the color green (which is the color reflected to be seen with our eyes.)

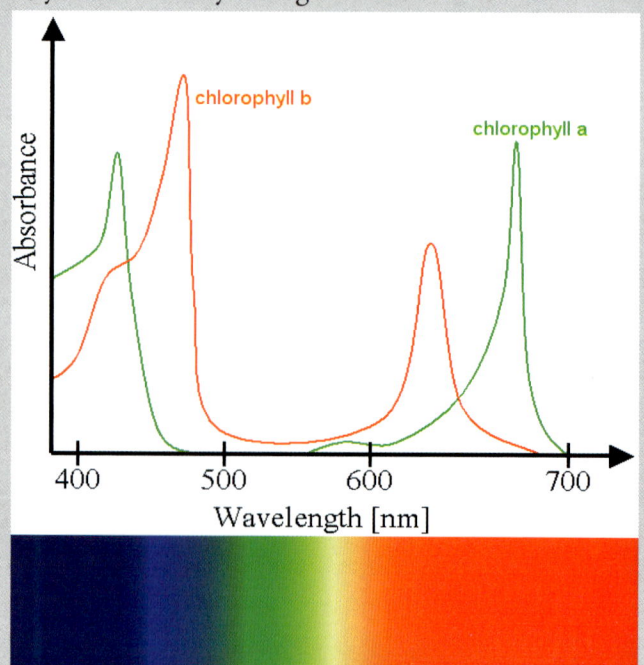

```
#include <TCS3200.h> // Add the library to the
code
#define S0 12        // Declare the variables
#define S1 11
#define S2 9
#define S3 8
#define sensorOut 10
int frequency = 0;
void setup() {
  pinMode(S0, OUTPUT); // Microcontroller
pins connections
  pinMode(S1, OUTPUT);
  pinMode(S2, OUTPUT);
  pinMode(S3, OUTPUT);
  pinMode(sensorOut, INPUT);
  digitalWrite(S0,HIGH);
  digitalWrite(S1,LOW);
  Serial.begin(9600); // Displays information on
serial plotter or monitor
}
void loop() {
  digitalWrite(S2,LOW);
  digitalWrite(S3,LOW);
  frequency = pulseIn( sensorOut, LOW);
  Serial.print("RED= "); // Displays the amount
of red color present
  Serial.print(frequency);
  Serial.print(" ");
  delay(50);
  digitalWrite(S2,HIGH);
  digitalWrite(S3,HIGH);
  frequency = pulseIn( sensorOut, LOW);
  Serial.print("GREEN= "); // Displays the
amount of green color present
  Serial.print(frequency);
  Serial.print(" ");
  delay(50);    // time of the wave in milliseconds
  digitalWrite(S2,LOW);
  digitalWrite(S3,HIGH);
  frequency = pulseIn( sensorOut, LOW);
  Serial.print("BLUE= "); // Displays the amount
of blue color present
  Serial.print(frequency);
  Serial.println(" ");
  delay(50);
}
```

Coding - Digital Prototype

13 Optional. Enter the code in Tinkercad® to match the code on this page.

14 Start the simulation. If the code has been entered correctly, the code will run without an error, but nothing will happen on the stage. ✓

Engineering Design - Project Build

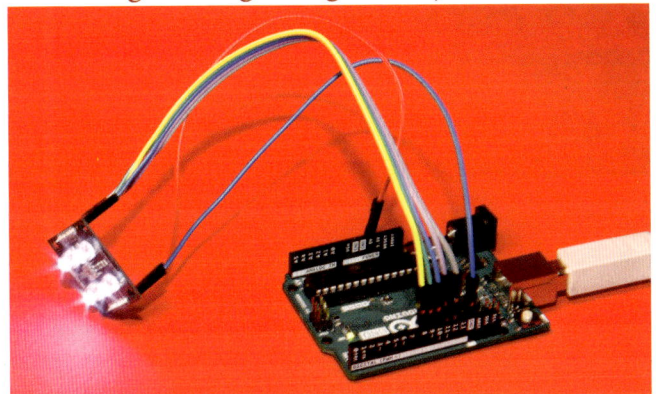

15 Connect the UNO board to the TCS3200 color sensor with jumper wires the following:
- the 5v pin to the Vcc /Vdd pin on the color detector
- a ground (GND) pin on the UNO board with the ground (GND) pin on the color detector
- pin eight (8) on the UNO board to the "S3" pin
- pin nine (9) on the UNO board to the "S2" pin
- pin 10 on the UNO board to the "OUT" pin
- pin 11 on the UNO board to the "S1" pin
- pin 12 on the UNO board to the "S0" pin

16 Study the circuit diagram for this project.

17 Connect the UNO board and computer using the USB cable.

18 Launch the Arduino® IDE software (pages 11-13) to make sure the board is communicating with the computer.

Go to the Tools menu and verify that the correct board is selected from the Board Manager menu. If not, select the correct board from the options listed.

Then confirm that the right Port is selected. If not, select the port that lists the UNO board from the options listed.

19 Download and install the Color Sensor Library from the book "Links" web page. Register for access.

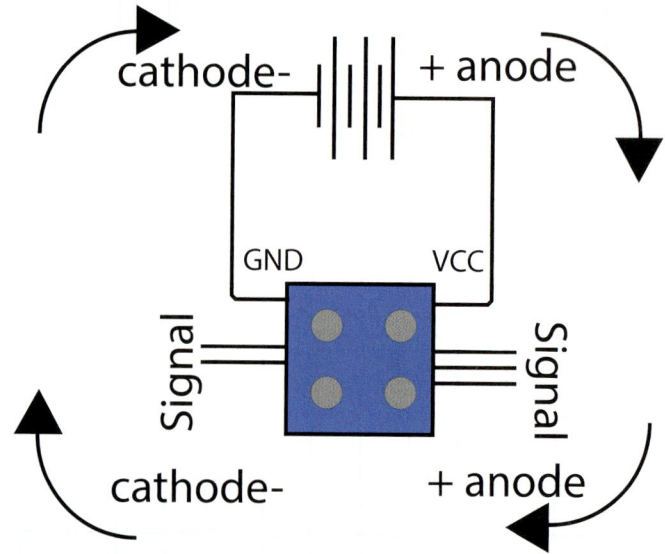

cathode-   + anode

GND   VCC

Signal   Signal

cathode-   + anode

Connection - Some complex math formulas generate color and shape patterns to form fractals. These patterns often show up in nature generated by the laws of physics or as designed by genetics.

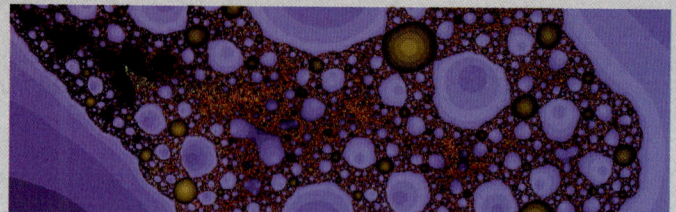

Coding - Project Build

20 Either copy and paste the code from the Tinkercad® prototype or type the code from page 147 into the Arduino® IDE software (pages 11-13).

21 Save the sketch (rename as needed).

22 Verify the sketch.

23 Upload the sketch to the UNO board.

24 Launch the Serial Plotter (pages 58-58).

25 Select several different colored objects and run the color detector over each while observing the Serial Plotter.

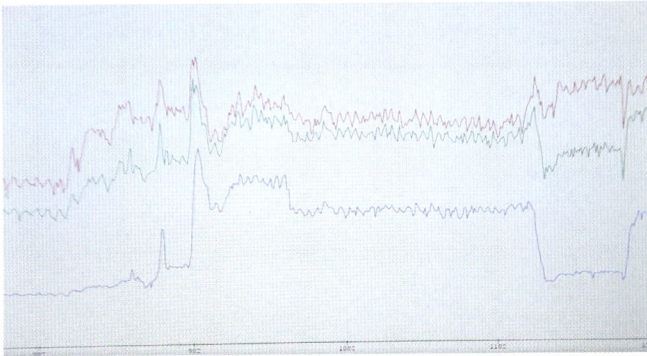

26 Document this project, discuss the project, and complete the extensions assigned from the next page.

```
//Project 13
//Color detector using a TCS3200 sensor
//thearduinoclassroom.com
//Copyright 2019, Isabel Mendiola and Peter Haydock

#include <TCS3200.h> // Add the library to the code
#define S0 12        // Declare the variables
#define S1 11
#define S2 9
#define S3 8
#define sensorOut 10
int frequency = 0;
void setup() {
  pinMode(S0, OUTPUT); // Microcontroller pins connections
  pinMode(S1, OUTPUT);
  pinMode(S2, OUTPUT);
  pinMode(S3, OUTPUT);
  pinMode(sensorOut, INPUT);

  digitalWrite(S0,HIGH);
  digitalWrite(S1,LOW);
  Serial.begin(9600); // Displays information on serial plotter or monitor
}
void loop() {
  digitalWrite(S2,LOW);
  digitalWrite(S3,LOW);
  frequency = pulseIn(sensorOut, LOW);
  Serial.print("RED= "); // Displays the amount of red color present
  Serial.print(frequency);
  Serial.print("  ");
  delay(50);
  digitalWrite(S2,HIGH);
  digitalWrite(S3,HIGH);
  frequency = pulseIn(sensorOut, LOW);
  Serial.print("GREEN= "); // Displays the amount of green color present
  Serial.print(frequency);
  Serial.print("  ");
  delay(50);     // time of the wave in milliseconds
  digitalWrite(S2,LOW);
  digitalWrite(S3,HIGH);
  frequency = pulseIn(sensorOut, LOW);
  Serial.print("BLUE= "); // Displays the amount of blue color present
  Serial.print(frequency);
  Serial.println("  ");
  delay(50);
}
```

Done Saving.

Connection - Most deciduous trees in the middle latitudes of Earth stop producing chlorophyll (which reflects green light) in late fall. The other photosynthetic pigments remain and depending on the species, the tree's leaves reveal new and different colors ranging from red to orange to yellow that the green had obscured. This color change allows the trees to continue photosynthesis using the light of fall more efficiently as the day grow shorter and the colors penetrating the atmosphere of sunlight change. Carotenes and xanthophylls remain which reflect these colors while absorbing blues and greens to continue feeding the trees and reflecting yellow, red and orange light.

A = Essential

Reflection/Summative Activities:

- Have each team member document in their project/classroom journal who was on their team, what went well, what they could improve upon, what they would do differently if they were to do the project again, and verify that every person in the group can do the project.

- In the journal, color a picture. Exchange the drawings for analysis by the other teams for the colors you used. Can you use colors they cannot identify?

For Discussion:

- What RGB values were expected from the color sensor as you tested different objects? What values were measured? Why would the results be different from those expected?

- Were there any colors not detected?

- How might industry use a color sensor?

Extensions:

- Compare results from different sensors using the same object. Explain the differences, if any.

- Determine the impact of ambient light on measuring colors.

- Find an online color chart. Print it. Use it to determine color components of objects and compare the results to the sensor.

B = Recommended

Industry Connection:

- Research and write short description of an industry or industrial application that uses color identification. Be sure to include the activities performed and how they are done.

Extensions:

- Compare and contrast leaves from different plants or leaves from the same plant at different times of the year.

- Redesign the project to do one of the following:
  1. have an RGB LED match the color(s) measured
  2. flash the RGB LED when a particular color value is measured or exceeded for each constituent color.
  3. make a sound when a particular color is measured.

- Submit your project to the book page on the website. Register and Login to submit at: thearduinoclassroom.com

From the homepage:
Click
  Books
Click
  UNO Edition Vol. 1

C = Optional

Extensions:

- Make the project portable with the LCD and the 9v battery barrel connector.

- Examine paint samples from a hardware or paint store with the sensor.

- Determine a way to calibrate color sensors such that regardless of which sensor used the readings will be the same.

Problem:

- Identify a problem that would be solved by reading color values like the one you have been working with in this project. How could this project be a solution? Include in your report the following:

Description of the problem.

Description of how the problem is solved by this project.

What other resources you would need to solve the problem.

Design and produce the solution.

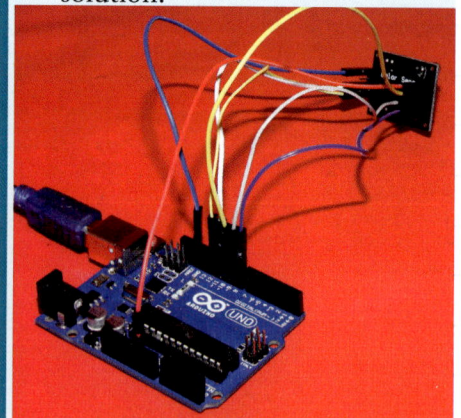

A project rubric can be found on page 236.

Anticipatory Sets
(from page 144)

Anticipatory Sets:
- Where would you see a color detector used? List at least three different places or situations where color detectors are used, describe the devices, and explain the purpose of the devices.
Analyzing paint samples at the hardware store, sorting fruit by color, examining the quality of fabric dyes. Examining the quality of candy.

A

Reflection/Summative Activities:
Responses Will Vary

For Discussion:
- What RGB values were expected from the color sensor as you tested different objects? What values were measured? Why would the results be different from those expected?
Responses Will Vary

- Were there any colors not detected?
All colors should be detected.

- How might industry use a color sensor?
Responses Will Vary

Extensions:
- Compare results from different sensors using the same object.

Explain the differences if any.
Responses Will Vary

- Determine the impact of ambient light on measuring colors.
Responses Will Vary

B

Industry Connection:
Responses Will Vary

Extensions:
- Compare and contrast leaves from different plants or leaves from the same plant at different times of the year.
Responses Will Vary

- Add or edit lines of code to the sketch to make a different repeating pattern of colors.
Responses Will Vary

- Submit your project to the book page on the website.
Responses Will Vary

C

Extensions:
- Make the project portable with the LCD and the 9v battery barrel connector.
Responses Will Vary

- Examine paint samples from a hardware or paint store with the sensor.
Responses Will Vary

- Determine a way to calibrate color sensors such that regardless of which sensor used the readings will be the same.
Responses Will Vary

Problem:
Responses Will Vary

Getting Started

Lesson Integration:	Groupings:	Level:	Time to Complete:
Physics - Ohm's Law, Circuits, Voltage, Resistance, Sound, Measurement Mathematics - Measurement, Ratios, Algebra	1 - 2	Intermediate	45 min. for the project and 45 min. for extensions.

Objectives:
- Investigate how circuits and electronic components interact with electrical energy.
- Investigate distance measurement with an ultrasonic device.

Prerequisite Skills:
- Time measurement (milliseconds)
- Metric length measurement (cm)
- Arduino® IDE software (pages 11-13)
- Tinkercad® (pages 15-16)

STEAM Connections:
Science - Circuits, Resistance, Ohm's law, Measurement (time and distance)
Technology - Code, Program Settings, Simulators, Digital Design, Electronic Components, Analog Data, Conditional Statements
Engineering - Designing, Building, and Using a Machine, Prototyping, Applied Physics
Allied Arts -
Math - Ratios, Conversions, Measurement

Purpose and Skills:
- Measure distance using a sensor

- Control an LED light and piezo using a digital input

Key Vocabulary:
Sonar - A process by which a sound is emitted underwater from a device that measures how long the echo from that sound takes to return to it as the sound reflects from an object. In turn a distance can be calculated based on the total trip time of the sound. Biological based Sonar is called echolocation.

Project Introduction:
- Introduce the groups to the purpose of the project, the skills developed, the standards met, and the goal of the project.

Anticipatory Sets:
- Where would you see distance detecting devices used? List at least three different places or situations where distance detecting technologies are used, describe the devices, and explain the purpose of the devices.

Educational Standards

ISTE Standards for Students
- Empowered Learner 1a, 1b, 1d
- Knowledge Constructor 3a, 3b, 3c, 3d
- Innovative Designer 4a, 4b, 4c, 4d
- Computational Thinker 5a, 5b, 5c, 5d
- Creative Communicator 6a, 6b, 6c, 6d
- Global Collaborator 7c, 7d

US Computer Science Standards
- Project correlations can be found on the book's web page.

US NGSS - Middle School
DCIs
- MS-PS4-3 PS4.C
- MS-PS4-1 PS4.A

Cross Cutting Concepts
- Cause and Effect
- Scale, Proportion and Quantity
- Structure and Function

Science and Engineering Practices
- Planning and Carrying Out Investigations
- Obtaining, Evaluating, and Communicating Information
- Analyzing and Interpreting Data
- Scientific Knowledge is Based on Empirical Evidence
- Using Mathematics and Computational Thinking

US NGSS - High School
- Correlations can be found on the book's web page.

US Common Core Language Arts and Mathematics
- Correlations can be found on the book's web page.

Materials List:
- Computer with IDE software
- Connection to the Internet
- USB Cable
- UNO or UNO Compatible Microcontroller
- Long Breadboard
- HC-SR04 Sensor
- Piezoelectric Buzzer
- 220 Ohm Resistor
- LED
- Nine (9) Jumper Wires (Male to Male)

Engineering Design - Digital Prototype

1 Launch and log into Tinkercad® (pages 15-16).

2 Start a new Circuits project.

3 Drag an UNO board to the workspace.

4 Drag a breadboard to the workspace.

5 Drag an HC-SR04 sensor to the breadboard such that the leads span four rows in a column.

6 Place an LED across two rows of a column.

7 Place a 220 Ohm resistor connecting the cathode(-) column and the same row as the cathode(-) leg of the LED.

8 Place a piezoelectric buzzer across two rows of a column.

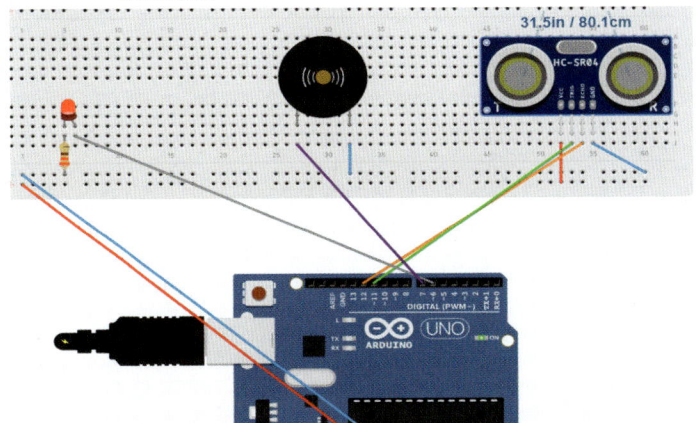

```
#define trigPin 11      // HCSR04 sensor connected to pin 11
#define echoPin 12      // HCSR04 sensor connected to pin 12
int Buzzer = 7 ;        // Buzzer is connected to pin 7
int ledPin= 6;          //LED connected to pin 6
int duration, distance; //distance-time taken
void setup() {
      Serial.begin (9600); // data can be read on serial monitor
      pinMode(11, OUTPUT);
      pinMode(12, INPUT);
      pinMode(Buzzer, OUTPUT);
      pinMode(ledPin, OUTPUT);
}
void loop() {
   digitalWrite(11, HIGH);
   delay(1000);
   digitalWrite(11, LOW);
   duration = pulseIn(12, HIGH);
   distance = (duration/2) / 29.1;//Distance greater than 100 or less than 0, buzzer and LED off
  if (distance >= 100 || distance <= 0)
      {
      Serial.println("no object detected");
      noTone(Buzzer);
      digitalWrite(ledPin,LOW);
      }
  else {
      Serial.println("cm object detected");
      Serial.print("distance= ");
      Serial.print(distance);        //displays distance between the range 0 to 100
      tone(Buzzer,600);              // 600Hz for 500 ms
      digitalWrite(ledPin,HIGH);
  }
}
```

9 Connect with jumper wires the following:
- the five volt (5v) pin to the anode(+) column on the breadboard
- a ground (GND) pin on the UNO board to the cathode(-) column on the breadboard.
- digital pin six (6) to the same row as the anode(+) lead of the LED.
- digital pin seven (7) to the same row as the anode(+) lead of the piezoelectric buzzer.
- the same row as the cathode(-) lead of the piezoelectric buzzer to the cathode(-) row
- the same row as connected to the ground (GND) lead on HCSR04 sensor to the cathode(-) column on the breadboard.
- digital pin 11 to the same row as the "TRIG" lead lead of the HCSR04 sensor.
- digital pin 12 to the same row as the "ECHO" lead of the HCSR04 sensor.
- the same row as the VCC lead on the HCSR04 sensor to the anode(+) column on the breadboard

Coding - Digital Prototype

10 Enter the code in Tinkercad® to match the code on this page. Notice the use of the conditional statements.

11 Test the distance sensor by clicking on "Code" to open the code then click the "Serial Monitor" icon at the bottom. Click "Start Simulation."

A light green cone with a darker green circle will appear with distance demarcations in cm and inches. Drag the circle such that the distances are within 100 cm. Then drag the circle away from the sensor past the 100 cm mark. Scale the stage as needed.

Observe the prototype and Serial Monitor as you move the circle.

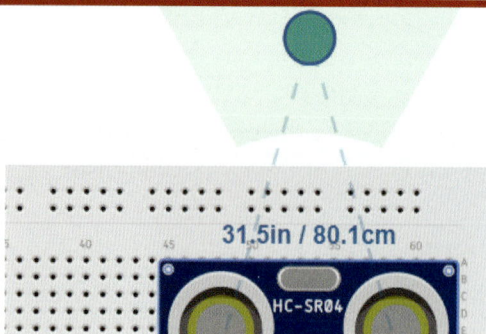

Connection - Being able to measure mass, volume, temperature, time, or length and obtain the same measurement no matter the location or conditions is foundational to mass production of goods and predictable services in a modern society. It is also essential for fair commerce between individuals, corporations, and nations. Everyone wants to know with confidence how much denim, avocados, or milk are purchased or what is the right time or temperature.

The world has for the past 200 years agreed to use metric units of measurement for length, mass, volume, and time for comparing and measuring. Even in the United States official measurements are done with the metric system.

The United States federal government agency responsible for the accuracy and precision of measurements is called NIST, the National Institute for Standards and Technology. NIST regulates and certifies almost every measurement and measuring device from rulers and tape measures to scales to clocks and many more things. It also certifies the speeds of computers, purity of materials, performance of devices, GPS measurements and even some aspects of cybersecurity. Perhaps the most famous device they maintain is the U.S. is the atomic clock. NIST is a part of the Department of Commerce.

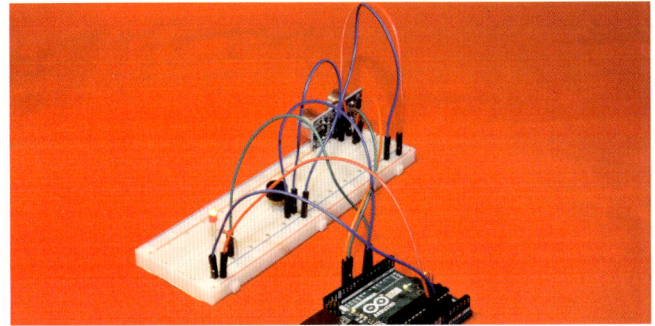

Engineering Design - Project Build

12 Place the US-100 sensor on the breadboard with all the leads in the same column. Remove the jumper on the back.

13 Place an LED across two rows of a column.

14 Place a 220 Ohm resistor connecting the cathode(-) column and the same row as the cathode(-) leg of the LED.

15 Place a piezoelectric buzzer across two rows of a column.

16 Connect with jumper wires the following:
- the five volt (5v) pin to the anode(+) column on the breadboard
- a ground (GND) pin on the UNO board to the cathode(-) column on the breadboard.
- digital pin six (6) to the same row as the anode(+) lead of the LED.
- digital pin seven (7) to the same row as the anode(+) lead of the piezoelectric buzzer.
- the same row as the cathode(-) lead of the piezoelectric buzzer to the cathode(-) row
- the same row as connected to the ground (GND) lead on HCSR04 sensor to the cathode(-) column on the breadboard.
- digital pin 11 to the same row as the "TRIG" lead lead of the HCSR04 sensor.
- digital pin 12 to the same row as the "ECHO" lead of the HCSR04 sensor.
- the same row as the VCC lead on the HCSR04 sensor to the anode(+) column on the breadboard

17 Connect the UNO board and computer using the USB cable.

18 Launch the Arduino® IDE software (pages 11-13) to make sure the board is communicating with the computer.

Go to the Tools menu and verify that the correct board is selected from the Board Manager menu. If not, select the correct board from the options listed.

Then confirm that the right Port is selected. If not, select the port that lists the UNO board from the options listed.

Coding - Project Build

19 Either copy and paste the code from the Tinkercad® prototype or type the code from page 154 into the Arduino® IDE software (pages 11-13).

20 Save the sketch (rename as needed).

21 Verify the sketch.

```
//Project 16
//Arduino Uno Digital Ruler
//thearduinoclassroom.com
//Copyright 2015, Isabel Mendiola and Peter Haydock

#define trigPin 11        // HCSR04 sensor connected to pin 11
#define echoPin 12        // HCSR04 sensor connected to pin 12
int Buzzer = 7 ;          // Buzzer is connected to pin 7
int ledPin= 6;            //LED connected to pin 6
int duration, distance; //distance-time taken
void setup() {
        Serial.begin (9600); // data can be read on serial monitor
        pinMode(11, OUTPUT);
        pinMode(12, INPUT);
        pinMode(Buzzer, OUTPUT);
        pinMode(ledPin, OUTPUT);
}
void loop() {
    digitalWrite(11, HIGH);
    delayMicroseconds(1000);
    digitalWrite(11, LOW);
    duration = pulseIn(12, HIGH);
    distance = (duration/2) / 29.1;//Distance greater than 100 or less than 0, buzzer and LED off
  if (distance >= 100 || distance <= 0)
        {
        Serial.println("no object detected");
        noTone(Buzzer);
        digitalWrite(ledPin,LOW);
        }
    else {
        Serial.println("cm object detected");
        Serial.print("distance= ");
        Serial.print(distance);          //displays distance between the range 0 to 100
        tone(Buzzer,600);                // 600Hz
        digitalWrite(ledPin,HIGH);
        }
}
```

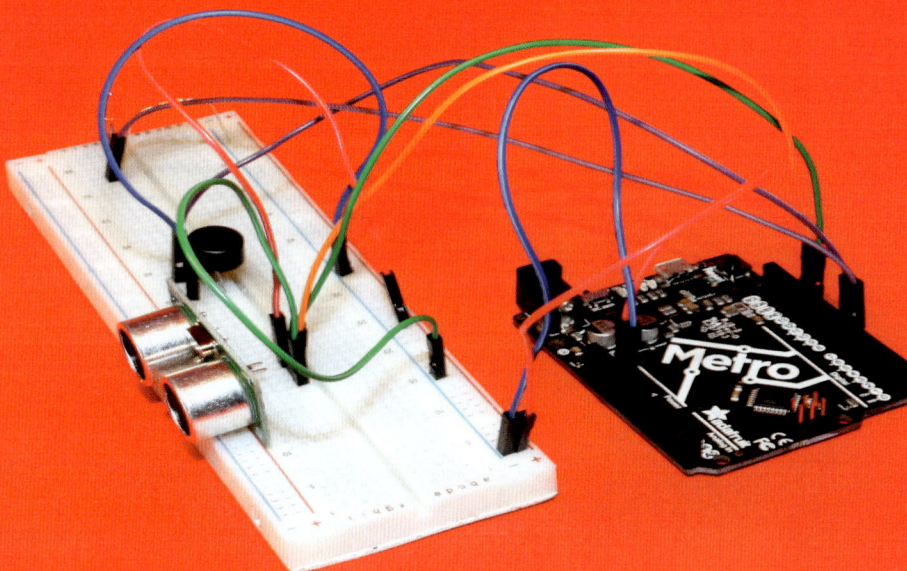

22 Upload the sketch to the UNO board.

23 Test the sensor with a variety of objects at different distances.

24 Document this project, discuss the project, and complete the extensions assigned from the next page.

Connection - Measurement requires specialized tools and skills. This photo, by Cruz Sanchez from about 1910-1919, shows a group of land surveyors in Yautepec, Morelos, Mexico with their measuring tools and equipment. At this time in Mexican history, the Mexican revolution was just starting. Different political, social, and economic groups were fighting about the future of Mexico after having been under the rule of a dictator for over 30 years.

One of the major issues of the Mexican Revolution was land ownership. At the start of the Revolution, 95% of the agricultural land was owned by large landowners or corporations. This led to numerous revolts and uprisings by landless and largely less educated workers who could only find employment with these entities. The workers often fought for increases to their wages. Starting in 1914 and continuing to this day land reform is an important political consideration in Mexican governance. With the start of land reform, land surveying became an important tool to accurately implement the social reform the government designed. Larger tracts of land were broken up into smaller private and community owned farms with the new land titles awarded to the working classes.

Land ownership claims rely on precise measurements to establish who own which parcel of land. Land then can be divided, combined, used as collateral for loans, or sold with certainty. Buildings and fences can be placed correctly on the property and mineral and water rights can enforced especially in times of social upheaval or legal disputes.

The tools here are similar to the one built in this project where the distance to an object is determined from a fixed point. In the photo, the measurement is done optically where the sensor uses a ultrasonic sound pulse.

Used with permission from the DeGolyer Library, Southern Methodist University

A = Essential

Reflection/Summative Activity:
- Have each team member document in their project/classroom journal who was on their team, what went well, what they could improve upon, what they would do differently if they were to do the project again, and verify that every person in the group can do the project.

For Discussion:
- What are the limitations and uncertainty of measurement of the sensor? How would this limit what you would do with it?

- What applications would this sensor be acceptable for?

Extensions:
- Test and verify the distances determined by the sensor.

- Change the code to detect objects further than 100 cm.

B = Recommended

Social Studies Connection:
- Assign each group a historical era and culture (to match what they are studying in their history/social studies class) to research the measurements used in this period of time and location. Include the origin of the measurements, who controlled the enforcement of the measurement, and why it was developed.

Connection:
- Research and write a short description of a modern tool or machine that uses a distance senor to measure length or distance as part of its operation.

Extension:
- Replace the LED with an RGB LED and edit the code for different colors to appear depending on the distance of the object in front of the sensor.

- Submit your project to the book page on the website. Register and Login to submit at: thearduinoclassroom.com

From the homepage:
Click
   Books
Click
   UNO Edition Vol. 1

C = Optional

Extensions:
- Add the LCD to the project to show the distance without using the Serial Monitor.

- Identify and research an Olympic sport(s) that their country participated in at the last Olympics where a measurement plays a key role in winning a medal.

- Add a pushbutton switch to start and stop measuring.

Problem:
- Identify a problem that would be solved having a distance sensor. How could this project be a solution? Include in your report the following:

Description of the problem.

Description of how the problem is solved by this project.

What other resources you would need to solve the problem.

Design and produce the solution.

A project rubric can be found on page 236.

Anticipatory Sets
(from page 152)

Anticipatory Sets:
- Where would you see distance detecting devices used? List at least three different places or situations where distance detecting devices are used, describe the devices, and explain the purpose of the devices.
Responses Will Vary

A

Reflection/Summative Activity:
Responses Will Vary

For Discussion:
- What are the limitations and uncertainty of measurement of the sensor?
The sensor is limited to about 400 cm, and accurate at +/- 3 mm.

- How would this limit what you would do with it?
Only objects within the 400 cm range could be detected.

- What applications would this sensor be acceptable for?
Responses Will Vary

Extensions:
- Test and verify the distances determined by the sensor.
Students should show that they have measured the distance with the sensor

and a tape measure, meter stick or other measuring device and compared the two measurements.
More advanced students should show the percent difference.
- Change the code to detect objects further than 100 cm.
Responses Will Vary

B

Social Studies Connection:
Responses Will Vary

Connections:
Responses Will Vary

Extension:
- Replace the LED with an RGB LED and code for different colors to appear depending on the distance of the object in front of the sensor.
Responses Will Vary

- Submit your project to the book page on the website.

Responses Will Vary

C

Extensions:
- Add the LCD to the project to show the distance without using the Serial Monitor.
Responses Will Vary

- Add a pushbutton switch to start and stop measuring.
Responses Will Vary

Problem:
Responses Will Vary

Project 17
Turbidity Sensor

Lesson Integration:	Groupings:	Level:	Time to Complete:
Physics - Ohm's Law, Circuits, Voltage, Resistance Biology - Ecology Mathematics - Arithmetic	1 - 2	Intermediate	45 min. for the project and 45 min. for extensions.

Objectives:
- Investigate how circuits and electronic components interact with electrical energy.

Prerequisite Skills:
- Time measurement (milliseconds)
- Arduino® IDE software (pages 11-13)
- Fritzing (pages 64-66)
- Serial Monitor and Plotter (pages 58-59)

Purpose and Skills:
- Analyze water quality by measuring turbidity

STEAM Connections:
Science - Circuits, Ecology, Light, Water Quality
Technology - Code, Program Settings, Simulators, Digital Design, Electronic Components, Analog Data, Conditional Statements
Engineering - Designing, Building, and Using a Machine, Prototyping, Applied Physics
Allied Arts - Aesthetics

Math - Ratios, Conversions

Key Vocabulary:
Turbidity - The cloudiness or opaqueness of a liquid from material suspended in the liquid.

Project Introduction:
- Introduce the groups to the purpose of the project, the skills developed, the standards met, and the goal of the project.

Anticipatory Sets:
- Investigate what turbidity means in terms of water quality. Determine what turbidity indicates about the water.

For this project you will need to access a file on the book's website. Register (free) and then download the file using the instructions in the project.

Project 17
Turbidity Sensor

Educational Standards

ISTE Standards for Students
- Empowered Learner 1a, 1b, 1d
- Knowledge Constructor 3a, 3b, 3c, 3d
- Innovative Designer 4a, 4b, 4c, 4d
- Computational Thinker 5a, 5b, 5c, 5d
- Creative Communicator 6a, 6b, 6c, 6d
- Global Collaborator 7c, 7d

US Computer Science Standards
- Project correlations can be found on the book's web page.

US NGSS - Middle School
DCIs
- MS-PS4-3 PS4.C
- MS-PS4-1 PS4.A
- MS-LS1-5 LS1.B
- MS-LS2-1 LS2.A
- MS-LS2-2 LS2.A
- MS-LS2-4 LS2.C
- MS-ESS2-1 ESS2.A
- MS-ESS2-2 ESS2.A
- MS-ESS2-4 ESS2.C

Cross Cutting Concepts
- Cause and Effect
- Scale, Proportion and Quantity
- Structure and Function

Science and Engineering Practices
- Planning and Carrying Out Investigations
- Obtaining, Evaluating, and Communicating Information
- Analyzing and Interpreting Data
- Scientific Knowledge is Based on Empirical Evidence Scale, Proportion and Quantity
- Using Mathematics and Computational Thinking

US NGSS - High School
- Correlations can be found on the book's web page.

Step-by-Step 1-18

Materials List:
- Computer with IDE software
- Connection to the Internet
- USB Cable
- UNO or UNO Compatible Microcontroller
- Turbidity Sensor
- Turbidity Sensor Module
- Turbidity Sensor Cable
- Turbidity Sensor Adapter Cable
- Three (3) Jumper Wires (Male to Male)
- At least three clear containers with water and varying amounts of soil to test turbidity.

Engineering Design - Digital Prototype

1 On a computer connected to the Internet go to thearduinoclassroom.com

2 Login to the site. If you do not have an account create an account.

3 Click the following in order
   Books (main menu)
   UNO Edition Vol. 1 (under books)
   Volume 1 Links (right sidebar menu)

4 Download the Turbidity Sensor Digital Prototype listed under Project 17.

5 Review the diagram provided. Print if needed.

6 Make a list of parts and connections for this project.

7 Review the code for the project found on the next page.

US Common Core Language Arts and Mathematics
- Correlations can be found on the book's web page.

Engineering Design - Project Build

8. Connect the turbidity sensor to the sensor adapter using the sensor cable provided.

9. Connect the sensor adapter cable to the sensor adapter.

10. Connect with jumper wires the following:
    - analog pin zero (A0) to the blue cable connected to the sensor adapter
    - 5v pin to the red cable connected to the center pin on the sensor adapter
    - GND to the black cable connected to the sensor adapter

11. Connect the UNO board and computer using the USB cable.

```
int sensorPin = A0;
float volt;
float NTU;
void setup() {
  Serial.begin(9600); // To use the Serial Monitor
that reads NTU (Nephelometric Turbidity Units) of
the water
}
void loop()
{
  for (int i=0; i<800; i++)
{
    int sensorvalue = analogRead(A0);
    volt += sensorvalue * (5.0 / 1023.0);  //Analog
reading is changed to 0-1023 to voltage 0-5v.
  }
  volt = volt/800;
  volt = round_to_dp(volt,1); //Volts are rounded.
  if (volt >4.2){
NTU = 0;
}
 else if(volt < 2.5){   //Prototype works only with
values between 2.5v ~ 4.2v
  NTU = 3000; //readings below 2.5v = 3000NTU
  }else{
    NTU =
-1120.4*square(volt)+5742.3*volt-4352.9;  //
Calculate current NTU
  }
  Serial.print(volt);  //Display voltage and NTU on
the Serial Monitor
  Serial.print(" v");
  Serial.print("\n");
  Serial.print(NTU);
  Serial.print(" NTU");
  Serial.println("\n");
  delay(2000);
}
float round_to_dp( float in_value, int decimal_place
)
{
  float multiplier = powf( 10.0f, decimal_place );
  in_value = roundf( in_value * multiplier ) /
multiplier;
  return in_value;
}
```

Connection - Measuring water turbidity is an important scientific indicator of the quality of water.

Turbid water is not necessarily dirty (as shown in this photo of a turbid river) nor clear water an indication of clean or healthy water.

Scientists will study the water regularly over a longer period of time (usually several times a year over decades) to understand how the water turbidity changes or does not change. In some cases water with a lot of sediment is good for the ecosystem because it is bringing new soil and nutrients down stream. In other cases high turbidity indicates erosion of soils that damage an ecosystem.

Even a clear blue lake like Blackfish Lake in Alaska does not necessarily indicate clean water. Scientists would have to study it over time to understand the health of the ecosystem with the turbidity of the water being one indicator.

12 Launch the Arduino® IDE software (pages 11-13) to make sure the board is communicating with the computer.

Go to the Tools menu and verify that the correct board is selected from the Board Manager menu. If not, select the correct board from the options listed.

Then confirm that the right Port is selected. If not, select the port that lists the UNO board from the options listed.

NTU - An NTU is a unit of measure for turbidity. It refers to how the light transmission is measured. An NTU measures light transmission at a 90 degree angle from the source. NTUs are the standard of measurement in the US for turbidity. Europe measures at the same angle but on a different scale.

Coding - Project Build

13 Type the code from page 162 into the Arduino® IDE software (pages 11-13).

14 Save the sketch (rename as needed).

15 Verify the sketch.

16 Upload the sketch to the UNO board.

17 Test the turbidity sensor in several glass containers of water mixed with various amounts of soil.

Do not submerge the sensor completely in water.

Open the Serial Monitor and observe the values recorded.

18 Document this project, discuss the project, and complete the extensions assigned from the next page.

```
TAC_UE_V1_P17_Turbidity | Arduino 1.8.9 (Windows Store 1.8.21.0)
File Edit Sketch Tools Help

TAC_UE_V1_P17_Turbidity
int sensorPin = A0;
float volt;
float NTU;

void setup() {
  Serial.begin(9600); // To use the Serial Monitor that reads NTU (Nephelometric Turbidity Units) of the water
}

void loop()
{
  for (int i=0; i<800; i++)
  {
      int sensorvalue = analogRead(A0);
      volt += sensorvalue * (5.0 / 1023.0);  //Analog reading is changed to 0~1023 to voltage 0-5V.
  }
  volt = volt/800;
  volt = round_to_dp(volt,1);       //Volts are rounded.
  if(volt < 2.5){                   //Prototype works only with values between 2.5V ~ 4.2V,
  NTU = 3000;                       //readings below 2.5V = 3000NTU
  }else{
    NTU = -1120.4*square(volt)+5742.3*volt-4352.9;  // Calculate current NTU
  }
  Serial.print(volt);              //Display voltage and NTU on the Serial Monitor
  Serial.print(" v");
  Serial.print("\n");
  Serial.print(NTU);
  Serial.print(" NTU");
  Serial.print("\n");
  delay(2000);
}
float round_to_dp( float in_value, int decimal_place )
{
  float multiplier = powf( 10.0f, decimal_place );
  in_value = roundf( in_value * multiplier ) / multiplier;
  return in_value;
}

Done compiling.
Sketch uses 3868 bytes (11%) of program storage space. Maximum is 32256 bytes.
Global variables use 214 bytes (10%) of dynamic memory, leaving 1834 bytes for local variables. Maximum is 2048

                                                                    Arduino/Genuino Uno on COM4
```

Connection - Two examples of bodies of water that have had observable levels of turbidity changes due to human impacts are Lake Michigan and Chesapeake Bay in the United States.

Before European colonization of the Americas, Chesapeake Bay was an ecosystem with oysters, sturgeon, crabs and many other species dominating the Bay. The first colonists remarked on oysters up to 30 cm (11. 8 inches) across, sturgeon up to 5 meters (16.4 feet) and soft shell crabs too numerous to count.

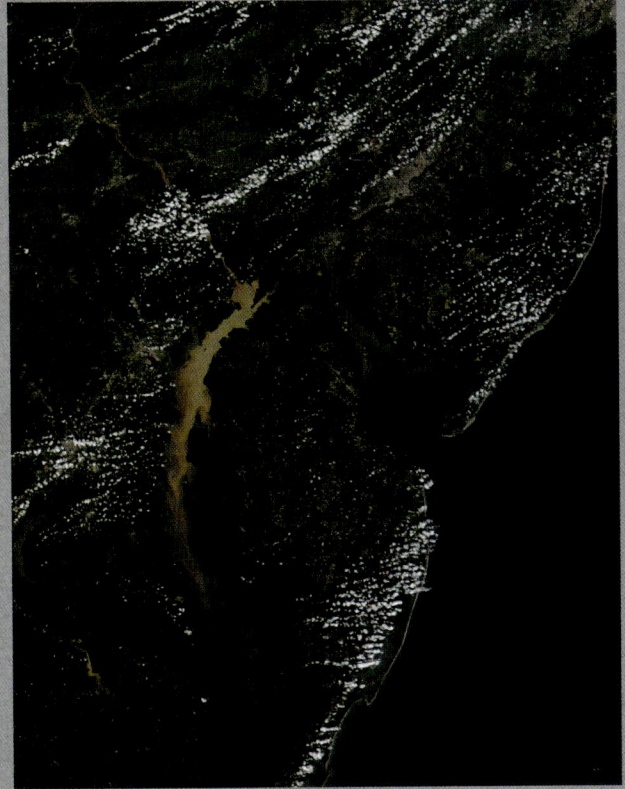

The water of the Bay was also extremely clear because there was no large-scale farming and the oysters that rested on the bottom of the bay filtered vast volumes of water each day. The colonists and their descendants quickly cleared large tracts of land and ate oysters, crabs and sturgeon to the point that over 90% of the animals were removed from the ecosystem. The land clearing allowed soil and silt into the water, suffocating many fish and crabs. Nutrients in the runoff fed toxic algae creating poisonous blooms. In the satellite image (Courtesy MODIS Rapid Response Team at NASA GSFC), the Chesapeake Bay is muddy brown due to run off from the rivers that feed into it after a hurricane. The surrounding states have spent billions of dollars in a coordinated attempt to clean and restore Chesapeake Bay.

More recently two invasive species from Europe have infiltrated the Great Lakes and many smaller lakes in the United States. The first invader was the Zebra mussel (left individual and bottom-right a colony), a small clam like animal that filters water for its food. Soon after, a slightly larger mussel, the Quagga mussel arrived. The two mussels, with no natural predators in North America, spread quickly attached to the bottom of boats or from bilge water from the ships that was emptied in to lakes. The mussels consumed the native algae at such fat rates that it is now common to see 8 to 10 meters deep in the Great Lakes when before the invasion one could only see a meter or two into the water. The removal of the algae increased penetration of light into the lakes allowing the rapid growth of larger water plants that could not previously live in the turbid waters. All of these changes to the ecosystem of the lakes has changed which fish are successful and which are not. Native small species no longer have the algae to eat which limits the food for native larger species like lake trout and perch.

Project Reflection and Summative Activities, Discussion Starters, Extensions, and Problem

A = Essential

Reflection/Summative Activities:

- Have each team member document in their project/classroom journal who was on their team, what went well, what they could improve upon, what they would do differently if they were to do the project again, and verify that every person in the group can do the project.

- List the bodies of water in your area.

For Discussion:

- Why is it important to measure the turbidity of body of water many times a year over many years?

- What does turbidity demonstrate about the ecosystem?

Extensions:

- Research online the bodies of water in your area to find if turbidity values are documented.

- Design a monitoring project for a body of water in your area using the turbidity sensor. Create a list of other measurements that would add value to the study.

- Build a Secchi Disk and use it in a local body of water along with the turbidity sensor.

- Compare and contrast the data collected with the turbidity sensor and the Secchi Disk.

B = Recommended

Professional Connection:

- Ask an ecologist to visit your classroom or visit one at a state, county, or national park. Request that they discuss how they monitor the ecosystem including the water quality of the bodies of water in their care. Summarize the discussion.

Extensions:

- As a class design and implement a study of bodies of water in your area that would include measuring turbidity.

- Create an info-graphic of your results.

- Submit your project to the book page on the website. Register and Login to submit at: thearduinoclassroom.com

From the homepage:
Click
    Books
Click
    UNO Edition Vol. 1

C = Optional

Extension:

- Make the turbidity sensor portable with a 9v battery, barrel battery connector, and LCD.

Research Extension:

- Research an invasive species in your area that may change or have already changed the water quality of local bodies of water. Detail the issue and explain what remediations plans are in place to deal with the species. Create a radio public service announcement that explains the issue in 30 seconds or less.

A project rubric can be found on page 236.

Anticipatory Sets
(from page 160)

Anticipatory Sets:
- Investigate what turbidity means in terms of water quality. Determine what turbidity indicates about the water.
Responses Will Vary

A

Reflection/Summative Activities:
Responses Will Vary

For Discussion:
- Why is it important to measure the turbidity of body of water many times a year over many years?
Responses Will Vary

- What does turbidity demonstrate about the ecosystem?
By itself very little. But in conjunction with a longterm study and other measures, the turbidity will shoe

Extensions:
- Design a monitoring project for a body of water in your area using the turbidity sensor. Create a list of other measurements that would add value to the study.
Responses Will Vary

B

Professional Connection:
- Ask an ecologist to visit your classroom or visit one at a state, county, or national park. Request that they discuss how they monitor the ecosystem including the water quality of the bodies of water in their care. Summarize the discussion.
Responses Will Vary

Extensions:
- As a class design and implement a study of bodies of water in your area that would include measuring turbidity.

- Create an info-graphic of your results.
Responses Will Vary

- Submit your project to the book page on the website.
Responses Will Vary

C

Extension:
- Make the turbidity sensor portable with a 9v battery, barrel battery connector, and LCD.
Responses Will Vary

Research Extension:
- Research another invasive species in your area that may change or have already changed the water quality of local bodies of water. Detail the issue and explain what remediations plans are in place to deal with the species. Create a radio public service announcement that explains the issue in 30 seconds or less.
Responses Will Vary

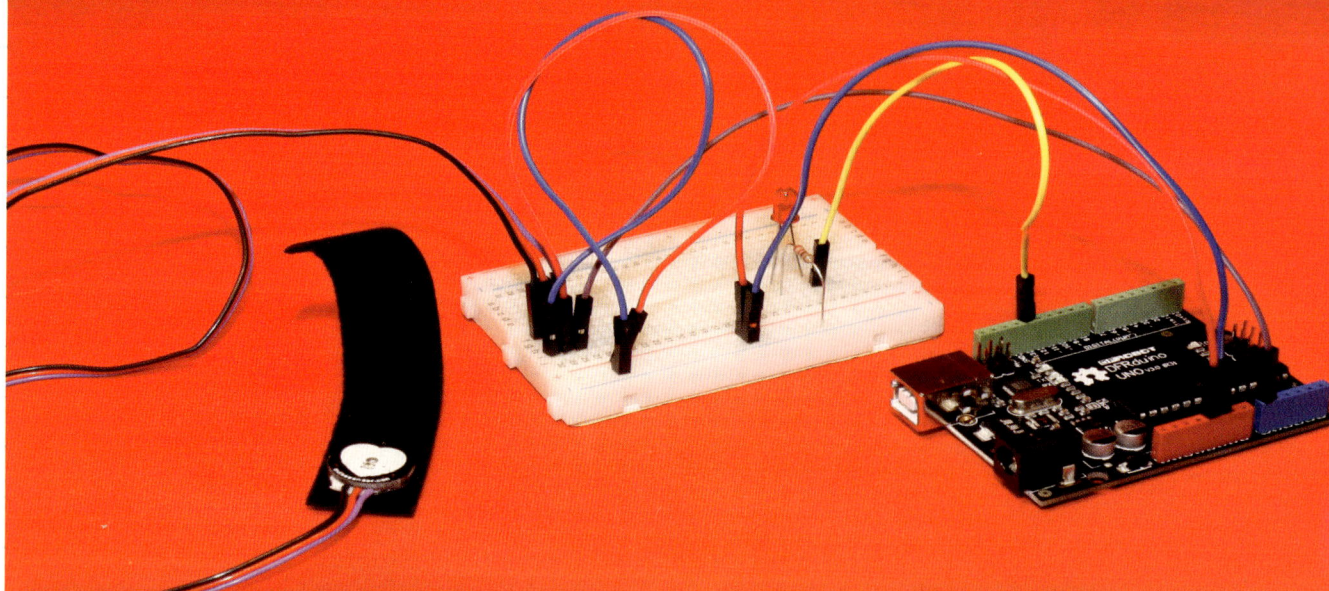

Getting Started

Lesson Integration:	Groupings:	Level:	Time to Complete:
Physics - Ohm's Law, Circuits, Voltage, Resistance Biology - Anatomy and Physiology: Circulatory System Mathematics - Arithmetic	1 - 2	Intermediate	45 min. of for the project and 30 min. for extensions.

Objectives:
- Investigate how circuits and electronic components interact with electrical energy.
- Record input from a sensor.

Prerequisite Skills:
- Time measurement (milliseconds)
- Arduino® IDE software (pages 11-13)
- Fritzing (pages 64-66)
- Serial Monitor and Plotter (pages 58-59)

Purpose and Skills:
- Record heartbeats

STEAM Connections:
Science - Circuits, Resistance, Ohm's law, Circulatory System
Technology - Code, Simulators, Digital Design, Electronic Components, Analog Data
Engineering - Designing, Building, and Using a Machine, Prototyping, Applied Physics
Allied Arts - Rhythm, Beats
Math - Data Graphing

Key Vocabulary:
Heart rate - the number of heart contractions over a period of time. Usually as beats per minute.
Diastolic Phase - The time the heart is using the ventricles to pump blood to the body and lungs.
Systolic Phase - The time the heart is **not** using the ventricles to pump blood. The heart muscles may be relaxed or the atria are pumping blood.

Project Introduction:
- Introduce the groups to the purpose of the project, the skills developed, the standards met, and the goal of the project.

Anticipatory Sets:
- Have each student measure their heartbeat using a clock.
- Investigate how blood circulates in the body.
- Investigate what heart rate indicates about a person's general health.
- Investigate how blood pressure and heart rate compliment each other.

Project 18
My Heartbeat

Educational Standards

ISTE Standards for Students
- Empowered Learner 1a, 1b, 1d
- Knowledge Constructor 3a, 3b, 3c, 3d
- Innovative Designer 4a, 4b, 4c, 4d
- Computational Thinker 5a, 5b, 5c, 5d
- Creative Communicator 6a, 6b, 6c, 6d
- Global Collaborator 7c, 7d

US Computer Science Standards
- Project correlations can be found on the book's web page.

US NGSS - Middle School
DCIs
- MS-PS4-3 PS4.C
- MS-PS4-1 PS4.A

Cross Cutting Concepts
- Cause and Effect
- Scale, Proportion and Quantity
- Structure and Function
- Patterns

Science and Engineering Practices
- Planning and Carrying Out Investigations
- Obtaining, Evaluating, and Communicating Information
- Analyzing and Interpreting Data
- Scientific Knowledge is Based on Empirical Evidence
- Using Mathematics and Computational Thinking

US NGSS - High School
- Correlations can be found on the book's web page.

US Common Core Language Arts and Mathematics
- Correlations can be found on the book's web page.

Materials List:
- Computer with IDE software and Fritzing
- Connection to the Internet
- USB Cable
- UNO or UNO Compatible Microcontroller
- Short Breadboard
- LED
- Seven (7) Jumper Wires (Male to Male)
- 220 Ohm Resistor
- Heartbeat Sensor

Engineering Design - Digital Prototype

1 Download, install, and launch (as needed) Fritzing (pages 64-66) to build a digital prototype.

2 Visit The Arduino® Classroom website and register for the website.

3 Navigate to the "UNO Edition Vol. 1" page by hovering over "Books" on the main menu and clicking on the "UNO Edition Vol. 1" link.

4 On the right of the page find and then click on the "Volume 1 Links" link.

5 Click on the "Fritzing Part Download" link for Project 18 which will take you to the page to download the file.

6 Extract the file.

7 Install the file (page 66).

8 Start a new sketch in Fritzing.

9 Drag an UNO board to the workspace.

10 Place an LED across two rows of a column on the breadboard.

11 Place a 220 Ohm resistor on the breadboard connecting the cathode(-) row with the same row as the cathode(-) lead of the LED

12) Place the Heartbeat sensor across three rows of a column on the breadboard.

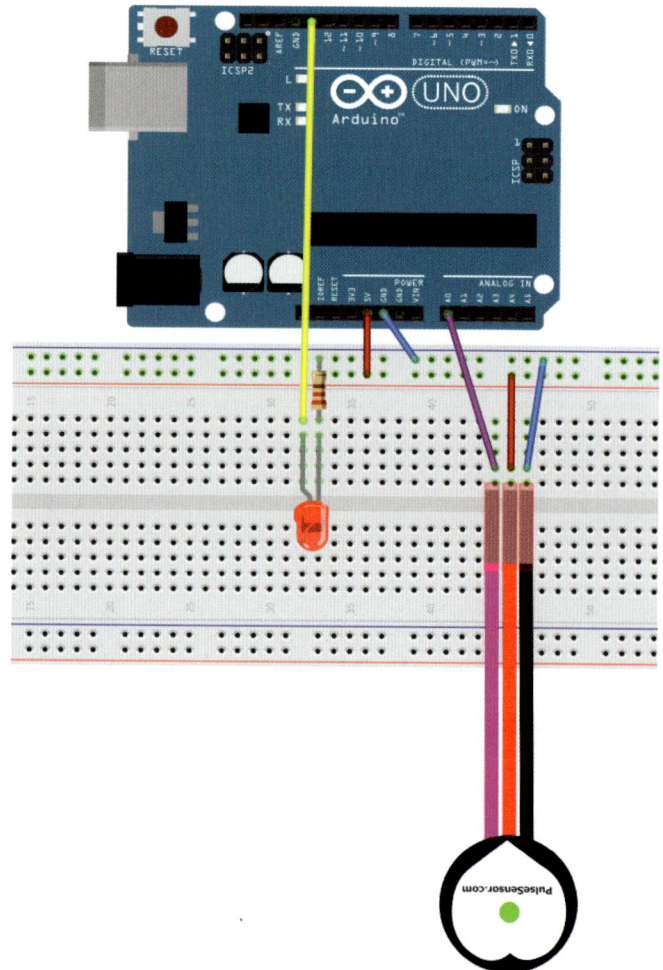

13) Connect with jumper wires the following:
- the same row as the cathode(-) lead of the heartbeat sensor with the cathode(-) column
- the same row as the anode(+) lead of the heartbeat sensor with the anode(+) column
- analog pin zero (A0) to the same row as the input lead of the heartbeat sensor.
- the five volt (5v) pin to the anode(+) column.
- a ground (GND) pin on the UNO board to the cathode(-) column on the breadboard.
- digital pin 13 to the same row as the anode(+) lead of the LED

Coding - Digital Prototype

14) Optional. Enter the code in Tinkercad® to match the code on the next page.

15) Verify the code.

Engineering Design - Project Build

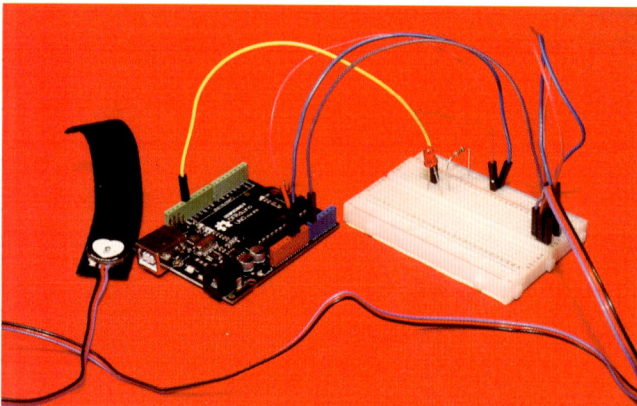

16) Place an LED across two rows of a column on the breadboard on one end of the breadboard.

17) Place a 220 Ohm resistor on the breadboard such that it connects the cathode(-) row with the same row as the cathode(-) lead of the LED and the same row.

18) Place the Heartbeat sensor across three rows of a column on the breadboard.

Assembling the Heartbeat Sensor

To assemble the heartbeat sensor apply one of the Velcro® cutouts from the kit to the back of the sensor. The back has the circuits. On the front, place one of the clear plastic cutouts. Attach the sensor to the Velcro® strip at one end.

To use the sensor place the side with the clear plastic facing the pointer finger and snugly, but not constricting wrap the strip around the finger to hold the sensor in place. Run the code, watch the LED and make sure to observe the values in the Serial Monitor.

```cpp
#define USE_ARDUINO_INTERRUPTS true    // For accurate BPM math.
#include <PulseSensorPlayground.h>     // PulseSensorPlayground Library.

//  Variables
const int PulseWire = 0;       // Sensor connected to Analog 0
const int LED13 = 13;          // LED pin 13.
int Threshold = 550;           // Choose which Signal to count as a beat and which to ignore.
    // Use the "Gettting Started Project" to fine-tune Threshold Value.
PulseSensorPlayground pulseSensor;

void setup() {
Serial.begin(9600);        // Reads Serial Monitor
pulseSensor.analogInput(PulseWire);
pulseSensor.blinkOnPulse(LED13);     //Blink Arduino's LED with heartbeat.
pulseSensor.setThreshold(Threshold);
if (pulseSensor.begin()) {
Serial.println("PulseSensor detector!");
}
}
void loop() {
 int myBPM = pulseSensor.getBeatsPerMinute();
if (pulseSensor.sawStartOfBeat()) {
Serial.println("♥ HeartBeat Detected!");
Serial.print("BPM: ");             // Displays "BPM: " on screen.
Serial.println(myBPM);             // Print the value inside of myBPM.
}
  delay(20);
}
```

19 Connect with jumper wires the following:
- the same row as the cathode(-) lead of the heartbeat sensor with the cathode(-) column
- the same row as the anode(+) lead of the heartbeat sensor with the anode(+) column
- analog pin zero (A0) to the same row as the input lead of the heartbeat sensor.
- the five volt (5v) pin to the anode(+) column.
- a ground (GND) pin on the UNO board to the cathode(-) column on the breadboard.
- digital pin 13 to the same row as the anode(+) lead of the LED

20 Connect the UNO board and computer using the USB cable.

Connection - When a health professional listens to a heartbeat with a stethoscope, they are only able to hear two distinct phases (systole and diastole) as the heart contracts and relaxes. When an EKG is used, five distinct heart actions are measured. The five actions correspond to the atria and ventricles contracting and relaxing in the heart. The "R" phase is the contraction of the ventricles moving blood to the body and lungs.

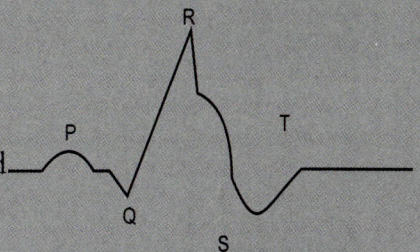

21 Launch the Arduino® IDE software (pages 11-13) to make sure the board is communicating with the computer.

Go to the Tools menu and verify that the correct board is selected from the Board Manager menu. If not, select the correct board from the options listed.

Then confirm that the right Port is selected. If not, select the port that lists the UNO board from the options listed.

Coding - Project Build

22 Either copy and paste the code from the Tinkercad® prototype or type the code on the previous page into the Arduino® IDE software (pages 11-13).

23 Install the sensor library (page 62) using the link on the "Links" page from the book web page.

24 Save the sketch (rename as needed).

25 Verify the sketch.

26 Upload the sketch to the UNO board.

27 Test the Heartbeat sensor and observe the LED on the breadboard. Open the Serial Monitor (ex to the right). Close the Serial Monitor and open and observe the Serial Plotter (ex. shown below).

28 Document this project, discuss the project, and complete the extensions assigned from the next page.

```
TAC_UE_V1_P18_Heartbeat | Arduino 1.8.9 (Windows Store 1.8.21.0)

File Edit Sketch Tools Help

TAC_UE_V1_P18_Heartbeat§

#define USE_ARDUINO_INTERRUPTS true   // For acurate BPM math.
#include <PulseSensorPlayground.h>    // PulseSensorPlayground Library.

// Variables

const int PulseWire = 0;      // Sensor connected to Analog 0
const int LED13 = 13;         // LED pin 13.
int Threshold = 550;          // Sets Signal to count as a beat and which to ignore.
    // Read "Gettting Started Project" to fine-tune Threshold Value.

PulseSensorPlayground pulseSensor;

void setup() {

Serial.begin(9600);           // Serial Monitor
pulseSensor.analogInput(PulseWire);
pulseSensor.blinkOnPulse(LED13);     //Blink Arduino's LED with heartbeat.
pulseSensor.setThreshold(Threshold);

if (pulseSensor.begin()) {
Serial.println("PulseSensor detector!");
}

}
void loop() {
  int myBPM = pulseSensor.getBeatsPerMinute();
if (pulseSensor.sawStartOfBeat()) {
Serial.println("♥ HeartBeat Detected!");
Serial.print("BPM: ");                       // Displays "BPM: " on screen.
Serial.println(myBPM);                        // Print the value inside of myBPM.
}
  delay(20);
}
```

```
Sketch uses 4784 bytes (14%) of program storage space. Maximum is 32256 bytes.
Global variables use 264 bytes (12%) of dynamic memory, leaving 1784 bytes for local variables. Max
```

♥ HeartBeat Detected!
BPM: 63
♥ HeartBeat Detected!
BPM: 63
♥ HeartBeat Detected!
BPM: 62
♥ HeartBeat Detected!
BPM: 62
♥ HeartBeat Detected!
BPM: 63
♥ HeartBeat Detected!

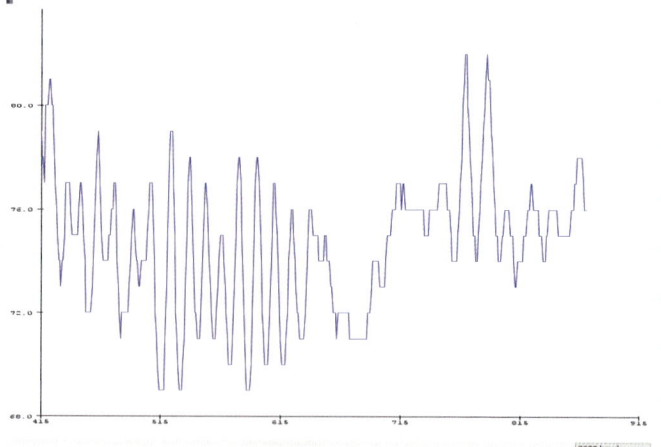

Connection - For thousands of years, humans have tried to understand the importance of the internal organs in the human body. For most of that time knowledge about the inner workings was often based on assumptions, superstition, or religious beliefs, and not on any science.

4000 years ago when the ancient Egyptians started to remove the internal organs and the brain to prepare the Pharaoh's body for mummification they chose to save five organs for burial with the body. The stomach, liver, intestines, and lungs were removed and then buried outside the body inside special jars called Canopic jars (shown here). The heart was believed to be the seat of the soul, the repository of memories, and place from which human wisdom resided. The heart was also needed for judgment in the afterlife and deemed so important that it was left inside the body. The remaining organs including the brain were considered unimportant and thrown away.

At the time of judgment the heart was put on a balance with the feather of Ma'at, the symbol for universal truth. If the heart weighed more than the feather, the soul was considered impure and the person to whom the heart belonged was denied life after death.

To tip the balance in the person's favor the body was buried with a scarab beetle amulet that prevented the heart from revealing anything impure about the person.

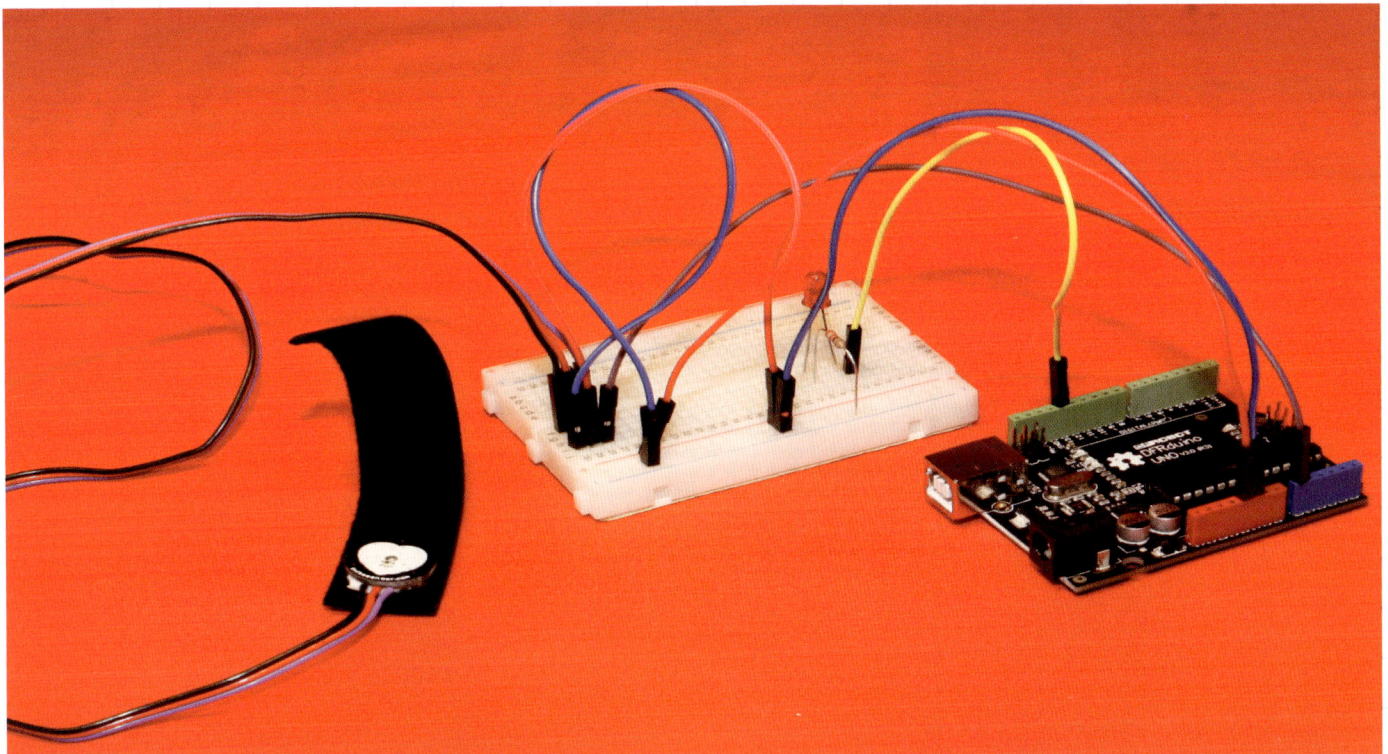

Project 18
My Heartbeat

Project Reflection and Summative Activities, Discussion Starters, Extensions, and Problem

A = Essential

Reflection/Summative Activity:
- Have each team member document in their project/classroom journal who was on their team, what went well, what they could improve upon, what they would do differently if they were to do the project again, and verify that every person in the group can do the project.

For Discussion:
- List the limitations and advantages of the heartbeat sensor.

Extensions:
- Design an implement a less complicated version of the heartbeat sensor.

- Replace the LED with a buzzer and implement the device.

B = Recommended

Professional Connections:
- Invite a fitness expert to explain the things that are monitored in a fitness program including diet and heart rate.

- Research and write a short description of a health profession that measures heart rates. Be sure to include the education the profession requires, any laws or regulations that guide their use, and what activities are performed.

Technology Connections:
- Many public places are installing heart defibrillator. Why? For what purpose? What does a defibrillator do?

- Compare and contrast this heartbeat sensor with a personal device (ex. Fitbit®).

Extension:
- Design and implement the project to use the 16x2 LCD.

- Submit your project to the book page on the website. Register and Login to submit at: thearduinoclassroom.com

From the homepage:
Click
 Books
Click
 UNO Edition Vol. 1

C = Optional

Extension:
- Using the 16x2 LCD, the barrel connection and a 9v battery, build a portable heartbeat sensor.

Problem:
- Identify a problem that would be solved by the heartbeat sensor, but not measuring a heartbeat. How could this project be a solution? Include in your report the following:

Description of the problem.

Description of how the problem is solved by this project

What other resources you would need to solve the problem.

Design and produce the solution.

A project rubric can be found on page 236.

Anticipatory Sets
(from page 168)

Anticipatory Sets:
- Have each student measure their heartbeat using a clock.
Responses Will Vary

- Investigate how blood circulates in the body.
Responses Will Vary

A

Reflection/Summative Activity:
Responses Will Vary

For Discussion:
- List the limitations and advantages of the heartbeat sensor.
The heartbeat sensor requires a precise fitting to measure heart rate accurately. It does not measure and average heart rate only estimates the heart rate by determining the time between beats.

Extensions:
- Design an implement a less complicated version of the heartbeat sensor.
Connect the heartbeat sensor directly to the UNO board which would remove the breadboard and LED on the breadboard.

- Replace the LED with a buzzer and implement the device.
Responses Will Vary

B

Professional Connections:
Responses Will Vary

Technology Connections:
- Many public places are installing heart defibrillator. Why? For what purpose? What does a defibrillator do?
To provide a life saving machine for heart attack victims. It shocks the heart to restart a regular beat.

- Compare and contrast this heartbeat sensor with a personal device (ex. Fitbit®).
Responses Will Vary

Extension:
- Design and implement the project to use the 16x2 LCD.
Responses Will Vary

- Submit your project to the book page on the website.
Responses Will Vary

C

Extension:
- Using the 16x2 LCD, the barrel connection and a 9v battery, build a portable heartbeat sensor.
Responses Will Vary

Problem:
Responses Will Vary

Project 19
pH Properties

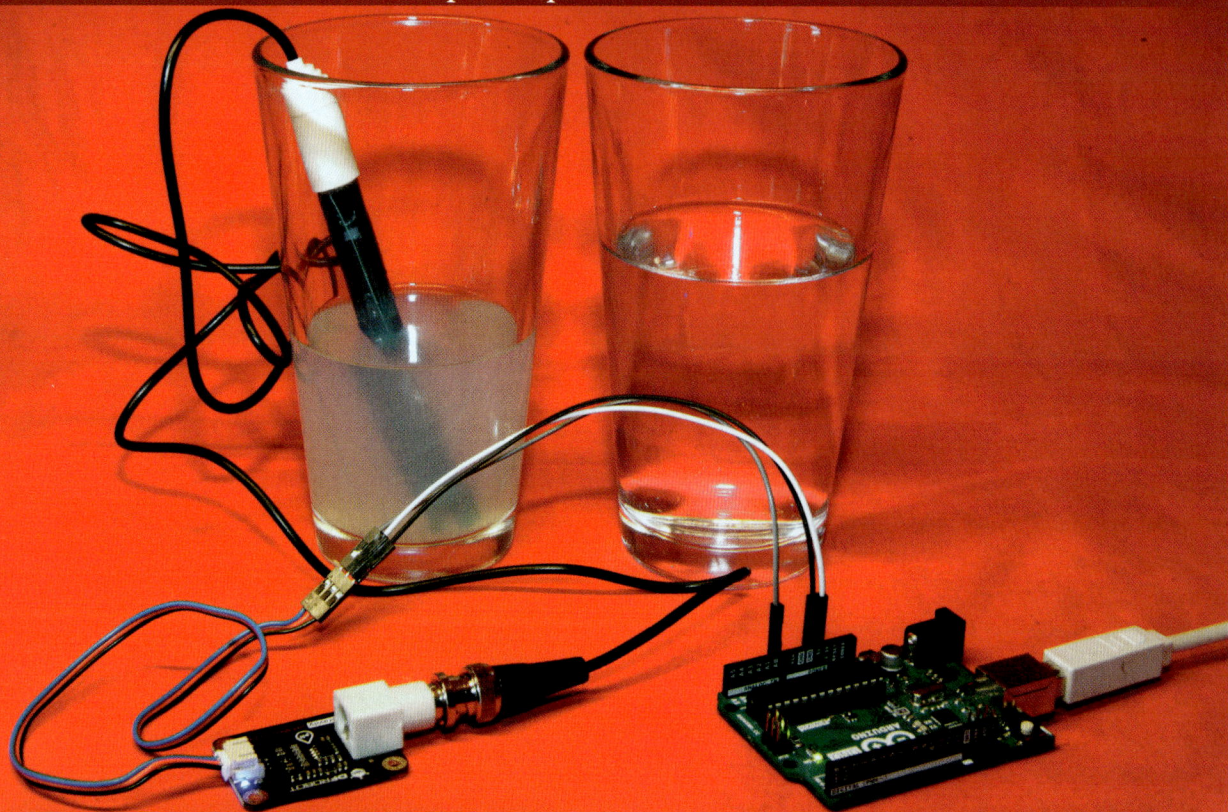

Lesson Integration:	Groupings:	Level:	Time to Complete:
Physics - Ohm's Law, Circuits, Voltage, Resistance Chemistry - pH, Ions, Acids, Bases, Buffers Biology - Digestive System Mathematics - Arithmetic, Algebra, Log Scales	1 - 2	Intermediate	45 min. for the project and 45 min. for extensions.

Objectives:
- Investigate how circuits and electronic components interact with electrical energy.
- Understand pH.

Prerequisite Skills:
- Molecules, Atoms, Ions, Acids, Bases, pH
- Arduino® IDE software (pages 11-13)
- Fritzing (pages 64-66)
- Serial Monitor and Plotter (pages 58-59)

STEAM Connections:
Science - Acids, Bases, pH, Digestion, Circuits,
Technology - Code, Analog Data, Calibration
Engineering - Designing, Building, and Using a Machine, Prototyping, Applied Physics
Allied Arts - Culinary Arts
Math - Ratios, Conversions, Log Scales

Purpose and Skills:
- Determine the pH of solutions

Key Vocabulary:
Acid - a molecule that releases H^+ ions in solution.
Base - a molecule that releases OH^- or accepts H^+ ions in solution.
Buffer - a solution that does not change pH easily.
Calibration - adjustment of an input sensor's data to match expected values.
pH - the measurement of OH^- or H^+ ions in a solution on a scale of 0-14 with 7 being neutral with equal amounts of OH^- and H^+ ions.

Project Introduction:
- Introduce the groups to the purpose of the project, the skills developed, the standards met, and the goal of the project.

Anticipatory Sets:
- List three locations you would see pH measured.

Educational Standards

ISTE Standards for Students
- Empowered Learner 1a, 1b, 1d
- Knowledge Constructor 3a, 3b, 3c, 3d
- Innovative Designer 4a, 4b, 4c, 4d
- Computational Thinker 5a, 5b, 5c, 5d
- Creative Communicator 6a, 6b, 6c, 6d
- Global Collaborator 7c, 7d

US Computer Science Standards
- Project correlations can be found on the book's web page.

US NGSS
DCIs
- MS-PS1-1. PS1.A
- MS-PS1-2 PS1.A & PS1.B
- MS-PS4-3 PS4.C
- MS-LS1-7 LS1.C
- MS-LS2-1 LS2.A

Cross Cutting Concepts
- Cause and Effect
- Scale, Proportion and Quantity
- Structure and Function

Science and Engineering Practices
- Planning and Carrying Out Investigations
- Obtaining, Evaluating, and Communicating Information
- Analyzing and Interpreting Data
- Scientific Knowledge is Based on Empirical Evidence
- Using Mathematics and Computational Thinking

US NGSS - High School
- Correlations can be found on the book's web page.

US Common Core Language Arts and Mathematics
- Correlations can be found on the book's web page.

Step-by-Step 1-36

Materials List:
- Computer with IDE software and Fritzing
- Connection to the Internet
- USB Cable
- UNO or UNO Compatible Microcontroller
- pH Sensor (with intermediate board and sensor cable)
- 3 Jumper Wires (Male to Male)
- Distilled Water - DO NOT USE TAP WATER
- Buffer solutions (pH 4.0, pH 7.0, pH 10.0)
- Testing Samples

Engineering Design - Digital Prototype

1. Download, install, and launch (as needed) Fritzing (pages 64-66) to build a digital prototype.

2. Visit The Arduino® Classroom website and register for the website.

3. Navigate to the "UNO Edition Vol. 1" page by hovering over "Books" on the main menu and clicking on the "UNO Edition Vol. 1" link.

4. On the right of the page find and then click on the "Volume 1 Links" link.

5. Click on the "Fritzing Part Download" link for Project 19 which will take you to the page to download the file.

6. Extract the file.

7. Install the file (page 66).

8. Start a new sketch in Fritzing.

9. Drag an UNO board to the workspace.

10. Drag the pH sensor board to the workspace.

Always store the pH sensor in fresh distilled water. Use proper safety protocols and equipment when handling chemicals including eye safety wear.

11 Connect with jumper wires the following:
- analog pin zero (A0) on the UNO board to the data pin of the pH sensor board.
- the five volt (5v) pin on the UNO board to the PWR pin of the pH sensor board.
- the GND pin on the UNO board to the GND pin of the pH sensor board.

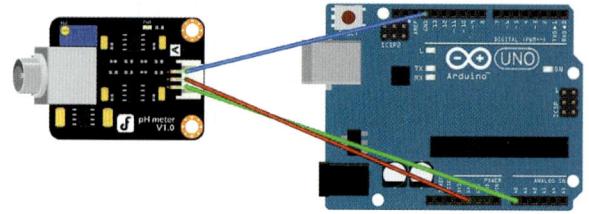

Engineering Design - Project Build

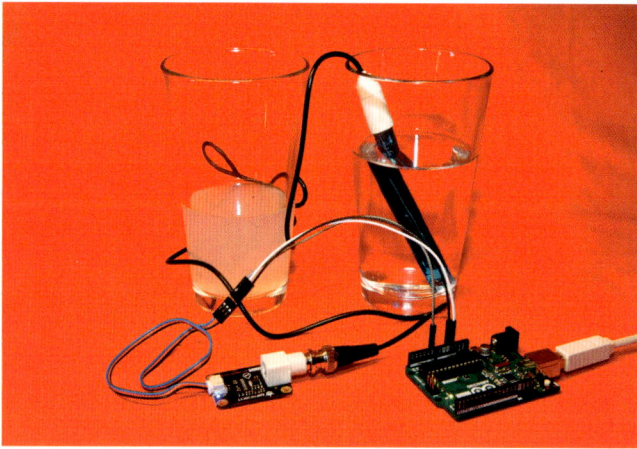

12 Connect with jumper wires the following:
- analog pin zero (A0) on the UNO board to the data pin of the pH sensor cable.
- the five volt (5v) pin t on the UNO board to the PWR pin of the pH sensor cable.
- the GND pin on the UNO board to the GND pin of the pH sensor cable.

13 Connect the pH sensor cable to the pH data board.

14 Connect the single coaxial pH cable the pH data board.

15 Connect the single coaxial pH cable the pH sensor.

16 Launch the Arduino® IDE software (pages 11-13) to make sure the board is communicating with the computer.

Go to the Tools menu and verify that the correct board is selected from the Board Manager menu.

pH	Examples
0	Battery Acid
1	Hydrochloric Acid
2	Lemon Juice / Stomach Acid
3	Grapefruit Juice / Soda
4	Tomato Juice / Acid Rain
5	Coffee / Sparkling Water
6	Saliva
7	Pure Water
8	Ocean Water
9	Baking Soda in Water
10	Milk of Magnesia
11	Ammonia Solution
12	Soap Water
13	Bleach / Oven Cleaner
14	Drain Cleaner (Liquid)

```
float calibration = 0.0; //Add value to calibrate
const int analogPin = A0;
int pHsensorValue = 0;
unsigned long int averageValue;
float b;
int buf[10],temp;
void setup() {
 Serial.begin(9600);
}
void loop() {
 for(int i=0;i<10;i++)
 {
 buf[i]=analogRead(analogPin);
 delay(30);
 }
 for(int i=0;i<9;i++)
 {
 for(int j=i+1;j<10;j++)
 {
 if(buf[i]>buf[j])
 {
 temp=buf[i];
 buf[i]=buf[j];
 buf[j]=temp;
 }
 }
 }
 averageValue=0;
 for(int i=2;i<8;i++)
 averageValue+=buf[i];
 float pHVol=(float)averageValue*5.0/1023/6;
 float pHsensorValue = 5.0 * pHVol + calibration;
 Serial.print("pH reading = ");
 Serial.println(pHsensorValue);
 delay(700);
}
```

If not, select the correct board from the options listed.

Then confirm that the right Port is selected. If not, select the port that lists the UNO board from the options listed.

Coding - Project Build

17 Type the code from this page into the Arduino® IDE software (pages 11-13).

18 Save the sketch (rename as needed).

19 Verify the sketch.

20 Upload the sketch to the UNO board.

21 Remove the protective covering off the end of the pH sensor.

22 Wash the pH sensor with distilled water.

23 Immerse the pH sensor in 7.0 buffer solution for at least two (2) minutes. Open the Serial Monitor and observe the pH output values. Record the value and difference from 7.00.

```
TAC_UE_V1_P19c
//Project 19
//pH Sensor
//thearduinoclassroom.com
//Copyright 2019, Isabel Mendiola and Peter Haydock

float calibration = 0.0; //Add value to calibrate
const int analogPin = A0;
int pHsensorValue = 0;
unsigned long int averageValue;
float b;
int buf[10],temp;
void setup() {
 Serial.begin(9600);
}
void loop() {
 for(int i=0;i<10;i++)
 {
 buf[i]=analogRead(analogPin);
 delay(30);
 }
 for(int i=0;i<9;i++)
 {
 for(int j=i+1;j<10;j++)
 {
 if(buf[i]>buf[j])
 {
 temp=buf[i];
 buf[i]=buf[j];
 buf[j]=temp;
 }
 }
 }
 averageValue=0;
 for(int i=2;i<8;i++)
 averageValue+=buf[i];
 float pHVol=(float)averageValue*5.0/1023/6;
 float pHsensorValue = 5.0 * pHVol + calibration;
 Serial.print("pH reading = ");
 Serial.println(pHsensorValue);
 delay(700);
}
```

Connection - The human body's digestive system uses acids and bases to process the food we eat. Starting in the mouth where chewing and enzymes begin the breaking down of food in a fairly neutral pH 6.2 - 7.2 solution.

After the food passes down the esophagus it is drenched in an acid solution of pH 1.5 to 3.0 in the stomach. The acid is so strong that the body must replace the lining of the stomach every five to seven days and coat it with a protective layer of mucus. So much acid is released into the stomach that when the byproducts of the acid production reach the blood, the pH in bloodstream sees an significant decrease in pH. If acid backs up into the esophagus, it is called heartburn.

Food then passes into the small intestines where it continues to be digested in a solution with a pH of 6. The pH slowly rises as the food passes through. By the time the food is ready for the large intestines the pH of the food solution can he as high as 7.4.

In the large intestines, the pH of the food and digestive juices drops to 5.7 and then climbs again to 6.7 before being expelled by the body.

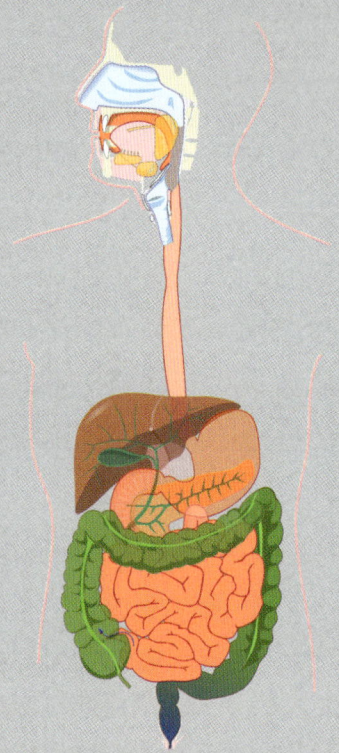

Project 19
pH Properties

24 Remove the pH sensor from the buffer solution and rinse with distilled water.

25 Immerse the pH sensor in 4.0 buffer solution for at least two (2) minutes.

26 Record the pH value and difference from 4.00 shown in the Serial Monitor.

27 Remove the pH sensor from the buffer solution and rinse with distilled water.

28 Immerse the pH sensor in 10.0 buffer solution for at least two (2) minutes.

29 Record the pH value and difference from 10.00 shown in the Serial Monitor.

30 Remove the pH sensor from the buffer solution and rinse with distilled water.

31 Immerse the pH sensor in an unknown solution for at least two (2) minutes. Compare the value to the three buffer solution readings. Edit the first line of code with the difference recorded from the buffer solution with the closest match to the unknown. Ex. buffer solution 7.0 returns a value of 5.5 and is closest to the unknown with a 5.4. Edit the code with a +1.5 and retest the unknown to obtain a more accurate pH value.

32 Observe the pH output values from Serial Monitor for at least two (2) minutes.

33 Remove the pH sensor from the buffer solution and rinse with distilled water.

34 Edit the code so the calibration number to 0.0.

35 Test other unknown solutions using the same process (step 31) of measurement and calibration.

36 Document this project, discuss the project, and complete the extensions assigned from the next page.

Connection - Chemical reactions using acids and bases are very energetic. These reactions can release light, heat, gas, salt, or even absorb heat to make an object very cold.

In the case of a first aid cold pack, the chemical ammonium nitrate is mixed with water which creates an acidic solution of a pH 5.25. When the NH_4^+ ions separate from the NO_3^- ions in the presence of water the mixture absorbs heat from the surroundings quickly with the solution becoming very cold. The cold pack then is placed on the body to lower the temperature of the injured part which in turn reduces swelling and blocks pain signals

When baking a cake, sodium bicarbonate (a base) is mixed with an acid (that varies based on the cake being made). The acid and sodium bicarbonate react and at the end of the chain of the reaction carbon dioxide (a gas) is released which forms bubbles in the cake mix. The bubbles then are baked into the cake making it fluffy.

Project 19
pH Properties

Project Reflection and Summative Activities, Discussion Starters, Extensions, and Problem

A = Essential

Reflection/Summative Activity:
- Have each team member document in their project/classroom journal who was on their team, what went well, what they could improve upon, what they would do differently if they were to do the project again, and verify that every person in the group can do the project.

For Discussion:
- Create a list of acids and bases found in your home. Place these on a single pH scale chart like the one on page 178.

- Why do people take antacids? What is the chemical/pH justification for this medicine?

Extensions:
- Add an LCD to the project and edit the code to support the project readings being displayed on the LCD.

- Add three LEDs to the project and edit the code to blink one LED when the solution being tested is lower than a pH of 3.5, another LED when the solution is above a pH of 11, and the third LED when the solution pH is between 3.5 and 11.

B = Recommended

Health Connections:
- Examine the ingredients in your food and determine which are acids and which are bases. What is the function of each ingredient and determine how removing the ingredient would impact the quality and taste of the food.

- Create a presentation based on your findings.

Culinary Connection:
- Video conference, invite a chef or cook to your classroom, or watch a show to learn about the chemistry of cooking.

Extension:
- Add a breadboard to the project and have different tones made depending on the pH of the solution being tested. Demonstrate the project.

- Submit your project to the book page on the website. Register and Login to submit at: thearduinoclassroom.com

From the homepage:
Click
   Books
Click
   UNO Edition Vol. 1

C = Optional

Extension:
- Create a chart comparing the accuracy of the pH sensor against another pH measuring device.

Problem:
- Identify a problem that would be solved by the pH sensor. How could this project be a solution? Include in your report the following:

Description of the problem.

Description of how the problem is solved by this project.

What other resources you would need to solve the problem.

Design and produce the solution.

A project rubric can be found on page 236.

Anticipatory Sets
(from page 176)

Anticipatory Sets:
- List three locations you would see pH measured.
Responses Will Vary

A

Reflection/Summative Activity:
- Have each team member document in their project/ classroom journal who was on their team, what went well, what they could improve upon, what they would do differently if they were to do the project again, and verify that every person in the group can do the project.
Responses Will Vary

For Discussion:
- Create a list of acids and bases found in your home. Place these on a single pH scale chart like the one on page 178.
Responses Will Vary

- Why do people take ant-acids? What is the chemical/pH justification for this medicine?
Responses Will Vary

Extensions:
- Add an LCD to the project and edit the code to support the project readings being displayed on the LCD.
Responses Will Vary

- Add three LEDs to the project and edit the code to blink one LED when the solution being tested is lower than a pH of 3.5, another LED when the solution is above a pH of 11, and the third LED when the solution pH is between 3.5 and 11.
Responses Will Vary

B

Health Connections:
- Examine the ingredients in your food and determine which are acids and which are bases. What is the function of each ingredient and determine how removing the ingredient would impact the quality and taste of the food.

- Create a presentation based on your findings.
Responses Will Vary

Culinary Connection:
- Video conference, invite a chef or cook to your classroom, or watch a show to learn about the chemistry of cooking.
Responses Will Vary

Extension:
- Add a breadboard to the project and have different tones made depending on the pH of the solution being tested. Demonstrate the project.
Responses Will Vary

- Submit your project to the book page on the website.
Responses Will Vary

C

Extension:
- Create a chart comparing the accuracy of the pH sensor against another pH measuring device.
Responses Will Vary

Problem:
Responses Will Vary

Project 20
Soil Moisture

Lesson Integration:	Groupings:	Level:	Time to Complete:
Physics - Ohm's Law, Circuits, Voltage, Resistance Biology - Botany Chemistry - Water, Conductivity Earth Science - Soils, Hydrology, Water Cycle Mathematics - Arithmetic	1 - 2	Intermediate	45 min. for the project and 45 min. for extensions.

Objectives:
- Investigate how circuits and electronic components interact with electrical energy.
- Create a sensor to measure soil moisture content

Prerequisite Skills:
- Arduino® IDE software (pages 11-13)
- Fritzing (pages 64-66)
- Serial Monitor and Plotter (pages 58-59)

Purpose and Skills:
- Use a sensor for water content of a soil.

STEAM Connections:
Science - Circuits, Resistance, Ohm's law, Soils, Conductivity, Water
Technology - Code, Simulators, Digital Design, Electronic Components, Analog Data
Engineering - Designing, Building, and Using a Machine, Prototyping, Applied Physics
Allied Arts -

Math - Numbers, Algebra

Key Vocabulary:
Conductivity - The ability of a material to allow or not allow electricity to pass through it.

Project Introduction:
- Introduce the groups to the purpose of the project, the skills developed, the standards met, and the goal of the project.

Anticipatory Sets:
- Identify where soil moisture is measured routinely. List at least three different places or situations where soil moisture is measured, describe the locations, and explain the purpose of the measurement.
- Assign each student a different location around the world to investigate the predominant type of soil (and its characteristics) that is found there.

Project 20
Soil Moisture

Educational Standards

ISTE Standards for Students
- Empowered Learner 1a, 1b, 1d
- Knowledge Constructor 3a, 3b, 3c, 3d
- Innovative Designer 4a, 4b, 4c, 4d
- Computational Thinker 5a, 5b, 5c, 5d
- Creative Communicator 6a, 6b, 6c, 6d
- Global Collaborator 7c, 7d

US Computer Science Standards
- Project correlations can be found on the book's web page.

US NGSS
DCIs
- MS-PS1-1 PS1.A
- MS-PS1-2 PS1.A
- MS-PS4-3 PS4.C
- MS-LS1-5 LS1.B
- MS-LS2-1 LS2.A
- MS-LS2-4 LS2.C
- MS-ESS2-4 ESS2.C

Cross Cutting Concepts
- Cause and Effect
- Scale, Proportion and Quantity
- Structure and Function

Science and Engineering Practices
- Planning and Carrying Out Investigations
- Obtaining, Evaluating, and Communicating Information
- Analyzing and Interpreting Data
- Scientific Knowledge is Based on Empirical Evidence
- Using Mathematics and Computational Thinking

US NGSS - High School
- Correlations can be found on the book's web page.

US Common Core Language Arts and Mathematics
- Correlations can be found on the book's web page.

Materials List:
- Computer with IDE and Fritzing software
- Connection to the Internet
- USB Cable
- UNO or UNO Compatible Microcontroller
- Short Breadboard
- Soil Moisture Sensor
- Sensor Connection Cable
- Eight (8) Jumper Wires
- 220 Ohm Resistor
- Piezoelectric Buzzer
- Several pots of soil
- Water

Engineering Design - Digital Prototype

1. Download, install, and launch (as needed) Fritzing (pages 64-66) to build a digital prototype.

2. Visit The Arduino® Classroom website and register for the website.

3. Navigate to the "UNO Edition Vol. 1" page by hovering over "Books" on the main menu and clicking on the "UNO Edition Vol. 1" link.

4. On the right of the page find and then click on the "Volume 1 Links" link.

5. Click on the "Fritzing Part Download" link for Project 20 which will take you to the page to download the file.

6. Extract the file.

7. Install the file (page 66).

8. Start a new sketch in Fritzing.

9. Drag an UNO board to the workspace.

10. Drag the soil moisture sensor to the workspace.

11. Drag the soil moisture sensor board to the workspace.

12. Place an LED on the breadboard.

13. Place a Piezoelectric buzzer on the breadboard.

14. Place a 220 Ohm resistor on the breadboard such that it connects the cathode(-) column and the row that has the cathode(-) lead of the LED.

15. Connect with jumper wires the following:
    - the five volt (5v) pin on the UNO board to the anode(+) column on the breadboard
    - a ground (GND) pin on the UNO board to the cathode(-) column on the breadboard.
    - analog pin zero (A0) to the data pin on the soil moisture data board
    - the anode(+) column on the breadboard to the anode(+) pin on the soil moisture data board.
    - the cathode(-) column on the breadboard to the cathode(-) pin on the soil moisture data board.
    - the cathode(-) column on the breadboard to the same row as the cathode(-) pin on the piezo.
    - digital pin three (3) to the same row as the anode(+) lead of the LED
    - digital pin two (2) to the same row as the anode(+) lead of the piezo

Coding - Digital Prototype

16. Optional. Enter the code in Tinkercad® to match the code on this page.

```
int sensor = A0;
int LED = 3;
int buzzer = 2;
int thresholdValue = 700; // can be adjusted
void setup(){
  pinMode(A0, INPUT);
  pinMode(LED, OUTPUT);
  pinMode(buzzer, OUTPUT);
  digitalWrite(LED, LOW);
  noTone(buzzer);
  Serial.begin(9600);
}
void loop() {
  digitalWrite(LED, LOW);
  noTone(buzzer);
  delay(5000);
  int sensorValue = analogRead(A0);// Reads sensor
  Serial.print(sensorValue);
  if(sensorValue > thresholdValue){
   Serial.println( " Your plant does not need water" );
   delay(1000);
  }
  else {
   Serial.println(" Time to add water to your plant");
   digitalWrite(LED, HIGH);
   tone(buzzer, 550);
   delay(1000);
  }
}
```

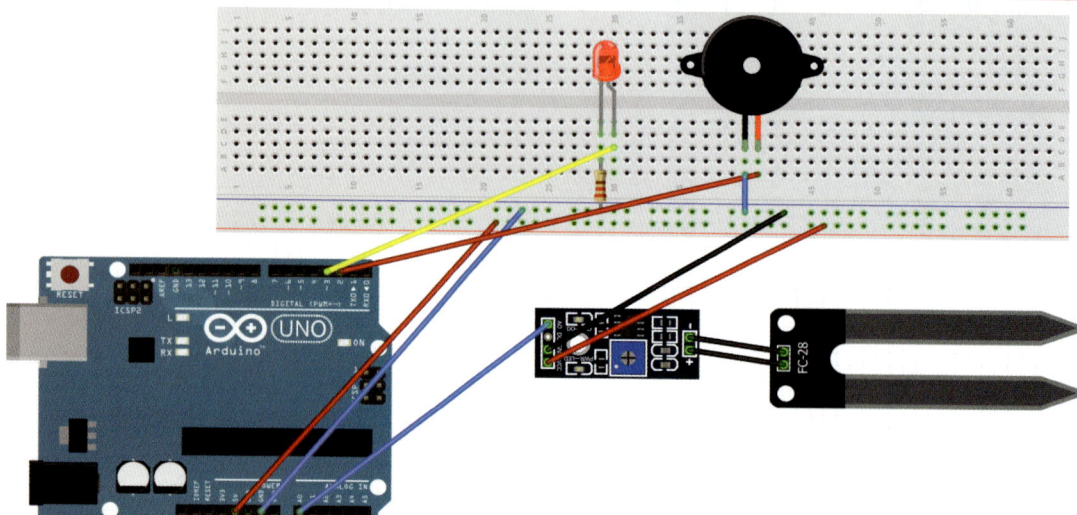

Connection - Scientists who study rain forests have shown that the soil holds little nutrient value to the plants. When it rains, the water washes most decaying plant and animal material and other soil nutrients quickly downstream along with much of the water.

To overcome this, most plants have evolved to use their roots to collect needed water and nutrients as quickly as possible. The water moves upwards through the plant and is transpired out of the leaves. The nutrients are carried upwards with the water from the roots to the leaves efficiently and feed the plant. The water that escapes into the atmosphere will fall as rain and be used to carry more nutrients up into the rain forest.

Some plants in the rain forest have evolved to not root in the soil, but in the air. The roots cling to other tress and draw nutrients straight from the air. In this photo the sloth climbs among plants with their roots hanging off other plants in their search for food an water.

Engineering Design - Project Build

17　Connect the soil moisture sensor to soil moisture sensor board.

18　Place an LED on the breadboard.

19　Place a Piezoelectric buzzer on the breadboard.

20　Place a 220 Ohm resistor on the breadboard such that it connects the cathode(-) column and the row that has the cathode(-) lead of the LED.

21　Connect with jumper wires the following:
- the five volt (5v) pin on the UNO board to the anode(+) column on the breadboard
- a ground (GND) pin on the UNO board to the cathode(-) column on the breadboard.
- analog pin zero (A0) to the data pin on the soil moisture data board
- the anode(+) column on the breadboard to the anode(+) pin on the soil moisture data board.
- the cathode(-) column on the breadboard to the cathode(-) pin on the soil moisture data board.
- the cathode(-) column on the breadboard to the same row as the cathode(-) pin on the piezo.
- digital pin three (3) to the same row as the anode(+) lead of the LED
- digital pin two (2) to the same row as the anode(+) lead of the piezo

22 Connect the UNO board and computer using the USB cable.

23 Launch the Arduino® IDE software (pages 11-13) to make sure the board is communicating with the computer.

Go to the Tools menu and verify that the correct board is selected from the Board Manager menu. If not, select the correct board from the options listed.

Then confirm that the right Port is selected. If not, select the port that lists the UNO board from the options listed.

Coding - Project Build

24 Either copy and paste the code from the Tinkercad® prototype or type the code from page 186 into the Arduino® IDE software (pages 11-13).

25 Save the sketch (rename as needed).

26 Verify the sketch.

27 Upload the sketch to the UNO board.

```
TAC_UE_V1_P20 | Arduino 1.8.9 (Windows Store 1.8.21.0)
File Edit Sketch Tools Help

TAC_UE_V1_P20§

//Project 20
//Soil moisture hygrometer detection sensor
//thearduinoclassroom.com
//Copyright 2019, Isabel Mendiola and Peter Haydock
int sensor = A0;
int LED = 3;
int buzzer = 2;
int thresholdValue = 700; // can be adjusted

void setup(){
  pinMode(A0, INPUT);
  pinMode(LED, OUTPUT);
  pinMode(buzzer, OUTPUT);
  digitalWrite(LED, LOW);
  noTone(buzzer);
  Serial.begin(9600);
}
void loop() {
  digitalWrite(LED, LOW);
  noTone(buzzer);
  delay(5000);
  int sensorValue = analogRead(A0);// Reads sensor
  Serial.print(sensorValue);
  if(sensorValue > thresholdValue){
    Serial.println( " Your plant does not need water" );
    delay(1000);
  }
  else {
    Serial.println(" Time to add water to your plant");
    digitalWrite(LED, HIGH);
    tone(buzzer, 550);
    delay(1000);
  }
}

Done uploading.
Sketch uses 3190 bytes (9%) of program storage space. Maximum is 32256 byt
Global variables use 271 bytes (13%) of dynamic memory, leaving 1777 bytes

34                                          Arduino/Genuino Uno on COM4
```

28 Open the Serial Monitor and observe the sensor under different soil conditions.

29 Document this project, discuss the project, and complete the extensions assigned from the next page. ✓

Connection - As seen from this photo (top right) taken by NASA from a Earth monitoring satellite, pivot irrigation is a common but inefficient practice used on many farms in the western plains states of the United States from Wyoming to Texas. In the upper photo the larger circles and semi-circles are a mile in diameter with the smaller circles approximately a half mile in diameter. Farmers dug wells that tapped into aquifers for water and then used an overhead sprinklers that traveled in a circle to water their crops (middle right). Pivot systems work best in flat or nearly flat landscapes with annual crops like wheat, barely and oats. These systems are being slowly abandoned because they waste a large percentage of the water.

Watering crops from above allows much of the water to evaporate before it can be used by the crops. Also, the systems did not water the corners leaving much valuable cropland unused. Additionally, the crops used so much water that the aquifers started to empty. Farmers first dug deeper wells, but that became too expensive and other solutions needed to be pursued.

Drip or ground based systems (as used on this pecan ranch in Candela, Mexico - bottom right) save water and money because they are more efficient.

Modern irrigation systems monitor the entire field and only water sections needing water and only at times of the day to maximize plant growth. The watering is timed with fertilizer application and pest management solution. The systems use satellite data, soil moisture sensors, drones, and other technologies to minimize water use and water crops only when needed.

Project 20
Soil Moisture

Project Reflection and Summative Activities, Discussion Starters, Extensions, and Problem

A = Essential

Reflection/Summative Activity:
- Have each team member document in their project/classroom journal who was on their team, what went well, what they could improve upon, what they would do differently if they were to do the project again, and verify that every person in the group can do the project.

- Students should create a graphic organizer of what they have learned by doing this project.

For Discussion:
- How does soil define what kinds of plants live in that type of soil.

- Create a class list of reasons for construction, agriculture, environmental sciences would test soil conditions.

Extension:
- Change the LED in the project to an RGB LED. Change the code to show different colors of light to indicate different levels of soil moisture.

B = Recommended

Geological Connections:
- Have students research soil conservation methods in different locations of the world.

- Research and write a summary of technologies used to monitor the soil.

Extension:
- Make the project portable with an LCD and the 9v battery barrel connector.

- Submit your project to the book page on the website. Register and Login to submit at: thearduinoclassroom.com

From the homepage:
Click
  Books
Click
  UNO Edition Vol. 1

C = Optional

Extensions:
- Redesign the project to sound the alarm when too much is present in the soil (think of a cactus).

- Design and implement a study on how quickly different soil types dry out and need watering to keep a plant alive.

Share the results in a poster.

Problem:
- Identify a problem that would be solved with a soil moisture sensor. How could this project be a solution? Include in your report the following:

Description of the problem.

Description of how the problem is solved by this project.

What other resources you would need to solve the problem.

Design and produce the solution.

A project rubric can be found on page 236.

Anticipatory Sets
(from page 184)

Anticipatory Sets:
- Identify where soil moisture is measured routinely. List at least three different places or situations where soil moisture is measured, describe the locations, and explain the purpose of the measurement.
Responses Will Vary

- Assign each student a different location of the world to investigate the predominant type of soil (and its characteristics) that is found there.
Responses Will Vary

A

Reflection/Summative Activity:
- Have each team member document in their project/classroom journal who was on their team, what went well, what they could improve upon, what they would do differently if they were to do the project again, and verify that every person in the group can do the project.
Responses Will Vary

- Students should create a graphic organizer of what they have learned by doing this project.
Responses Will Vary

For Discussion:
- How does soil define what

kinds of plants live in that type of soil.
Responses Will Vary

Create a class list of reasons for construction, agriculture, environmental sciences would test soil conditions.
Responses Will Vary

Extensions:
- Change the LED in the project to a RGB LED. Change the code to show different colors of light to indicate different levels of soil moisture.
Responses Will Vary

B

Geological Connections:
- Have students research soil conservation methods in different locations of the world.
Responses Will Vary

- Research and write a summary of technologies used to monitor the soil.
Responses Will Vary

Extensions:
- Make the project portable with an LCD and the 9v battery barrel connector.
Responses Will Vary

- Submit your project to the book page on the website.
Responses Will Vary

C

Extensions:
- Design and implement a study on how quickly different soil types dry out and need watering to keep a plant alive.
Responses Will Vary

- Share the results in a poster.
Responses Will Vary

Problem:
Responses Will Vary

Project 21
Gas Sniffer

Lesson Integration:	Groupings:	Level:	Time to Complete:
Physics - Ohm's Law, Circuits, Voltage, Resistance Chemistry - Organic Chemistry, Molecules Biology - Digestion Earth Science - Atmospheric Composition Mathematics - Arithmetic	1 - 2	Advanced	1:15 hr:min. for the project and 45 min. for extensions.

Objectives:
- Investigate how circuits and electronic components interact with electrical energy.
- Sense for different gas molecules.

Prerequisite Skills:
- Arduino® IDE software (pages 11-13)
- Fritzing (pages 64-66)
- Serial Monitor and Plotter (pages 58-59)
- Compounds and Molecules

Purpose and Skills:
Create a sensor for different gases.

STEAM Connections:
Science - Circuits, Resistance, Ohm's law, Molecules, Gases, Organic Compounds
Technology - Code, Simulators, Digital Design, Electronic Components, Analog Data, Calibration
Engineering - Designing, Building, and Using a Machine, Prototyping, Applied Physics
Allied Arts - Smell
Math - Ratios, Conversions

Key Vocabulary:
Gas - A state of matter that has no definite volume

or shape

Project Introduction:
- Introduce the groups to the purpose of the project, the skills developed, the standards met, and the goal of the project.

Anticipatory Sets:
- Identify where gas sensors are seen. List at least three different places or situations where temperatures are shown, describe the locations, and explain the purpose of the displays.

- How do gas delivery companies warn people of a gas leak without a sensor?

WARNING: This project is not meant to diagnose any gas leaks (CH_4 or otherwise) or replace commercially produced sensors. CH_4 and other gases may be explosive gases and should only be tested under supervision of an experienced adult.

The sensor heats up during use and touching the metal protective screen may cause a burn. Handle the sensor with caution.

Project 21
Gas Sniffer

Step-by-Step 1-27

Educational Standards

ISTE Standards for Students
- Empowered Learner 1a, 1b, 1d
- Knowledge Constructor 3a, 3b, 3c, 3d
- Innovative Designer 4a, 4b, 4c, 4d
- Computational Thinker 5a, 5b, 5c, 5d
- Creative Communicator 6a, 6b, 6c, 6d
- Global Collaborator 7c, 7d

US Computer Science Standards
- Project correlations can be found on the book's web page.

US NGSS - Middle School
DCIs
- MS-PS1-1 PS1.A
- MS-PS1-2 PS1.A
- MS-PS4-3 PS4.C
- MS-LS2-4 LS2.C
- MS-ESS3-5 ESS3.D

Cross Cutting Concepts
- Cause and Effect
- Scale, Proportion and Quantity
- Structure and Function

Science and Engineering Practices
- Planning and Carrying Out Investigations
- Obtaining, Evaluating, and Communicating Information
- Analyzing and Interpreting Data
- Scientific Knowledge is Based on Empirical Evidence
- Using Mathematics and Computational Thinking

US NGSS - High School
- Correlations can be found on the book's web page.

US Common Core Language Arts and Mathematics
- Correlations can be found on the book's web page.

Materials List:
- Computer with IDE and Fritzing software
- Connection to the Internet
- USB Cable
- UNO or UNO Compatible Microcontroller
- Short Breadboard
- MQ4 Gas Sensor
- LED
- 220 Ohm Resistor
- Six (6) Jumper Wires (Male to Male)

Engineering Design - Digital Prototype

1. Download, install, and launch (as needed) Fritzing (pages 64-66) to build a digital prototype.

2. Visit The Arduino® Classroom website and register for the website.

3. Navigate to the "UNO Edition Vol. 1" page by hovering over "Books" on the main menu and clicking on the "UNO Edition Vol. 1" link.

4. On the right of the page find and then click on the "Volume 1 Links" link.

5. Click on the "Fritzing Part Download" link for Project 21 which will take you to the page to download the file.

6. Extract the file.

7. Install the file (page 66).

8. Start a new sketch in Fritzing.

9. Drag an UNO board to the workspace.

10. Drag the MQ4 sensor to the workspace.

11. Place an LED across two rows of a column.

Project 21
Gas Sniffer

```
#include <GAS_MQ4.h> // Include Library
int LED = 3;        // Set Constants
int sensor = A0;
int MQ4SensorValue = 0;

void setup() {
 pinMode(LED, OUTPUT); //Assign Pins
 pinMode(sensor, INPUT);
 Serial.begin(9600); //Turn On Serial Monitor
 Serial.println("MQ4 Sensor Analog START");
 Serial.println(" ");
}
void loop() {
 delay(3000); // Sensing Loop
 MQ4SensorValue = analogRead(A0);
 Serial.print("MQ4 Sensor Analog Reading ");
 Serial.println(MQ4SensorValue);
if (MQ4SensorValue > 170) {
  Serial.println("Warning CH4 or other gas
detected.");
  Serial.println(" ");
  digitalWrite(LED, HIGH);
}
  else {
    Serial.println("NO WARNING");
  Serial.println(" ");
  digitalWrite(LED, LOW);
  }
}
```

12 Place a 220 Ohm resistor on the breadboard such that it connects the cathode(-) column and the row that has the cathode(-) lead of the LED.

13 Connect with jumper wires the following:
- the five volt (5v) pin on the UNO board to the anode(+) column on the breadboard.
- a ground (GND) pin on the UNO board to the cathode(-) column on the breadboard.
- analog pin zero (A0) to data pin on the MQ4 sensor.
- the anode(+) column on the breadboard to the anode(+) pin on the MQ4 sensor.
- the cathode(-) column on the breadboard to the cathode(-) pin on the MQ4 sensor.
- digital pin three (3) to the same row as the anode(+) lead of the LED

Coding - Digital Prototype

14 Optional. Enter the code in Tinkercad® to match the code on this page. Verify.

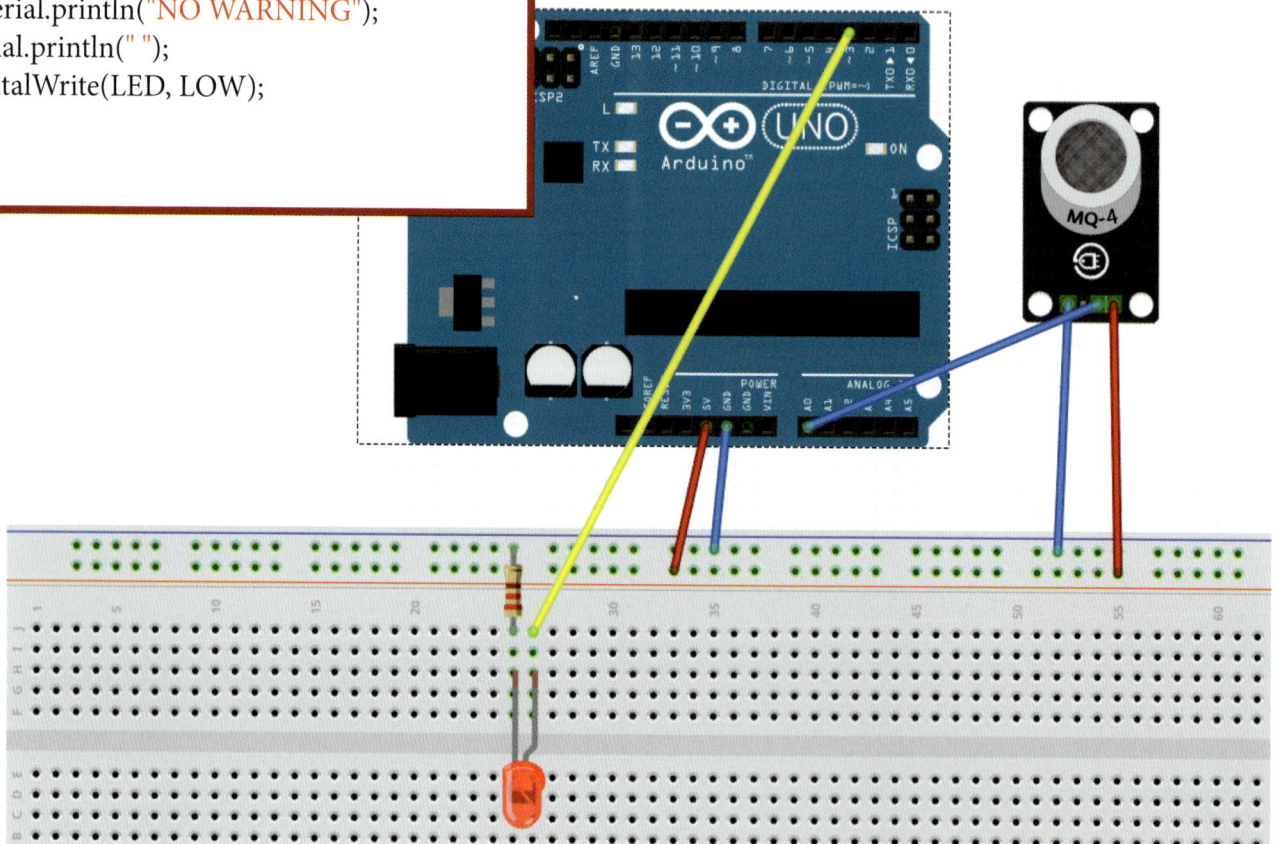

Project 21
Gas Sniffer

Connection - Scientists have been studying global climate change for many years now and primarily focused on the link between carbon dioxide (CO_2) emissions and temperature increases. CO_2 acts like the glass on a car parked in the open on a sunny day. The gases in the atmosphere including CO_2 let light pass though the atmosphere and strike Earth. When the light is turned into heat by the ground or water, the reflected heat is trapped inside the atmosphere more efficiently by the CO_2. The more CO_2, the easier it is for the heat to stay inside the atmosphere.

The added heat warms the oceans and atmosphere just a little bit more. On average the temperature rises enough to change the cycles of ice forming and melting at the poles, when plants grow where and when they bloom, when animals migrate, and how weather systems develop.

While CO_2 is only .04% of the composition of the atmosphere it can have a big impact. Considering it was only .033% just 50 years ago the growth in this time frame has scientists concerned. Methane is an even stronger molecule when it comes to keeping heat in the atmosphere. Methane is only .00017% of the atmosphere but is 30 times stronger than CO_2 in terms of heat retention.

Scientists have identified cows and other domesticated livestock raised for meat as a contributor of methane to the atmosphere.

Engineering Design - Project Build

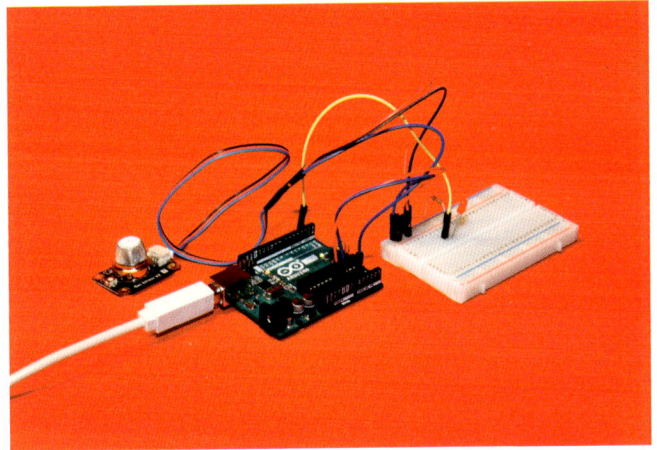

15 Place an LED across two rows of a column.

16 Place a 220 Ohm resistor on the breadboard such that it connects the cathode(-) column and the row that has the cathode(-) lead of the LED.

17 Connect with jumper wires the following:
- the five volt (5v) pin on the UNO board to the anode(+) column on the breadboard.
- a ground (GND) pin on the UNO board to the cathode(-) column on the breadboard.
- analog pin zero (A0) to data pin on the MQ4 sensor.
- the anode(+) column on the breadboard to the anode(+) pin on the MQ4 sensor.
- the cathode(-) column on the breadboard to the cathode(-) pin on the MQ4 sensor.
- digital pin three (3) to the same row as the anode(+) lead of the LED

18 Connect the UNO board and computer using the USB cable.

19 Launch the Arduino® IDE software (pages 11-13) to make sure the board is communicating with the computer.

Go to the Tools menu and verify that the correct board is selected from the Board Manager menu. If not then select the correct board from the options listed.

Then confirm that the right Port is selected. If not then select the port that lists the UNO board from the options listed.

Coding - Project Build

20 Either copy and paste the code from the Tinkercad® prototype or type the code from page 194 into the Arduino® IDE software (pages 11-13). Also install the library. A link is on the links page for the book.

21 Save the sketch (rename as needed).

22 Verify the sketch.

23 Upload the sketch to the UNO board.

24 Observe the LED on the breadboard.

25 Under supervision, test the sensor to detect CH_4 (Methane). Rubbing alcohol may be substituted.

26 Observe the data output in the Serial Monitor.

27 Document this project, discuss the project, and complete the extensions assigned from the next two pages.

```
TAC_UE_V1_P21 | Arduino 1.8.9 (Windows Store 1.8.21.0)
File Edit Sketch Tools Help

TAC_UE_V1_P21

// Project 21
// Methane Detector
//thearduinoclassroom.com
//Copyright 2019,  and Peter Haydock

#include <GAS_MQ4.h> // Include Library
int LED = 3;          // Set Constants
int sensor = A0;
int MQ4SensorValue = 0;

void setup() {
  pinMode(LED, OUTPUT); //Assign Pins
  pinMode(sensor, INPUT);
  Serial.begin(9600); //Turn On Serial Monitor
  Serial.println("MQ4 Sensor Analog START");
  Serial.println(" ");
}
void loop() {
  delay(3000); // Sensing Loop
  MQ4SensorValue = analogRead(A0);
  Serial.print("MQ4 Sensor Analog Reading ");
  Serial.println(MQ4SensorValue);
if (MQ4SensorValue > 170) {
    Serial.println("Warning CH4 or other gas detected.");
    Serial.println(" ");
    digitalWrite(LED, HIGH);
}
    else {
        Serial.println("NO WARNING");
      Serial.println(" ");
      digitalWrite(LED, LOW);
    }
}
```

```
Done Saving.

Sketch uses 2476 bytes (7%) of program storage space. Maximum is 32256 bytes.
Global variables use 286 bytes (13%) of dynamic memory, leaving 1762 bytes for loca

19                                              Arduino/Genuino Uno on COM4
```

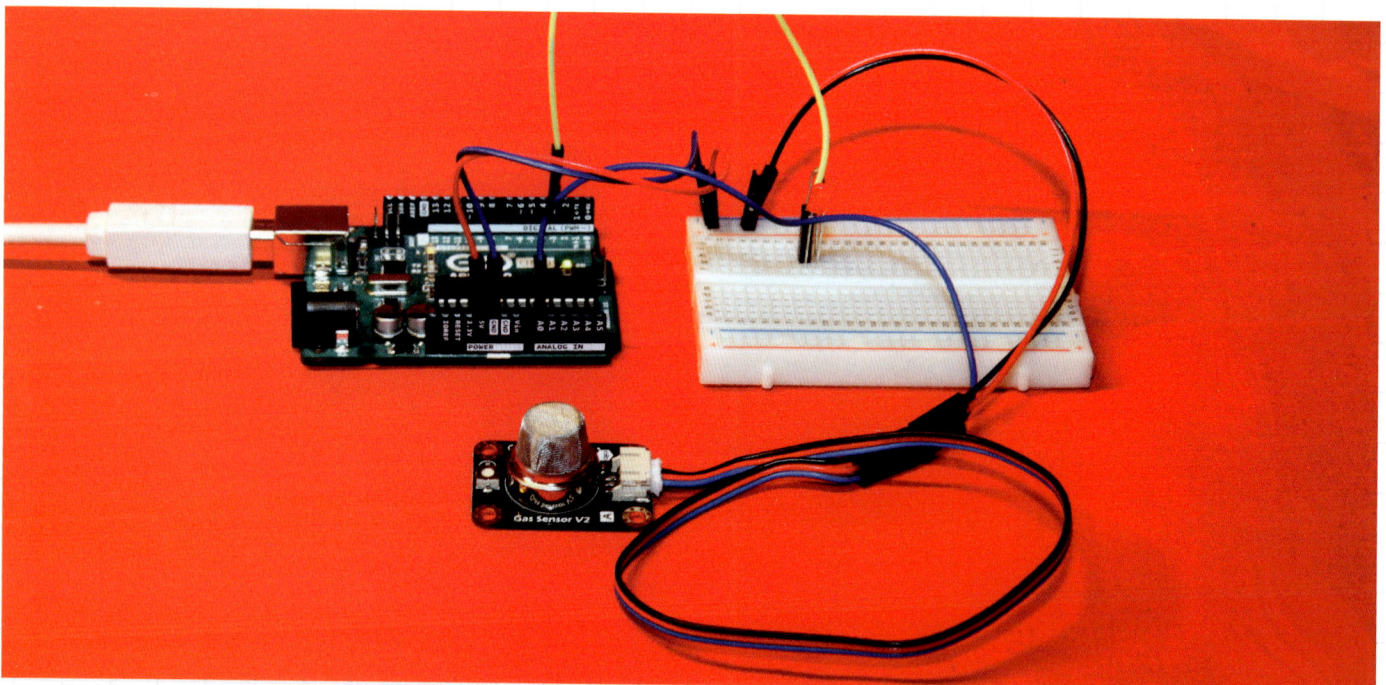

Project Build - Extension

Data accuracy for the MQ4 benefits from letting the sensor warm up for at least 30 minutes. Underneath the protective screen of the sensor is a metal membrane that works best when hot.

The MQ4 output is sensitive to air temperature and humidity which may require calibrating the sensor. The sensor in the book has a potentiometer built in that will change the calibration and sensitivity of the sensor. It works, but without knowing the physics of the calibration, it is not likely to return a scientific value. The data returned in the first part of the project is a relative number with no scientific value.

However there are ways to turn the MQ4 sensor into a more scientifically valid sensor.

On the book website there are links to a more complex MQ4 sensor project that will take two sketches and more time. Doing this will return a closer value to the actual amount of methane in the space being observed.

Step one: Download and run the calibration sketch after the MQ4 sensor has been on for at least an hour. The sketch will return a calibration number.

Step two: Insert the calibration number into the second sketch. The data output of the sketch will show methane levels with a parts per million value.

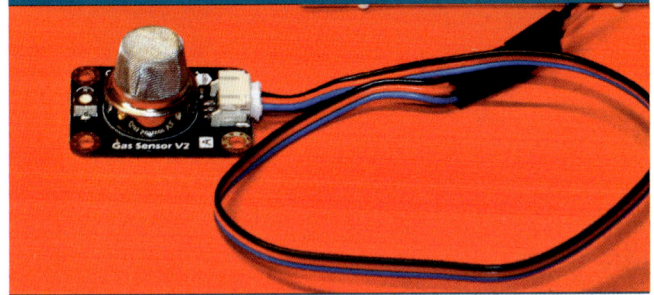

Connection - Methane is a biogas. It is an important source of energy for many people. It provides efficient heat, cooking fuel, and even can be used in vehicles for transportation. It only leaves carbon dioxide and water behind when ignited. Methane can be generated by waste organics and sewage, buried trash in landfills, and as a byproduct in the extraction of oil or manufacturing of gasoline. It can be compressed and transported by train, pipeline, and ship.

This bus in Germany is powered by biogas.

A = Essential

Reflection/Summative Activity:
- Have each team member document in their project/classroom journal who was on their team, what went well, what they could improve upon, what they would do differently if they were to do the project again, and verify that every person in the group can do the project.

Section Reflection/Summative Activities:
- Have the team create a graphic organizer summarizing the skills and knowledge they obtained in doing Projects Thirteen through Twenty One.

- Document in your journal key vocabulary, inputs and controls used, and skills developed in this section.

For Discussion:
- What are the advantages and disadvantages to using methane as a source of fuel?

- Why might detecting methane part of a safety system for a building?

Extensions:
- Change the conditional statements, warning language, and sensitivity levels to convey more information about the level of methane detected.

- Change the LED to an RGB LED and change the code to indicate different levels of methane.

B = Recommended

Professional Connection:
- Research and prepare a two slide presentation about a profession that is involved in energy production or delivery.

Research Connection:
- Investigate the methane output of different plants, animals, or biological processes (ex. decomposition or swamp gas).

- As a class compare your investigations.

Extension:
- Redesign the project portable with the 9v battery barrel connector and an LCD.

- Submit your project to the book page on the website. Register and Login to submit at: thearduinoclassroom.com

From the homepage:
Click
    Books
Click
    UNO Edition Vol. 1

C = Optional

Extension:
- Implement the extension on page 197.

Problem:
- Research and write about an energy issue facing your community.

Include in your report the following:

Description of the problem.

Description of how the problem is solved.

What other resources you would need to solve the problem.

Communicate your solution through a poster.

A project rubric can be found on page 236.

Anticipatory Sets
(from page 192)

Anticipatory Sets:
- Identify where gas sensors are seen. List at least three different places or situations where temperatures are shown, describe the locations, and explain the purpose of the displays.
Responses Will Vary

- How do gas delivery companies warn people of a gas leak without a sensor?
Responses Will Vary

A

Reflection/Summative Activity:
- Have each team member document in their project/classroom journal who was on their team, what went well, what they could improve upon, what they would do differently if they were to do the project again, and verify that every person in the group can do the project.
Responses Will Vary

Section Reflection/Summative Activities:
- Have the team create a graphic organizer summarizing the skills and knowledge they obtained in doing Projects Thirteen through Twenty One.
Responses Will Vary

- Document in your journal key vocabulary, inputs and controls used, and skills developed in this section.
Responses Will Vary

For Discussion:
- What are the advantages and disadvantages to using methane as a source of fuel?
Responses Will Vary

- Why might detecting methane part of a safety system for a building?
Responses Will Vary

Extensions:
- Change the conditional statements, warning language, and sensitivity levels to convey more information about the level of methane detected.
Responses Will Vary

- Change the LED to an RGB LED and change the code to indicate different levels of methane.
Responses Will Vary

B

Professional Connection:
- Research and prepare a two slide presentation about a profession that is involved in energy production or delivery.

Research Connection:
- Investigate the methane output of different plants, animals, or biological processes (ex. decomposition or swamp gas).

- As a class compare your investigations.
Responses Will Vary

Extension:
- Redesign the project portable with the 9v battery barrel connector and an LCD.

- Submit your project to the book page on the website. Register and Login to submit at: thearduinoclassroom.com

From the homepage:
Click
    Books
Click

    UNO Edition Vol. 1

C = Optional

Extension:
- Implement the extension on page 197.

Problem:
- Research and write about an energy issue facing your community.

Include in your report the following:

Description of the problem.

Description of how the problem is solved.

What other resources you would need to solve the problem.

Communicate your solution through a poster.
Responses Will Vary

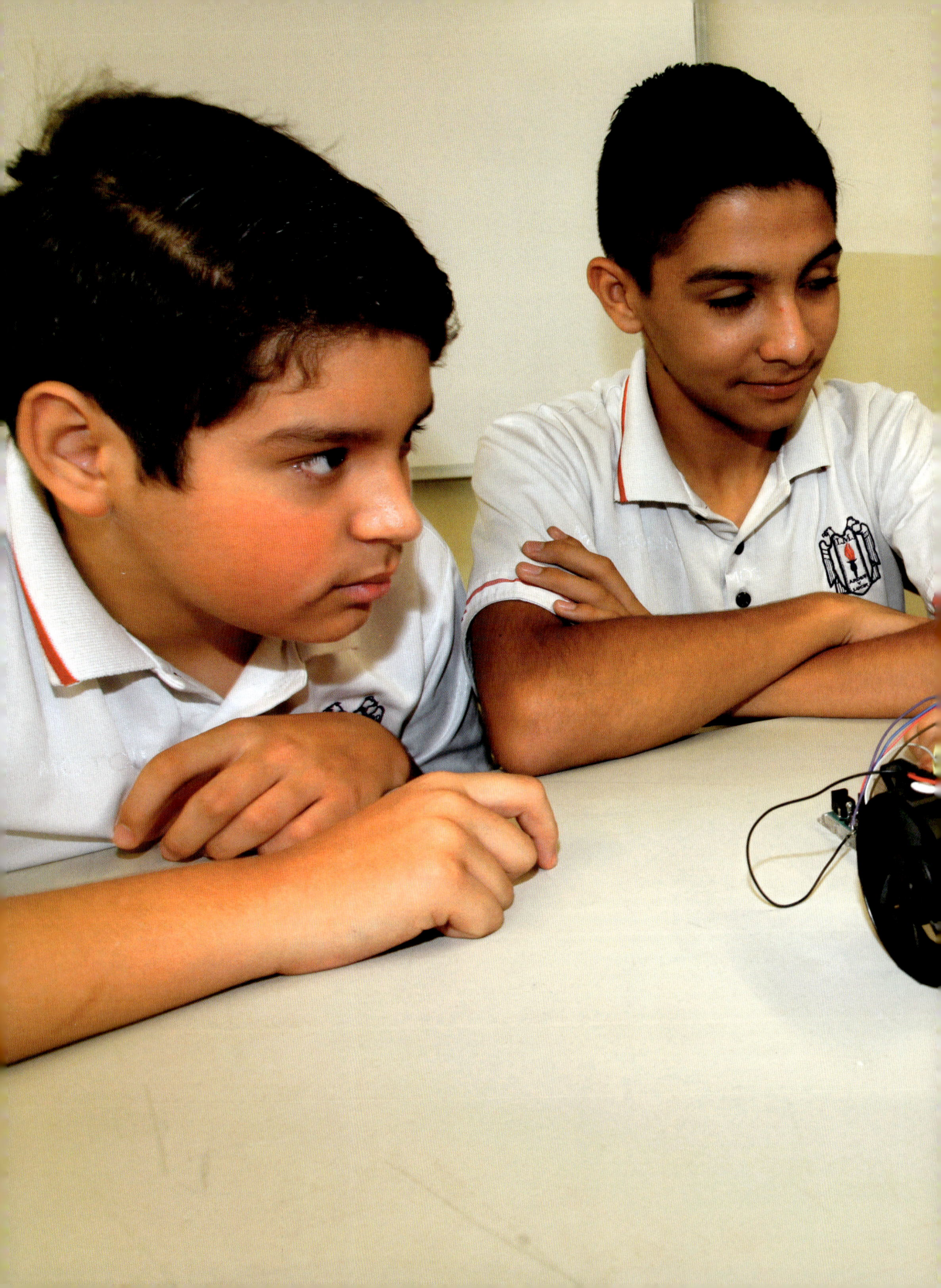

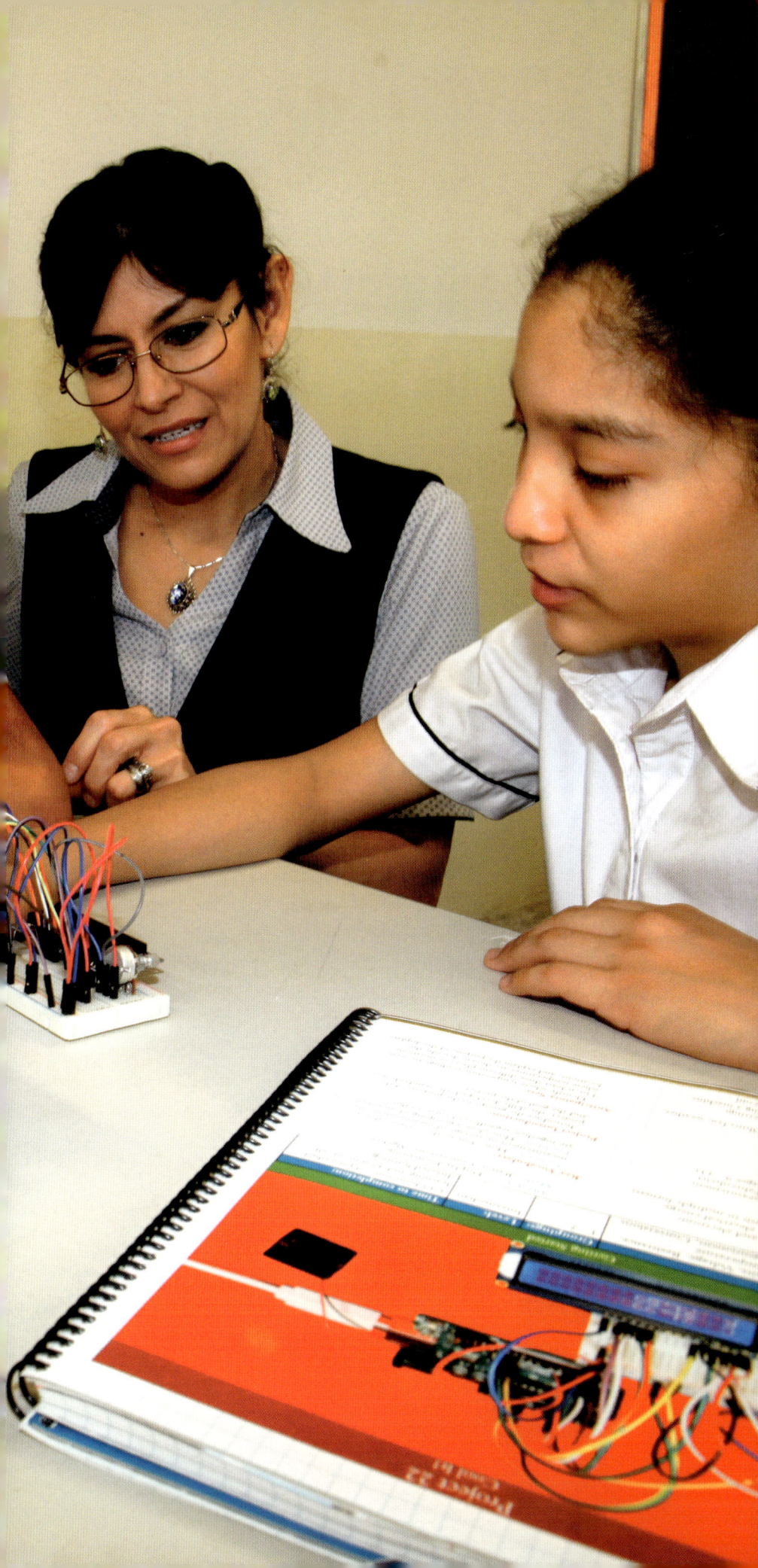

Section 4
Combining Inputs and Outputs

PROJECTS

Project 22
Cool It!

Lesson Integration:	Groupings:	Level:	Time to Complete:
Physics - Ohm's Law, Circuits, Voltage, Resistance, Earth Science - Weather, Temperature Mathematics - Algebra, Measurement, Conversions	1 - 2	Intermediate	1:15 hour:min. for the project and 45 min. for extensions.

Objectives:
- Investigate how circuits and electronic components interact with electrical energy.
- Display sensor information in multiple formats.
- Control multiple outputs with a sensor

Prerequisite Skills:
- Time and Temperature measurement (milliseconds, Celsius and Fahrenheit)
- Arduino® IDE software (pages 11-13)
- Libraries (page 62)

Purpose and Skills:
- Measure temperature
- Calibrate a sensor

STEAM Connections:
Science - Circuits, Ohm's law, Weather, Temperature
Technology - Code, Simulators, Digital Design, Electronic Components, Analog Data

Engineering - Designing, Building, and Using a Machine, Prototyping, Applied Physics
Allied Arts - Display Formatting
Math - Ratios, Conversions, Algebra

Key Vocabulary:
Accuracy - How close to the correct value a measurement or average of measurements are.
Precision - How close a grouping of measurements are together, regardless of correct value.

Project Introduction:
- Introduce the groups to the purpose of the project, the skills developed, the standards met, and the goal of the project.

Anticipatory Sets:
- Identify where temperature readings are seen. List at least three different places or situations where temperatures are shown, describe the locations, and explain the purpose of the displays.

Educational Standards

ISTE Standards for Students
- Empowered Learner 1a, 1b, 1d
- Knowledge Constructor 3a, 3b, 3c, 3d
- Innovative Designer 4a, 4b, 4c, 4d
- Computational Thinker 5a, 5b, 5c, 5d
- Creative Communicator 6a, 6b, 6c, 6d
- Global Collaborator 7c, 7d

US Computer Science Standards
- Project correlations can be found on the book's web page.

US NGSS - Middle School
DCIs
- MS-PS1-4 PS1.A, PS3.A
- MS-PS4-3 PS4.C
- MS-ESS2-6 ESS2.C

Cross Cutting Concepts
- Cause and Effect
- Scale, Proportion and Quantity
- Structure and Function

Science and Engineering Practices
- Planning and Carrying Out Investigations
- Obtaining, Evaluating, and Communicating Information
- Analyzing and Interpreting Data
- Scientific Knowledge is Based on Empirical Evidence
- Using Mathematics and Computational Thinking

US NGSS - High School
- Correlations can be found on the book's web page.

US Common Core Language Arts and Mathematics
- Correlations can be found on the book's web page.

Materials List:
- Computer with IDE software
- Connection to the Internet
- USB Cable
- UNO or UNO Compatible Microcontroller
- Short Breadboard
- 19 Jumper Wires (Male to Male)
- 16x2 LCD
- 10k Potentiometer
- 220 Ohm Resistor
- TMP36 Temperature Sensor
- 5v Computer Fan

Notes for Project 22

This is the most complicated project in the book in terms of the jumper wires. With 19 wires there are a lot of connections to manage. The TMP36 is very sensitive to voltage changes that often occur with loose wires or poor connections. If the temperature readings are high or low or fluctuate by more than a C degree, reseat the wires carefully to ensure tight connections. Moving the sensor, LCD, and/or potentiometer over by a row or column may help as well. Then move the jumper wires as needed. It may take a few moves too and some extra time to get the connections just right. Using the large breadboard is also an option to get the connections.

The other change that may be needed is recalibrating the TMP36. The TMP36 is calibrated at the factory and is set for a specific set of conditions. Once the temperature readings are stable, but do not match the known temperature, you may need to change one setting in the code to recalibrate. The Step-by-step directions will address that when it comes time to enter the code.

Engineering Design - Digital Prototype

1. Launch and log into Tinkercad® (pages 13 - 14).

2. Start a new Circuits project.
Use the diagram on this page for placement assistance.

3. Drag an UNO board to the workspace.

4. Drag a breadboard to the workspace.

5. Drag a 16x2 LCD to one end of the breadboard leaving at least two columns uncovered.

6. Drag one 10k potentiometer to the breadboard such that the leads span three rows in a column at the other end on the same half of the breadboard as the LCD.

7. Drag a 5v motor to the workspace. This will substitute for the fan.

8. Drag a LM35 (a substitute for TMP36) sensor to the left end of the breadboard to cross three rows in the same column.

9. Place a 220 Ohm resistor on the breadboard such that it connects same row with the pin 15 (LED) of the LCD to the row on the other half of the breadboard.

10. Connect with jumper wires the following:
   - one terminal on the motor to digital pin seven (7) on the UNO board and the other terminal to a GND pin on the UNO board. Two jumper wires.
   - the anode(+) column to the rows connected to the anode(+) lead of the LM35, the 10k potentiometer, the VCC pin of the LCD, the 220K Ohm resistor, and the 5v pin on the UNO board. Five jumper wires.
   - the cathode(-) column to the rows connected to the cathode(-) lead of the LM35, the 10k potentiometer, the GND pin (pin 1)of the LCD, the RW pin of the LCD, the LCD GND pin (pin 16) of the LCD, and the GND pin on the UNO board. Six jumper wires.
   - digital pin 12 to the same row as the RS pin of the LCD
   - digital pin 11 to the same row as the E pin of the LCD
   - digital pin five (5) to the same row as the DB4 pin of the LCD
   - digital pin four (4) to the same row as the DB5 pin of the LCD
   - digital pin three (3) to the same row as the DB6 pin of the LCD
   - digital pin two (2) to the same row as the DB7 pin of the LCD
   - The same row of the LM35 signal pin (center pin) to the Analog A0 pin
   - The same row of the potentiometer signal pin (center pin) to the same row of the V0 pin of the LCD

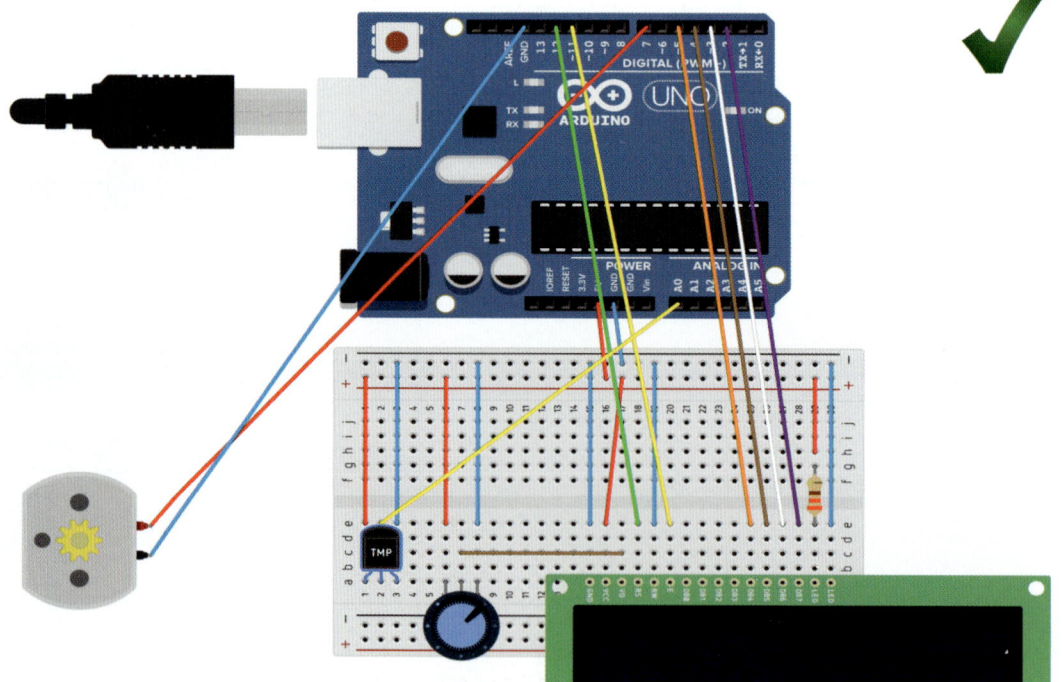

Project 22
Cool It!

```
#include<LiquidCrystal.h>
LiquidCrystal lcd(12, 11, 5, 4, 3, 2);
int fan = 7;
const int sensor = A0; // Assigning analog pin A0 to variable 'sensor'
float tempc;  //variable for temperature in degree Celsius
float tempf;  //variable for temperature in Fahrenheit
float volts;  //variable to hold sensor reading.

void setup()
{
  Serial.begin(9600);
  pinMode (7, OUTPUT);
  lcd.begin(16, 2);
  lcd.setCursor(0, 0);
  lcd.print("The Arduino    ");
  lcd.setCursor(0, 1);
  lcd.print(" Classroom   ");
  delay(3000);
  lcd.clear();
  lcd.print("Temperature ");
  lcd.setCursor(0, 1);
  lcd.print("Controlled Fan ");
  delay(3000);
  lcd.clear();
}
void loop()
{
  volts = analogRead(sensor);
  volts = (volts * 500) / 1023; //Change the "500" to calibrate if
temperatures are not accurate
  tempc = volts; // Storing value in Degree Celsius
  tempf = (volts * 1.8) + 32; // Converting to Fahrenheit
  lcd.setCursor(0, 0);
  lcd.print("C ");
  lcd.print(tempc);
  lcd.setCursor(0, 1);
  lcd.print("F ");
  lcd.print(tempf);
  delay(1000);
  if (volts < 20) {
    digitalWrite (7, LOW);
  }
  else
    digitalWrite (7, HIGH);
  fan = 0;
}
```

Coding - Digital Prototype

11 Enter the code in Tinkercad® to match the code on this page.

12 Run the code and observe the digital prototype.

13 Click once on the LM35 (TMP) sensor. It will open a slider to change the temperature.

At the time of printing the book, the LM35 sensor in Tinkercad® was not calculating the voltage correctly and therefore the temperature readings were not correct. ✓

Engineering Design - Project Build

14 Place the 16x2 LCD on one end of the breadboard leaving at least two columns uncovered.

15 Place the 10k potentiometer to the breadboard such that the leads span three rows in a column at the other end on the same half of the breadboard as the LCD.

16 Place the TMP36 at the left end of the breadboard across three rows of the same column.

17 Place a 220 Ohm resistor on the breadboard such that it connects same row with the pin 15 (LED) of the LCD to the row on the other half of the breadboard.

18 Connect the anode(+), red lead on the fan to digital pin seven (7) on the UNO board and the cathode(-), black lead to a GND pin on the UNO board.

19 Connect with jumper wires the following:
- the anode(+) column to the rows connected to the anode(+) lead of the TMP36, the 10k potentiometer, the VCC pin of the LCD, the 220K Ohm resistor, and the 5v pin on the UNO board. Five jumper wires.
- the cathode(-) column to the rows connected to the cathode(-) lead of the TMP36, the 10k potentiometer, the GND pin (pin 1) of the LCD, the RW pin of the LCD, the LCD GND pin (pin 16) of the LCD, and the GND pin on the UNO board. Six jumper wires.
- digital pin 12 to the same row as the RS pin of the LCD
- digital pin 11 to the same row as the E pin of the LCD
- digital pin five (5) to the same row as the DB4 pin of the LCD
- digital pin four (4) to the same row as the DB5 pin of the LCD
- digital pin three (3) to the same row as the DB6 pin of the LCD
- digital pin two (2) to the same row as the DB7 pin of the LCD
- The same row of the TMP36 signal pin (center pin) to the Analog A0 pin
- The same row of the potentiometer signal pin (center pin) to the same row of the V0 pin of the LCD

20 Connect the UNO board and computer using the USB cable.

21 Launch the Arduino® IDE software (pages 11-13) to make sure the board is communicating with the computer.

Go to the Tools menu and verify that the correct board is selected from the Board Manager menu. If not, select the correct board from the options listed.

Then confirm that the right Port is selected. If not, select the port that lists the UNO board from the options listed.

Coding - Project Build

22 Either copy and paste the code from the Tinkercad® prototype or type the code on the previous page into the Arduino® IDE software (pages 11-13). Install the library if needed.

23 Save the sketch (rename as needed).

24 Verify the sketch.

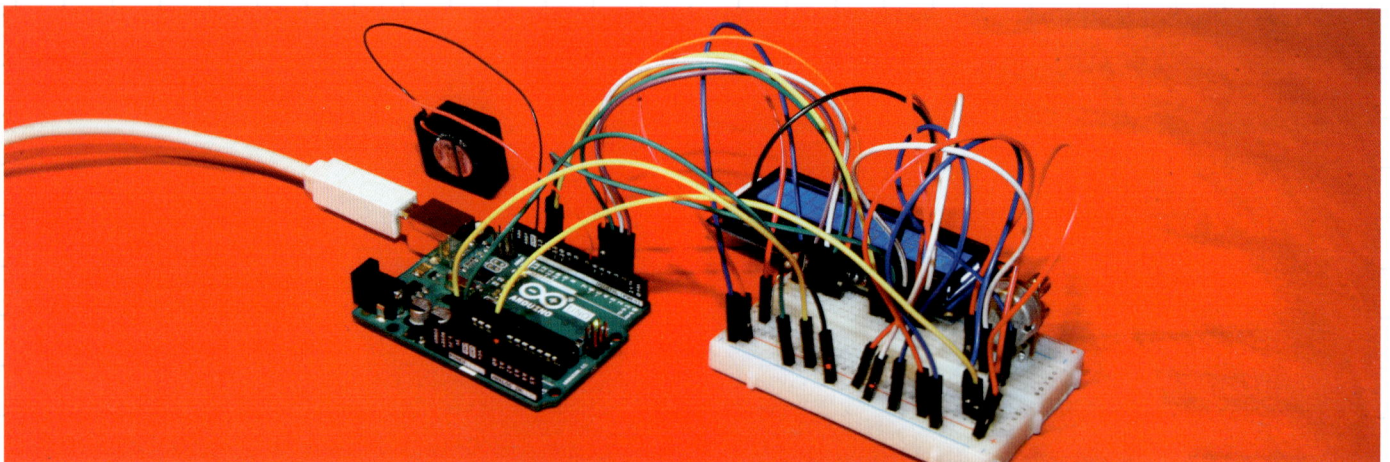

25 Upload the sketch to the UNO board.

26 Observe the 16x2 LCD.

If the temperature fluctuates by more than a one degree C, reseat the jumper wires. If needed move the sensor, LCD, and potentiometer by shifting them by at least one row or column. This may take several moves and reseatings of the wires. Once the temperature is stable, compare it to a known temperature from a thermometer or thermostat in the room.

If the temperature is reading too high or too low, the line in the code

volts = (volts * 500) / 1023; //Change the "500" to calibrate if temperatures are not accurate

will need calibration.

If the temperature is too high, reduce the "500"

by 10's and upload the new saved code to the UNO board until the temperature is accurate. If the temperature is too low, increase the "500" by 10's and upload the new saved code to the UNO board until the temperature is correct.

27 Test the thermometer and fan under varying conditions

28 Document this project, discuss the project, and complete any of the extensions assigned to you as found on the next page.

Connection - What does $325 million buy in terms of a computer? It can buy two supercomputers, one of which is open to public research called Summit. Summit is the newest supercomputer at Oak Ridge National Lab in Oak Ridge, Tennessee which can do 200,000 trillion calculations per second. It takes up to 13 megawatts (MW) of power to run Summit. That is enough power to supply nearly 900 homes with electricity.

In order to run the Summit at peak efficiency it needs a super cooling system of temperature controlled fans and cooling water. Together, this system keeps Summit, when running at full speed, at 24 C (75 F) or cooler.

When this cool, the supercomputer can work its fastest. Fans can circulate the entire volume of air in Summit's room in 20 seconds while thousands of gallons of water are pumped into cooling devices in each cabinet to cool the computers from the inside.

```
//Project 22
//LM35 Digital Thermometer
//thearduinoclassroom.com
//Copyright 2019, Isabel Mendiola and Peter Haydock

#include<LiquidCrystal.h>
LiquidCrystal lcd(12, 11, 5, 4, 3, 2);
int fan = 7;
const int sensor = A0; // Assigning analog pin A0 to variable 'sensor'
float tempc;  //variable for temperature in degree Celsius
float tempf;  //variable for temperature in Fahreinheit
float volts;  //variable to hold sensor reading.

void setup()
{
  Serial.begin(9600);
  pinMode (7, OUTPUT);
  lcd.begin(16, 2);
  lcd.setCursor(0, 0);
  lcd.print("The Arduino     ");
  lcd.setCursor(0, 1);
  lcd.print(" Classroom   ");
  delay(3000);
  lcd.clear();
  lcd.print("Temperature ");
  lcd.setCursor(0, 1);
  lcd.print("Controlled Fan ");
  delay(3000);
  lcd.clear();
}
void loop()
{
  volts = analogRead(sensor);
  volts = (volts * 500) / 1023; //Change the "500" to calibrate if temperatures are not accurate
  tempc = volts; // Storing value in Degree Celsius
  tempf = (volts * 1.8) + 32; // Converting to Fahrenheit
  lcd.setCursor(0, 0);
  lcd.print("C  ");
  lcd.print(tempc);
  lcd.setCursor(0, 1);
  lcd.print("F  ");
  lcd.print(tempf);
  delay(1000);
  if (volts < 20) {
    digitalWrite (7, LOW);
  }
  else
    digitalWrite (7, HIGH);
    fan = 0;
}
```

Done uploading.

Sketch uses 4970 bytes (15%) of program storage space. Maximum is 32256 bytes.
Global variables use 312 bytes (15%) of dynamic memory, leaving 1736 bytes for local variables. Max

Project 22
Cool It!

Project Reflection and Summative Activities, Discussion Starters, Extensions, and Problem

A = Essential

Reflection/Summative Activity:
- Have each team member document in their project/classroom journal who was on their team, what went well, what they could improve upon, what they would do differently if they were to do the project again, and verify that every person in the group can do the project.

For Discussion:
- Describe locations and situations that would benefit from or require a temperature controlled fan?

- What other conditions would use a sensor controlled fan?

- List other devices that might work with a temperature controlled fan.

Extensions:
- Add an RGB LED to the project that changes color depending on the temperature.

- Replace the fan with a buzzer to sound at extreme temperatures.

B = Recommended

Professional Connection:
- Research and write a short description of a profession that relies on controlling temperature. Be sure to include the education the profession requires, any laws or regulations that guide their use, and what activities are performed.

Communication Connection:
- Create an ad campaign for your Cool It! device.

Extensions:
- Make the project portable with the 9v battery and barrel connector.

- Edit the code to display the temperature in Kelvins.

- Change the controller of the fan from the TMP36 to another controller (ex. photoresistor, color sensor, 10k potentiometer etc..).

- Submit your project to the book page on the website. Register and Login to submit at: thearduinoclassroom.com

From the homepage:
Click
    Books
Click
    UNO Edition Vol. 1

C = Optional

Extension:
- Redesign the project to run the fan at specific time intervals.

Problem:
- Identify a problem that would be solved by a device like the one you have been working with in this project. How could this project be a solution? Include in your report the following:

Description of the problem

Description of how the problem is solved by this project

What other resources you would need to solve the problem.

Design and produce the solution.

A project rubric can be found on page 236.

Anticipatory Sets
(from page 202)

Anticipatory Sets:
- Identify where temperature readings are seen. List at least three different places or situations where temperatures are shown, describe the locations, and explain the purpose of the displays.
Responses Will Vary

A

Reflection/Summative Activity:
- Have each team member document in their project/classroom journal who was on their team, what went well, what they could improve upon, what they would do differently if they were to do the project again, and verify that every person in the group can do the project.
Responses Will Vary

For Discussion:
- Describe locations and situations that would benefit from or require a temperature controlled fan?
Responses Will Vary

- What other conditions would use a sensor controlled fan?
Responses Will Vary

- List other devices that might work with a temperature controlled fan.

Responses Will Vary

Extensions:
- Add an RGB LED to the project that changes color depending on the temperature.
Responses Will Vary

- Replace the fan with a buzzer to sound at extreme temperatures.
Responses Will Vary

B

Professional Connection:
- Research and write a short description of a profession that relies on controlling temperature. Be sure to include the education the profession requires, any laws or regulations that guide their use, and what activities are performed.
Responses Will Vary

Communication Connection:
- Create an ad campaign for your "Cool It!" device.
Responses Will Vary

Extensions:
- Make the project portable with the 9v battery and barrel connector.
Responses Will Vary

- Edit the code to display the temperature in Kelvins.
Responses Will Vary

- Change the controller of the fan from the TMP36 to another controller (ex. photoresistor, color sensor, 10k potentiometer

etc..).
Responses Will Vary

- Submit your project to the book page on the website.
Responses Will Vary

C

Extension:
- Redesign the project to run the fan at specific time intervals.
Responses Will Vary

Problem:
Responses Will Vary

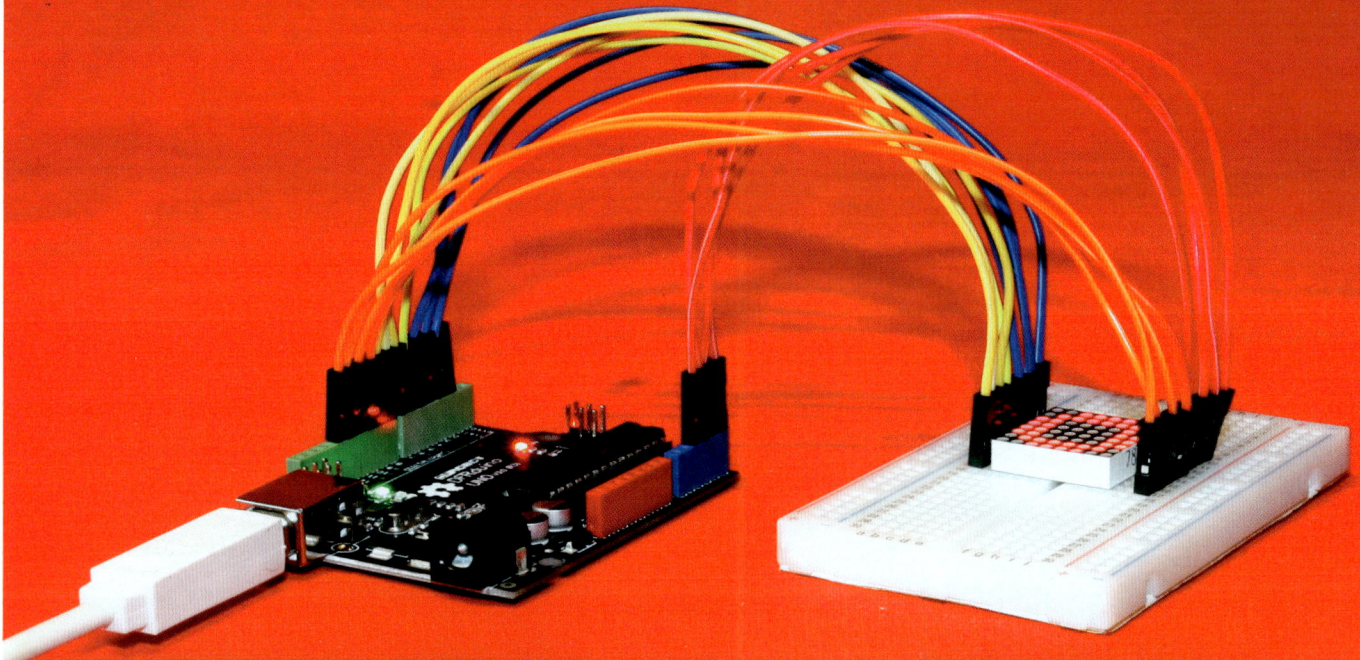

Project 23
Emoji Me

Lesson Integration:	Groupings:	Level:	Time to Complete:
Physics - Ohm's Law, Circuits, Voltage, Resistance Mathematics - Binary, Matrix	1 - 2	Advanced	1:15 hour:min. for the project and 45 min. for extensions.

Objectives:
- Investigate how circuits and electronic components interact with electrical energy.

Prerequisite Skills:
- Time measurement (milliseconds)
- Arduino® IDE software (pages 11-13)
- Fritzing (pages 64-66)

Purpose and Skills:
- Control an LED matrix.

STEAM Connections:
Science - Circuits, Resistance, Ohm's law
Technology - Code, Simulators, Digital Design, Electronic Components, Digital Data
Engineering - Designing, Building, and Using a Machine, Prototyping, Applied Physics
Allied Arts - Graphics, Displays, Binary Art
Math - Ratios, Conversions , Matrix

Key Vocabulary:
Data Array - Data defined by two ore more pieces of information. Examples include a date; July 4,

1968, positions on a chessboard Queen's Rook 3, or on the LED Matrix Row One (1), Column (3), and "On" defines the state of a particular LED.
Binary - A choice or variable with two options. It can be a "Yes" or "No", "1" or "0", or any other condition with two possibilities.
Boolean - A type of decision making that involves a structure of combining, separating, narrowing, expanding, or deciding choices. Uses terms like "AND," "OR," "NOT," "IF," "ELSE," etc..
Matrix - Something that is defined by two or more parts. The LED matrix has rows and columns that are used to define the location of each LED.

Project Introduction:
- Introduce the groups to the purpose of the project, the skills developed, the standards met, and the goal of the project.

Anticipatory Sets:
- Identify where LED signs are seen. List at least three different places or situations where LED technologies are used, describe the locations, and explain the purpose of the displays.

Project 23
Emoji Me

Educational Standards

ISTE Standards for Students
- Empowered Learner 1a, 1b, 1d
- Knowledge Constructor 3a, 3b, 3c, 3d
- Innovative Designer 4a, 4b, 4c, 4d
- Computational Thinker 5a, 5b, 5c, 5d
- Creative Communicator 6a, 6b, 6c, 6d
- Global Collaborator 7c, 7d

US Computer Science Standards
- Project correlations can be found on the book's web page.

US NGSS - Middle School
DCIs
- MS-PS4-3 PS4.C

Cross Cutting Concepts
- Cause and Effect
- Scale, Proportion and Quantity
- Structure and Function

Science and Engineering Practices
- Planning and Carrying Out Investigations
- Obtaining, Evaluating, and Communicating Information
- Analyzing and Interpreting Data
- Scientific Knowledge is Based on Empirical Evidence
- Using Mathematics and Computational Thinking

US NGSS - High School
- Correlations can be found on the book's web page.

US Common Core Language Arts and Mathematics
- Correlations can be found on the book's web page.

Materials List:
- Computer with IDE and Fritzing software
- Connection to the Internet
- USB Cable
- UNO or UNO Compatible Microcontroller
- Short Breadboard
- 20 mm 8x8 LED Matrix
- 16 Jumper Wires (Male to Male)

Engineering Design - Digital Prototype

1 Download, install, and launch (as needed) Fritzing (pages 64-66) to build a digital prototype.

2 Start a new sketch in Fritzing.

3 Drag an UNO board to the workspace.

4 Drag two short breadboards to the workspace.

5 Drag an 8x8 LED matrix to the breadboard and place it across the two breadboards.

6 Connect with jumper wires the following:
To determine the pin number on the LED matrix click on the matrix and then hover over the red dots on the matrix. If needed rotate the matrix.
- analog pin five (A5) to the same row as the lead of pin 8 of the LED matrix
- analog pin four (A4) to the same row as the lead of pin seven (7) of the LED matrix
- analog pin three (A3) to the same row as the lead of pin six (6) of the LED matrix
- analog pin two (A2) to the same row as the lead of pin five (5) of the LED matrix
- digital pin 10 to the same row as the lead of pin four (4) of the LED matrix
- digital pin 11 to the same row as the lead of pin three (3) of the LED matrix
- digital pin 12 to the same row as the lead of pin two (2) of the LED matrix
- digital pin 13 to the same row as the lead of pin one (1) of the LED matrix

- digital pin nine (9) to the same row as the lead of pin 16 of the LED matrix
- digital pin eight (8) to the same row as the lead of pin 15 of the LED matrix
- digital pin seven (7) to the same row as the lead of pin 14 of the LED matrix
- digital pin six (6) to the same row as the lead of pin 13 of the LED matrix
- digital pin five (5) to the same row as the lead of pin 12 of the LED matrix
- digital pin four (4) to the same row as the lead of pin 11 of the LED matrix
- digital pin three (3) to the same row as the lead of pin 10 of the LED matrix
- digital pin two (2) to the same row as the lead of pin nine (9) of the LED matrix

Coding - Digital Prototype

7 Enter the code in Tinkercad® to match the code on the next page.

8 Verify the code in Tinkercad®

Connection - Emojis are governed by an international organization called Unicode Consortium, a not-for-profit group run out of Silicon Valley in Northern California, U.S.A.

This group oversees the emojis and the characters that make up languages and ensures that technology companies are aware of additions, changes, or deletions from the character or emoji list. There are 12 people on the emoji sub-committee that considers new emojis or the retirement of old emojis.

The word emoji comes from two Japanese words, "e" for the word picture and "moji" for the word character. In Japanese, emoji looks like this. 絵文字

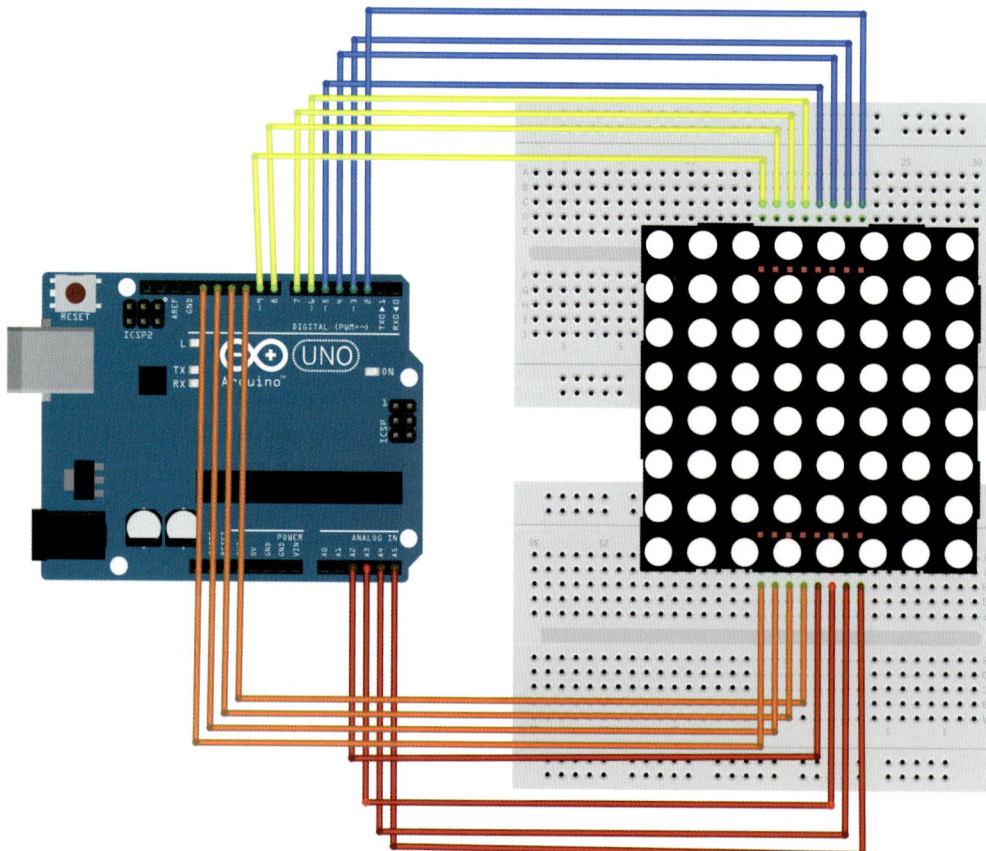

```cpp
const int row[8] = { //
Connection to rows in LED
Matrix
  6, 11, 10, 3, 17, 4, 8, 9 };
const int col[8] = {// Connection
to columns in LED Matrix
  16, 12, 18, 13, 5, 19, 7, 2 };
int pixels[8][8]; //8 pins x 8 pins
int count = 500;
char str[] = "FABCDEDCBA";
int strLen = sizeof(str);
int ptrChar = 0;
typedef bool charMapType[8][8];
const charMapType charBlank = {
  {0, 0, 0, 0, 0, 0, 0, 0},
  {0, 0, 0, 0, 0, 0, 0, 0},
  {0, 0, 0, 0, 0, 0, 0, 0},
  {0, 0, 0, 0, 0, 0, 0, 0},
  {0, 0, 0, 0, 0, 0, 0, 0},
  {0, 0, 0, 0, 0, 0, 0, 0},
  {0, 0, 0, 0, 0, 0, 0, 0},
  {0, 0, 0, 0, 0, 0, 0, 0} };
const charMapType character0
= {
  {0, 0, 0, 0, 0, 0, 0, 0},
  {0, 0, 0, 0, 0, 0, 0, 0},
  {0, 0, 0, 0, 0, 0, 0, 0},
  {0, 0, 0, 0, 0, 0, 0, 0},
  {0, 0, 0, 1, 1, 0, 0, 0},
  {0, 0, 0, 1, 1, 0, 0, 0},
  {0, 0, 0, 0, 0, 0, 0, 0},
  {0, 0, 0, 0, 0, 0, 0, 0} };
const charMapType character1
= {
  {0, 0, 0, 0, 0, 0, 0, 0},
  {0, 0, 0, 0, 0, 0, 0, 0},
  {0, 0, 0, 0, 0, 0, 0, 0},
  {0, 0, 0, 1, 1, 0, 0, 0},
  {0, 0, 1, 1, 1, 1, 0, 0},
  {0, 0, 1, 1, 1, 1, 0, 0},
  {0, 0, 0, 1, 1, 0, 0, 0},
  {0, 0, 0, 0, 0, 0, 0, 0} };
const charMapType character2
= {
  {0, 0, 0, 0, 0, 0, 0, 0},
  {0, 0, 0, 0, 0, 0, 0, 0},
  {0, 1, 1, 0, 0, 1, 1, 0},
  {1, 1, 1, 1, 1, 1, 1, 1},
  {1, 1, 1, 1, 1, 1, 1, 1},
  {0, 1, 1, 1, 1, 1, 1, 0},
  {0, 0, 1, 1, 1, 1, 0, 0},
  {0, 0, 0, 1, 1, 0, 0, 0} };
const charMapType character3
= {
  {0, 0, 0, 0, 0, 0, 0, 0},
  {0, 1, 1, 0, 0, 1, 1, 0},
  {1, 1, 1, 1, 1, 1, 1, 1},
  {1, 1, 1, 1, 1, 1, 1, 1},
  {0, 1, 1, 1, 1, 1, 1, 0},
  {0, 0, 1, 1, 1, 1, 0, 0},
  {0, 0, 0, 1, 1, 0, 0, 0},
  {0, 0, 0, 0, 0, 0, 0, 0} };
const charMapType character4
= {
  {0, 1, 1, 0, 0, 1, 1, 0},
  {1, 1, 1, 1, 1, 1, 1, 1},
  {1, 1, 1, 1, 1, 1, 1, 1},
  {0, 1, 1, 1, 1, 1, 1, 0},
  {0, 0, 1, 1, 1, 1, 0, 0},
  {0, 0, 0, 1, 1, 0, 0, 0},
  {0, 0, 0, 0, 0, 0, 0, 0},
  {0, 0, 0, 0, 0, 0, 0, 0} };
const charMapType *charMap[6]
= {&character0, &character1,
&character2, &character3,
&character4, &charBlank};
void setup() {
  for (int thisPin = 0; thisPin < 8;
thisPin++) {
    // initialize the output pins:
    pinMode(col[thisPin],
OUTPUT);
    pinMode(row[thisPin],
OUTPUT);
    digitalWrite(col[thisPin],
HIGH);
  }
  setupChar(); //screen
}
void loop() {
  refreshScreen();
  if(count-- == 0){
    count = 1000;
    setupChar();
  }
}
void setupChar(){
  char c = str[ptrChar];
  int offset = c - 'A';
  const charMapType *cMap =
charMap[offset];
  for (int x = 0; x < 8; x++) {
    for (int y = 0; y < 8; y++) {
      bool v = (*cMap)[x][y];
      if(v){
        pixels[x][y] = LOW;
      }else{
        pixels[x][y] = HIGH;
      }
    }
  }
  ptrChar++;
  if(ptrChar>=strLen-1){
    ptrChar = 0;
  }
}
void refreshScreen() {
  for (int thisCol = 0; thisCol < 8;
thisCol++) {
    digitalWrite(col[thisCol],
HIGH);
    for (int thisRow = 0; thisRow <
8; thisRow++) {
      int thisPixel = pixels[thisCol]
[thisRow];// Row is HIGH and
the column is LOW,
      digitalWrite(row[thisRow],
thisPixel);
      if (thisPixel == LOW) {
        digitalWrite(row[thisRow],
HIGH);
      }
    }
    digitalWrite(col[thisCol],
LOW);//turns off
  }
}
```

Project 23
Emoji Me

Engineering Design - Project Build

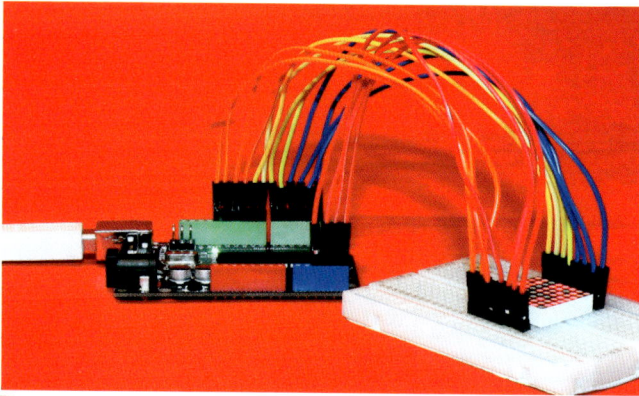

9 Place the 22 mm 8x8 LED across the center of the breadboard. Note the location of the part number printed on the side.

PIN 16   (far side)   PIN 9
Column 7 --->   ---LED (8,8)
--->
Column 1 --->   ---LED (8,1)
7888S
PIN 1---->   <----PIN 8

10 Connect with jumper wires the following:
- analog pin five (A5) to the same row as the lead of pin eight (8) of the LED matrix
- analog pin four (A4) to the same row as the lead of pin seven (7) of the LED matrix
- analog pin three (A3) to the same row as the lead of pin six (6) of the LED matrix
- analog pin two (A2) to the same row as the lead of pin five (5) of the LED matrix
- digital pin 10 to the same row as the lead of pin four (4) of the LED matrix
- digital pin 11 to the same row as the lead of pin three (3) of the LED matrix
- digital pin 12 to the same row as the lead of pin two (2) of the LED matrix
- digital pin 13 to the same row as the lead of pin one (1) of the LED matrix
- digital pin nine (9) to the same row as the

lead of pin 16 of the LED matrix
- digital pin eight (8) to the same row as the lead of pin 15 of the LED matrix
- digital pin seven (7) to the same row as the lead of pin 14 of the LED matrix
- digital pin six (6) to the same row as the lead of pin 13 of the LED matrix
- digital pin five (5) to the same row as the lead of pin 12 of the LED matrix
- digital pin four (4) to the same row as the lead of pin 11 of the LED matrix
- digital pin three (3) to the same row as the lead of pin 10 of the LED matrix
- digital pin two (2) to the same row as the lead of pin nine (9) of the LED matrix

11 Connect the UNO board and computer using the USB cable.

Coding - Project Build

12 Launch IDE software and either copy and paste the code from the Tinkercad® prototype or type the code on the previous page into the IDE software.

13 Save the sketch (rename as needed).

14 Verify the sketch.

15 Upload the sketch to the UNO board.

16 Observe the 20 mm 8x8 LED matrix.

17 Document this project, discuss the project, and complete any of the extensions assigned to you as found on the page 216.

If the LED does not work, use the guide on the next page to troubleshoot the problem.

Troubleshooting the 8x8 LED Matrix.

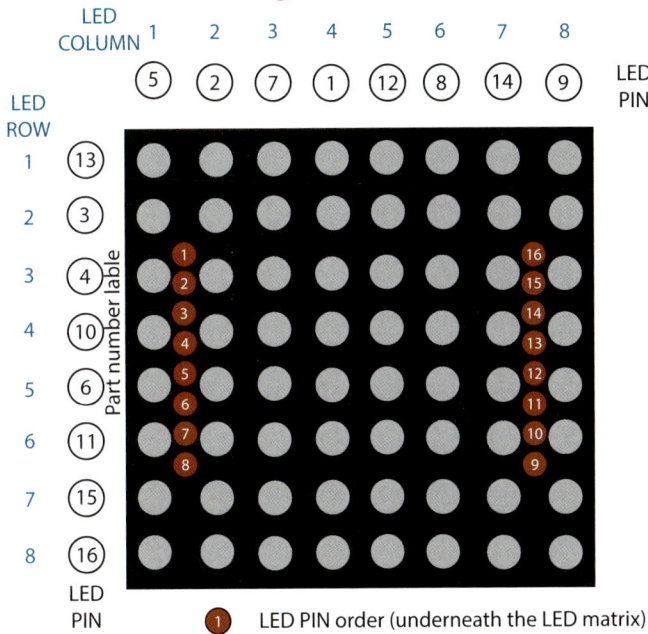

LED PIN order (underneath the LED matrix)

(column 1) and 13 (row 1). LED (4,8) is turned on by pin 16 (row 8) and pin 1 (column 4).

The data on the pin diagram for the rows and columns has two numbers, the circled number being the row or column number of the LED the second number being the corresponding pin number.

With one more translation, the pins are then listed in the code in order first row, then column.

```
const int row[8] = { // Connection to rows in LED Matrix
    6, 11, 10, 3, 17, 4, 8, 9 // UNO connections in row order by pin number};
    const int col[8] = {// Connection to columns in LED Matrix
16, 12, 18, 13, 5, 19, 7, 2 // UNO connections in column order by pin number };
```

The numbers entered in the code are the associated UNO board connection number and listed in pin order on the LED matrix.

The code for the project lists six for the first row value. This means that UNO digital pin 6 should connect with the pin on the LED matrix that controls row one (1) which is pin 13 on the bottom of the matrix.

Code values are as follows:
   UNO Digital pins 2 -13 are the same in the code listing (ex. Digital pin 12 is listed as 12 in the code).
   Analog pin A2 is listed as 16, Analog pin A3 is listed as 17, Analog pin A4 is listed as 18 and Analog pin A5 is listed as 19 in the code.

Use the Project 23 Pin Identification Chart to match the LED matrix positions with the pin connections on the book website. As needed edit the information in the code to fix the LED matrix output.

If this does not solve the issues, Use the Project 23 Pin Identification Step-By-Step document to match the LED matrix positions with the pin connections on the book website.

Look at the documentation of the LCD (if it was shipped with the LCD) for the pin diagram. The pin diagram will help you under stand which pins on the bottom of the LED control which rows and columns on the LED matrix.

Go to the book website and download the Project 23 Pin Identification Chart to fill in. This will guide matching the LED matrix values to UNO board.

The LEDs in the matrix are identified by their row number and column number. For example, the first LED is (1,1), the LED in row 5 column 8 is (5,8) and the last LED is (8,8). The part number printed on the LED is printed parallel to column one. Rows extend to the other side.

There are two sets of eight (8) pins on the bottom of the LED. When looking at the model number printed on the side of the LED matrix pin one (1) is on the left and pin eight (8) in on the right. The numbering wraps around with the second set starting with nine (9) and going up through 16.

Each pin is assigned to control a row or column. When two pins are combined, the pin numbers translate to an LED matrix value. For this project example LED (1,1) is turned on by pins number 5

Project Reflection and Summative Activities, Discussion Starters, Extensions, and Problem

A = Essential

Reflection/Summative Activities:

- Have each team member document in their project/classroom journal who was on their team, what went well, what they could improve upon, what they would do differently if they were to do the project again, and verify that every person in the group can do the project.

- Review each team member's documentation of this project using emojis.

For Discussion:

- What are the advantages and disadvantages to using emojis?

- What are the advantages and disadvantages to using an LED matrix?

Extensions:

- Edit the code to change the speed of the emoji graphic movement.

- Edit the code to change the direction of movement of the heart graphic.

- Edit the code to make a new series of graphics.

- Edit the code to add more graphics to the code (increase from 6 graphics generated to 8 or more graphics).

B = Recommended

Communications Connections:

- Create a proposal for a new emoji. Research the process, the requirements, and prepare a written proposal.

- Research the rules about using copyrighted work for your own purposes.

Extension:

-Go to the book website thearduinoclassroom.com

From the homepage:
Click
    Books
Click
    UNO Edition Vol.

Find the "Links" link on the right and locate the Project 23 link to the Character Generator.

Visit this site and generate a new message for the 8x8 LED Matrix.

Insert the new message into the code.

Submit your project to the book page on the website. Register and Login to submit at:
thearduinoclassroom.com

From the homepage:
Click
    Books
Click
    UNO Edition Vol. 1

C = Optional

Extensions:

- Add the 8x8 LED Matrix to a previously implemented project.

-From the Project 23 link page click on the Character Library link.

Install the .zip file and library (page 62).

Edit the code to use the library to generate characters for the 8x8 LED matrix instead of using binary code.

Problem:

Identify a problem that would be solved by combining a sensor with an LED matrix. How could this project be a solution? Include in your report the following:

Description of the problem.

Description of how the problem is solved by this project.

What other resources you would need to solve the problem.

Design and produce the solution.

A project rubric can be found on page 236.

Anticipatory Sets
(from page 210)

Anticipatory Sets:
Responses Will Vary

A

Reflection/Summative Activities:
- Have each team member document in their project/classroom journal who was on their team, what went well, what they could improve upon, what they would do differently if they were to do the project again, and verify that every person in the group can do the project.
Responses Will Vary

- Review each team member's documentation of this project using emojis.
Responses Will Vary

For Discussion:
- What are the advantages and disadvantages to using emojis?
Responses Will Vary

- What are the advantages and disadvantages to using an LED matrix?
Responses Will Vary

Extensions:
- Edit the code to change the speed of the emoji graphic movement.
Responses Will Vary

- Edit the code to change the direction of movement of the heart graphic.
Responses Will Vary

- Edit the code to make a new series of graphics.
Responses Will Vary

- Edit the code to add more graphics to the code (increase from 6 graphics generated to 8 or more graphics).
Responses Will Vary

B

Communications Connections:
- Create a proposal for a new emoji. Research the process, the requirements, and prepare a written proposal.
Responses Will Vary

- Research the rules about using copyrighted work for your own purposes.
Responses Will Vary

Extension:
- Go to the book website thearduinoclassroom.com

From the homepage:
Click
    Books
Click
    UNO Edition Vol.

Find the "Links" link on the right and locate the Project 23 link to the Character Generator. Visit this site and generate a new message for the 8x8 LED Matrix.

Insert the new message into the code.
Responses Will Vary

Submit your project to the book page on the website.
Responses Will Vary

C

Extensions:
- Add the 8x8 LED Matrix to a previously implemented project.
Responses Will Vary

-From the Project 23 link page click on the Character Library link.

Install the .zip file and library (page 62).

Edit the code to use the library to generate characters for the 8x8 LED matrix instead of using binary code.
Responses Will Vary

Problem:
Responses Will Vary

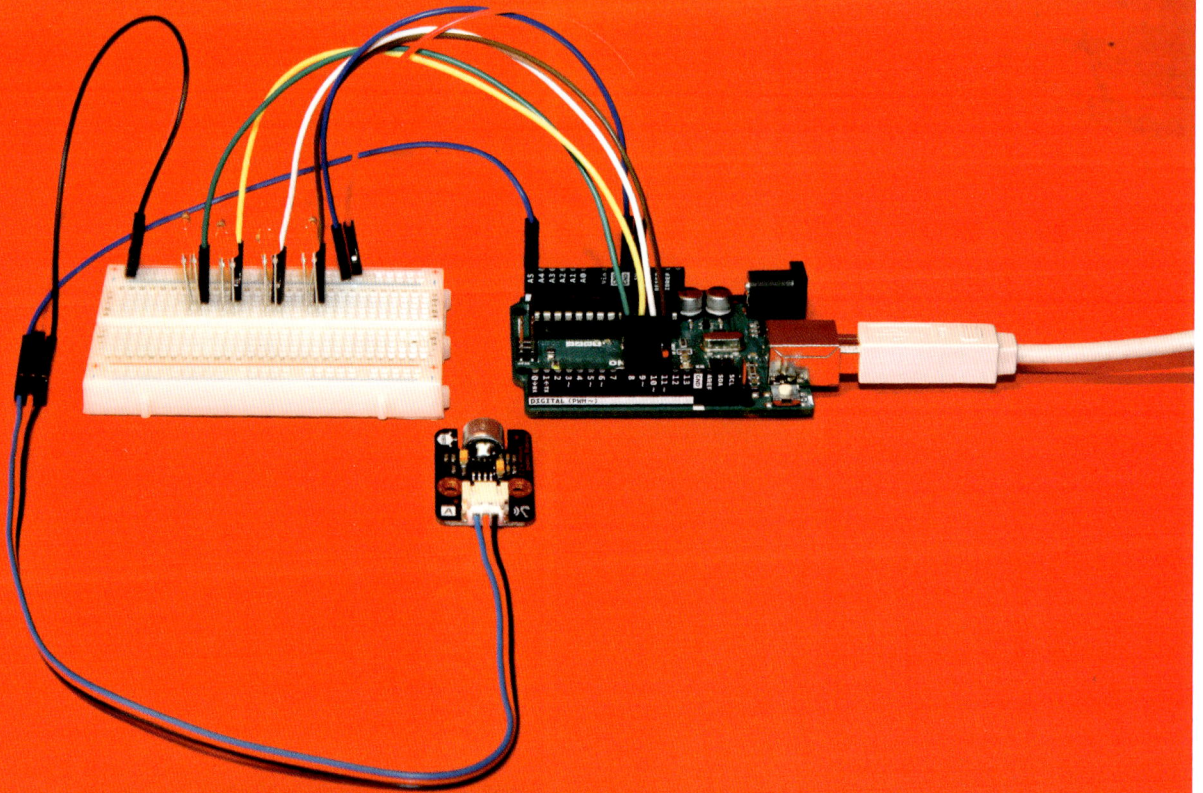

Getting Started

Lesson Integration:	Groupings:	Level:	Time to Complete:
Physics - Ohm's Law, Circuits, Voltage, Resistance, Sound, Intensity, Decibels Biology - Nervous System, Senses, Hearing Mathematics - Arithmetic	1 - 2	Intermediate	1:15 hour:min for the project and 45 min. for extensions.

Objectives:

- Investigate how circuits and electronic components interact with electrical energy.

Prerequisite Skills:

- Time measurement (milliseconds)
- Understanding of Sound and Sound Intensity
- Arduino® IDE software (pages 11-13)
- Fritzing (pages 64-66)

Purpose and Skills:

- Interpret sound sensor input.

STEAM Connections:

Science - Circuits, Resistance, Ohm's law, Sound, Intensity

Technology - Code, Simulators, Digital Design, Electronic Components, Analog Data

Engineering - Designing, Building, and Using a Machine, Prototyping, Applied Physics

Allied Arts - Sound

Math - Ratios, Conversions

Key Vocabulary:

Decibel - The unit of measure for loudness of a sound with respect to the listener.

Project Introduction:

- Introduce the groups to the purpose of the project, the skills developed, the standards met, and the goal of the project.

Anticipatory Sets:

- Where would you see a sound sensor used? List at least three different places or situations where sound sensor are used, describe the devices, and explain the purpose of the devices.

Project 24
Did You Hear Me?

Educational Standards

ISTE Standards for Students
- Empowered Learner 1a, 1b, 1d
- Knowledge Constructor 3a, 3b, 3c, 3d
- Innovative Designer 4a, 4b, 4c, 4d
- Computational Thinker 5a, 5b, 5c, 5d
- Creative Communicator 6a, 6b, 6c, 6d
- Global Collaborator 7c, 7d

US Computer Science Standards
- Project correlations can be found on the book's web page.

US NGSS - Middle School
DCIs
- MS-PS4-3 PS4.C
- MS-PS4-1 PS4.A
- MS-LS1-8 LS1.D

Cross Cutting Concepts
- Cause and Effect
- Scale, Proportion and Quantity
- Structure and Function

Science and Engineering Practices
- Planning and Carrying Out Investigations
- Obtaining, Evaluating, and Communicating Information
- Analyzing and Interpreting Data
- Scientific Knowledge is Based on Empirical Evidence
- Using Mathematics and Computational Thinking

US NGSS - High School
- Correlations can be found on the book's web page.

US Common Core Language Arts and Mathematics
- Correlations can be found on the book's web page.

Materials List:
- Computer with IDE and Fritzing software
- Connection to the Internet
- USB Cable
- UNO or UNO Compatible Microcontroller
- Short Breadboard
- Four (4) 220 Ohm Resistors
- Four (4) LEDs
- 10 Jumper Wires (Male to Male)
- MAX 9814 (Sound and Microphone)

Engineering Design - Digital Prototype

1. Download, install, and launch (as needed) Fritzing (pages 64-66) to build a digital prototype.

2. Visit The Arduino® Classroom website and register for the website.

3. Navigate to the "UNO Edition Vol. 1" page by hovering over "Books" on the main menu and clicking on the "UNO Edition Vol. 1" link.

4. On the right of the page find and then click on the "Volume 1 Links" link.

5. Click on the "Fritzing Part Download" link for Project 24 which will take you to the page to download the file.

6. Extract the file.

7. Install the file (page 66).

8. Start a new sketch in Fritzing.

9. Drag an UNO board to the workspace.

10. Drag a breadboard to the workspace.

11. Place four (4) LEDs on the breadboard across the same column.

12 Place four (4) 220 Ohm resistors on the breadboard connecting the cathode(-) row and the same row as the cathode(-) leg of each LED.

13 Connect with jumper wires the following:
- the five volt (5v) pin to the anode(+) column on the breadboard
- a ground (GND) pin on the UNO board to the cathode(-) column on the breadboard.
- digital pin 11 to the same row as the anode(+) lead of the LED.
- digital pin 11 with the same row connected to the anode(+) leg of the first LED
- digital pin 10 with the same row connected to the anode(+) leg of the second LED
- digital pin nine (9) with the same row connected to the anode(+) leg of the third LED
- digital pin eight (8) with the same row connected to the anode(+) leg of the last LED
- the GND lead on the sensor to the cathode(-) column on the breadboard.
- the VCC lead on the sensor to the anode(+) column on the breadboard
- the data lead on the sensor to the analog five (A5) pin on the UNO board

Coding - Digital Prototype

14 Optional. Enter the code in Tinkercad® to match the code on this page.

15 Verify the code.

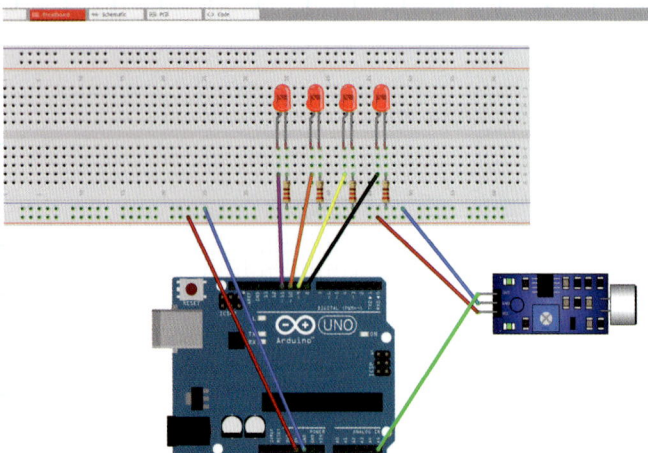

```
int DA = A5;        //  Connection to the Analog Input Sensor
int threshold = 90; //  Minimum threshold for LED
int sensorvalue = 0;
void setup()
{
  Serial.begin(9600);
  pinMode(8, OUTPUT);
  pinMode(9, OUTPUT);
  pinMode(10, OUTPUT);
  pinMode(11, OUTPUT);
}
  void loop ()
{
  sensorvalue = analogRead(DA); //Reading analog values
  Serial.println (" ");
  Serial.print("Analog Sound Reading: ");
  Serial.print(sensorvalue);    //Displays analog value
  Serial.println (" ");
  delay(250);
  if (sensorvalue >= threshold) //Compares analog values with threshold
  {
digitalWrite(8, HIGH);
digitalWrite(9, HIGH);
digitalWrite(10, HIGH);
digitalWrite(11, HIGH);
Serial.print("ALARM!!!!");
Serial.println (" ");
Serial.println (" ");
}
else {
  digitalWrite(8, LOW);
  digitalWrite(9, LOW);
  digitalWrite(10, LOW);
  digitalWrite(11, LOW);
  Serial.print("NO ALARM");
  Serial.println (" ");
Serial.println (" ");
}
}
```

zing

Project 24
Did You Hear Me?

Engineering Design - Project Build

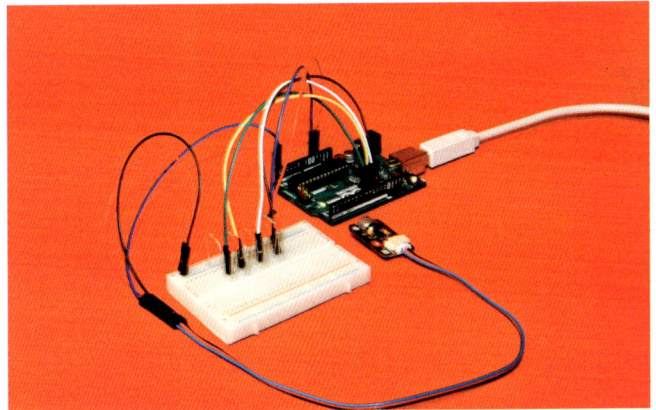

Connection - SONAR, an acronym for SOund NAvigation and Ranging, was developed as a technology in the early 20th century, which came to full use as a means of object detection and evasion in the water in the middle of that century. First used as a way to navigate as well as discover and evade warships and submarines, SONAR has evolved into a tool for searching and even sport fishing. Leonardo Da Vinci was the first person to note that sounds can travel through water and be used to identify nearby vessels. However, it was not until more than 400 years later that the technology was developed.

SONAR devices can be used in two modes, active and passive. Active mode sends sound out and then listens to the sound returned. The data of the returned echo is then interpreted with the to understand the shape, size and movement of another object. Passive SONAR only listens to sounds and while it returns less information to the listener, it does not reveal their position either.

Animals like bats, toothed whales (like Orca and Beluga whales), and dolphins use a naturally evolved form of SONAR that is called echolocation. Research with dolphins has shown that they can "see" with echolocation as well as humans see with their eyes. Here an Orca whale leaps over a pod mate.

Credit David Ellifrit (NOAA), Alaska Fisheries Science Center, NOAA Fisheries Service

Project 16 uses a sound emitter and sensor, very much like SONAR, to detect distances.

16 Place four (4) LEDs on the breadboard across the same column.

17 Place four (4) 220 Ohm resistors on the breadboard connecting the cathode(-) row and the same row as the cathode(-) leg of each LED.

18 Connect the sensor cable and then with jumper wires the following:
- the five volt (5v) pin to the anode(+) column on the breadboard
- a ground (GND) pin on the UNO board to the cathode(-) column on the breadboard.
- digital pin 11 to the same row as the anode(+) lead of the LED.
- digital pin 11 with the same row connected to the anode(+) leg of the first LED
- digital pin 10 with the same row connected to the anode(+) leg of the second LED
- digital pin nine (9) with the same row connected to the anode(+) leg of the third LED
- digital pin eight (8) with the same row connected to the anode(+) leg of the last LED
- the GND lead on the sensor to the cathode(-) column on the breadboard.
- the VCC lead on the sensor to the anode(+) column on the breadboard
- the data lead on the sensor to the analog five (A5) pin on the UNO board

19 Connect the UNO board and computer using the USB cable.

20 Launch the Arduino® IDE software (pages 11-13) to make sure the board is communicating with the computer.

Go to the Tools menu and verify that the correct board is selected from the Board Manager menu. If not, select the correct board from the options listed.

Then confirm that the right Port is selected. If not, select the port that lists the UNO board from the options listed.

Coding - Project Build

21 Either copy and paste the code from the Tinkercad® prototype or type the code from page 220 into the Arduino® IDE software (pages 11-13).

22 Save the sketch (rename as needed).

23 Verify the sketch.

24 Upload the sketch to the UNO board.

25 Test the sensor LED indicators with sounds of varying loudness.

26 Open the Serial Monitor and observe the data being collected by the sensor under varying levels of sound.

27 Open the Serial Plotter and observe the data being collected by the sensor under varying levels of sound. See the next page for a sample graph of the data.

28 Document this project, discuss the project, and complete the extensions assigned from the next page.

```
TAC_UE_V1_P24 | Arduino 1.8.9 (Windows Store 1.8.21.0)
File Edit Sketch Tools Help

TAC_UE_V1_P24
//Project 24
//Sound Detector with LED
//thearduinoclassroom.com
//Copyright 2019, Isabel Mendiola and Peter Haydock

int DA = A5;          //  Connection to the Analog Input Sensor
int threshold = 90; //  Minimum threshold for LED
int sensorvalue = 0;
void setup()
{
  Serial.begin(9600);
  pinMode(8, OUTPUT);
  pinMode(9, OUTPUT);
  pinMode(10, OUTPUT);
  pinMode(11, OUTPUT);
}
  void loop ()
{
  sensorvalue = analogRead(DA); //Reading analog values
  Serial.println (" ");
  Serial.print("Analog Sound Reading: ");
  Serial.print(sensorvalue);      //Displays analog value
  Serial.println (" ");
  delay(250);
  if (sensorvalue >= threshold) //Compares analog values with threshold
  {
digitalWrite(8, HIGH);
digitalWrite(9, HIGH);
digitalWrite(10, HIGH);
digitalWrite(11, HIGH);
Serial.print("ALARM!!!!");
Serial.println (" ");
Serial.println (" ");
}
else {
  digitalWrite(8, LOW);
  digitalWrite(9, LOW);
  digitalWrite(10, LOW);
  digitalWrite(11, LOW);
  Serial.print("NO ALARM");
  Serial.println (" ");
Serial.println (" ");
}
}

Sketch uses 2416 bytes (7%) of program storage space. Maximum is 32256 bytes.
Global variables use 232 bytes (11%) of dynamic memory, leaving 1816 bytes for local variables. Maximum is 2048

90                                                    Arduino/Genuino Uno on COM4
```

Connection - Ears are critical to the sense of hearing in most mammals. Other animals, like reptiles, fish, and birds can sense sounds (or the pressure of sound waves), but do not have external lobs that form outer ears like those of mammals.

The main purpose of ears in a mammal is to focus and amplify sounds. Having a pair of ears allows most mammals the ability to tell which direction the sound is coming from. This gives the mammal an advantage to hear softer sounds and determine better the direction of sounds. The shape, orientation, and ability to change the position of the ears indicate the pressures the mammal faces in its ecosystem.

The elephant has evolved the ability to use their ears as giant heat radiators. When their bodies need cooling, more blood is pumped to the ears while the elephant fans the ears to help more heat escape their body.

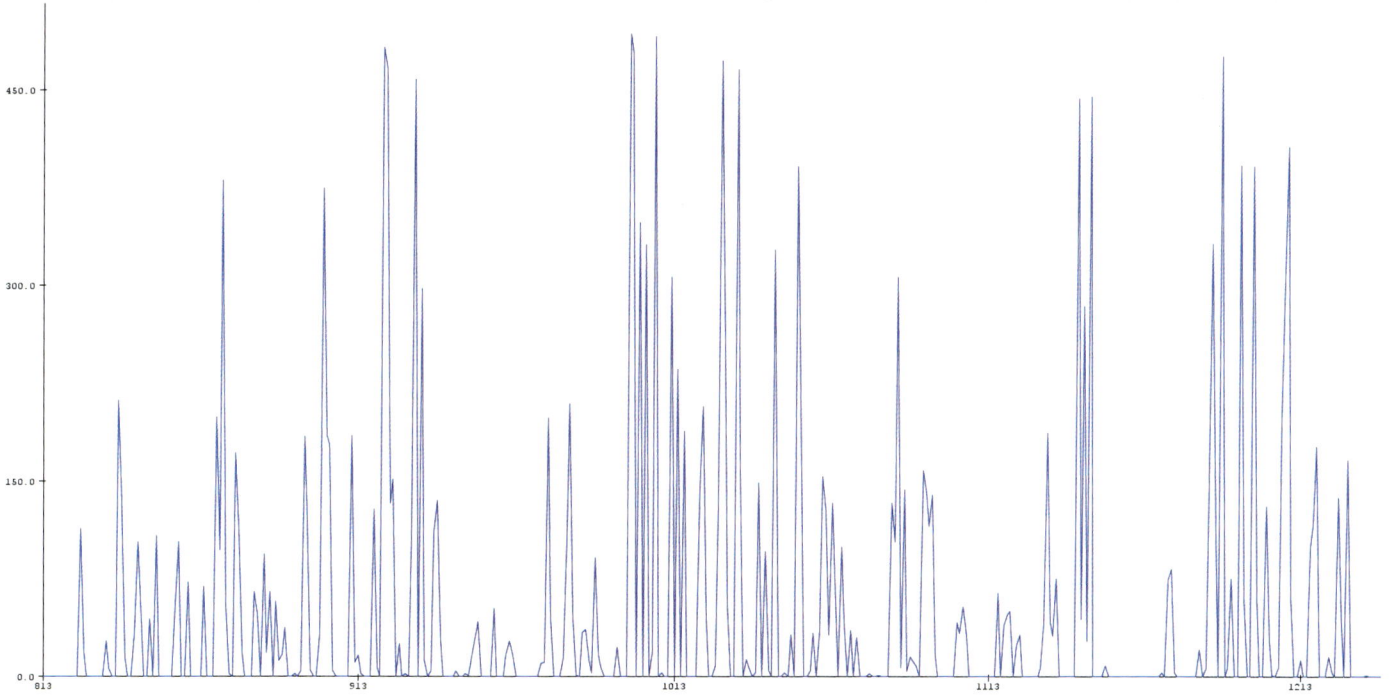

Project Reflection and Summative Activities, Discussion Starters, Extensions, and Problem

A = Essential

Reflection/Summative Activity:

- Have each team member document in their project/ classroom journal who was on their team, what went well, what they could improve upon, what they would do differently if they were to do the project again, and verify that every person in the group can do the project.

For Discussion:

- How do bats, toothed whales and dolphins use echolocation?

- Develop two lists that compare and contrast SONAR and echolocation.

Extensions:

- Adjust the code to be more or less sensitive to sound.

- Add an LED to the project to indicate that the sensor is "listening."

- Research and explain using an infographic how earth scientists listen to Earth and what information do they collect for studying.

B = Recommended

Biological Connection:

- Create a model of an mammal's ears and explain how they function, what advantages they give the mammal and limitations for the mammal that the ears might have.

Music Connection:

- Play some music for the sound sensor and observe the Serial Plotter. Note any correlations between what the Plotter graphs and what you hear.

Extensions:

- Develop an experiment to show how distance and orientation from a sound impacts the measurements and data generated by the sensor. Carry out that experiment.

- Edit the code to light the LEDs in an order to show the loudness of the sound being observed?

- Submit your project to the book page on the website. Register and Login to submit at: thearduinoclassroom.com

From the homepage:
Click
    Books
Click
    UNO Edition Vol. 1

C = Optional

Extensions:

- Use a decibel meter and change the code to calibrate the sensor signal to decibels.

- Go online to the book web page (make sure to register) and access the Project 24 extension. Photos, a Fritzing diagram and code can be found at this link.

Problem:

- Identify a problem that would be solved by a sound sensor like the one used in this project. How could this project be a solution? Include in your report the following:

Description of the problem.

Description of how the problem is solved by this project.

What other resources you would need to solve the problem.

Design and produce the solution.

A project rubric can be found on page 236.

Anticipatory Sets
(from page 218)

Anticipatory Sets:
Sound sensors are found in SONAR systems, security systems, and decibel meters that measure sound loudness to protect people's hearing.

A

Reflection/Summative Activity:
Responses Will Vary

For Discussion:
- How do bats, toothed whales and dolphins use echolocation?
Each species sends a high pitched clicking sound out and then listens to echo return.

- Develop two lists that compare and contrast SONAR and echolocation.
Responses Will Vary

Extensions:
- Adjust the code to be more or less sensitive to sound.
Responses Will Vary

- Add an LED to the project to indicate that the sensor is "listening."
Responses Will Vary

B

Biological Connections:
- Create a model of an mammal's ears and explain how they function, what advantages they give the mammal and limitations for the mammal that the ears might have.
Responses Will Vary

Music Connection:
- Play some music for the sound sensor and observe the Serial Plotter. Note any correlations between what the Plotter graphs and what you hear.
Responses Will Vary

Extensions:
- Develop an experiment to show how distance and orientation from a sound impacts the measurements and data generated by the sensor. Carry out that experiment.
Responses Will Vary

- Edit the code to light the LEDs in an order to show the loudness of the sound being observed?
Responses Will Vary

- Submit your project to the book page on the website.
Responses Will Vary

C

Extensions:
- Use a decibel meter and change the code to calibrate the sensor signal to decibels.

- Go online to the book web page (make sure to register) and access the Project 24 extension. Photos, a Fritzing diagram and code can be found at this link.
Responses Will Vary

Problem:
Responses Will Vary

Project 25
Robot Race

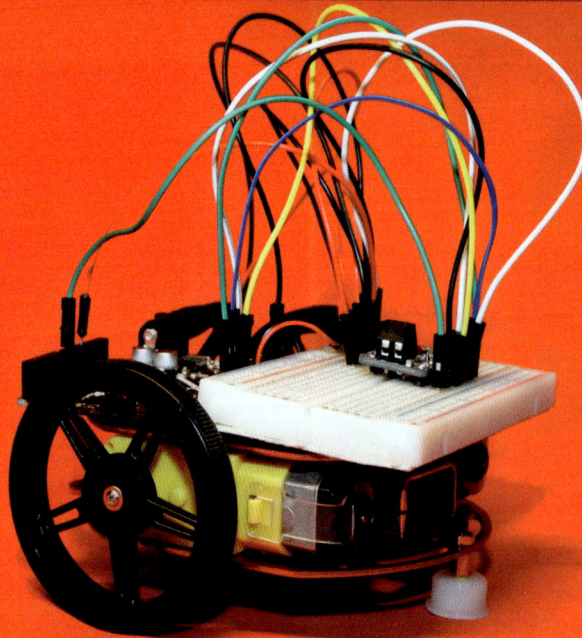

Lesson Integration:	Groupings:	Level:	Time to Complete:
Physics - Ohm's Law, Circuits, Voltage, Resistance, Motion, Time, Rate Mathematics - Arithmetic	1 - 2	Advanced	2 hours for the project and 2 hours for extensions.

Objectives:
- Investigate how circuits and electronic components interact with electrical energy.
- Control a robot

Prerequisite Skills:
- Time and Distance measurement (milliseconds)
- Arduino® IDE software (pages 11-13)
- Fritzing (pages 64-66)

Purpose and Skills:
- Build a robotic car guided by different sensors

STEAM Connections:
Science - Circuits, Resistance, Ohm's law
Technology - Code, Simulators, Digital Design, Electronic Components, Analog Data, Robotics
Engineering - Designing, Building, and Using a Machine, Prototyping, Applied Physics
Allied Arts - Motion
Math - Ratios, Conversions

Key Vocabulary:
Robot - A device that moves based on computer

controls that simulates or replaces human motions.

Project Introduction:
- Introduce the groups to the purpose of the project, the skills developed, the standards met, and the goal of the project.

Anticipatory Sets:
- Identify where robots are seen. List at least three different places or situations where robots are used, describe the robots and explain the purpose of the lights.

- Find videos of robots in use and create a playlist for the class.

- Prepare a list of pros and cons for using robots.

- Watch a robot competition on TV or the Internet and summarize what you saw with an emphasis on the design and capabilities of the robot.

Project 25
Robot Race

Educational Standards

ISTE Standards for Students
- Empowered Learner 1a, 1b, 1d
- Knowledge Constructor 3a, 3b, 3c, 3d
- Innovative Designer 4a, 4b, 4c, 4d
- Computational Thinker 5a, 5b, 5c, 5d
- Creative Communicator 6a, 6b, 6c, 6d
- Global Collaborator 7c, 7d

US Computer Science Standards
- Project correlations can be found on the book's web page.

US NGSS - Middle School
DCIs
- MS-PS2-2 PS2.A
- MS-PS3-5 PS3.B

Cross Cutting Concepts
- Cause and Effect
- Scale, Proportion and Quantity
- Structure and Function

Science and Engineering Practices
- Planning and Carrying Out Investigations
- Obtaining, Evaluating, and Communicating Information
- Analyzing and Interpreting Data
- Using Mathematics and Computational Thinking

US NGSS - High School
- Correlations can be found on the book's web page.

US Common Core Language Arts and Mathematics
- Correlations can be found on the book's web page.

The robotic car kit requires a #1 Phillips screwdriver and some soldering. Use caution when soldering as the tip of the iron is hot and can burn. The soldering should be done by or under the supervision of a responsible adult.

Step-by-Step 1-28

Materials List:
- Computer with IDE and Fritzing software
- Connection to the Internet
- USB Cable
- UNO or UNO Compatible Microcontroller
- Short Breadboard
- Robotic Car Kit
- IR Distance Sensor
- Stepper Motor
- 15 Jumper Wires (Male to Male)
- 9v Battery & Snap Barrel Connector
- Optional - Double-sided tape

Engineering Design - Digital Prototype

1. Download, install, and launch (as needed) Fritzing (pages 64-66) to build a digital prototype.

2. Visit The Arduino® Classroom website and register for the website.

3. Navigate to the "UNO Edition Vol. 1" page by hovering over "Books" on the main menu and clicking on the "UNO Edition Vol. 1" link.

4. On the right of the page find and then click on the "Volume 1 Links" link.

5. Click on the "Fritzing Part Download" link for Project 25 which will take you to the page to download the file.

6. Click on "View raw." This will download the file.

7. Install the file (page 66).

8. Start a new sketch in Fritzing.

9. Drag an UNO board to the workspace.

10. Drag two (2) M1 motors to the workspace.

11. Drag one (1) DRV8833 stepper motor to the workspace and place it on the breadboard.

12 Connect with jumper wires the following:
- the five volt (5v) pin to the anode(+) column of the breadboard
- a ground (GND) pin on the UNO board to the cathode(-) column on the breadboard.
- Vin on the UNO board to the VM on the DRV8833 stepper motor board
- GND on the DRV8833 stepper motor board to the cathode(-) column on the breadboard
- digital pin seven(7) to BIN1 on the DRV8833 stepper motor board
- digital pin six (6) to BIN2 on the DRV8833 stepper motor board
- SLP on the DRV8833 stepper motor board to the anode(+) column

- digital pin five (5) to AIN2 on the DRV8833 stepper motor board
- digital pin four (4) to AIN1 on the DRV8833 stepper motor board

13 Connect each motor to the DRV8833 stepper motor
- AOUT 1 on the DRV8833 stepper motor board to the GND (black) lead on motor 1
- AOUT 2 on the DRV8833 stepper motor board to the 5v (red) lead on motor 1
- BOUT 2 on the DRV8833 stepper motor board to the 5v lead (red) lead on motor 2
- BOUT 1 on the DRV8833 stepper motor board to the GND (black) on motor 2

Connection - An artist's concept of the next Mars rover to be built by the Jet Propulsion Laboratory for NASA and launched in July 2020. The rover has a coring drill to collect samples of rocks and soils that may be returned to Earth on a future mission. The rover will also demonstrate technologies that would be needed for human expeditions to Mars. An oxygen producing prototype machine will be on board and sensors to identify subsurface water and other necessary resources. The rover will also help improve landing techniques and collect data on weather, dust, and other environmental conditions. This data will help NASA plan for future missions that will include astronauts living and working on Mars.

Image Credit NASA/JPL-Caltech

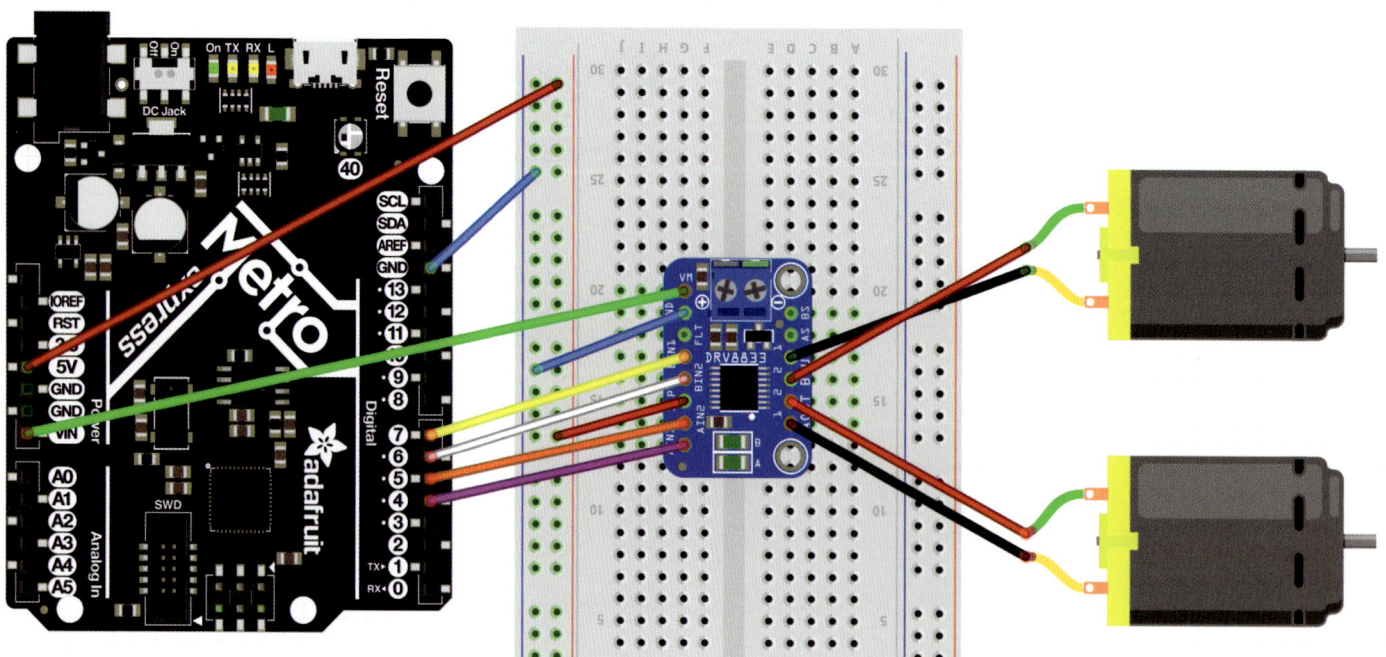

Engineering Design - Project Build

14 Build the robot chassis per the included instructions. Optional. Affix the breadboard and UNO to the chassis with double-sided tape.

15 Place the DRV8833 stepper motor (solder on the pins before connecting to the breadboard) on the breadboard spanning the two halves.

16 Connect with jumper wires the following:
- the five volt (5v) pin to the anode(+) column of the breadboard
- a ground (GND) pin on the UNO board to the cathode(-) column on the breadboard.
- Vin on the UNO board to the VM on the DRV8833 stepper motor board
- GND on the DRV8833 stepper motor board to the cathode(-) column on the breadboard
- digital pin seven(7) to BIN1 on the DRV8833 stepper motor board
- digital pin six (6) to BIN2 on the DRV8833 stepper motor board
- SLP on the DRV8833 stepper motor board to the anode(+) column
- digital pin five (5) to AIN2 on the DRV8833 stepper motor board
- digital pin four (4) to AIN1 on the DRV8833 stepper motor board

17 Connect each motor to the DRV8833 stepper motor
- AOUT 1 on the DRV8833 stepper motor board to the GND (black) lead on motor 1
- AOUT 2 on the DRV8833 stepper motor board to the 5v (red) lead on motor 1
- BOUT 2 on the DRV8833 stepper motor board to the 5v lead (red) lead on motor 2
- BOUT 1 on the DRV8833 stepper motor board to the GND (black) on motor 2

18 Connect the UNO board and computer using the USB cable for uploading the sketch.

19 Launch the Arduino® IDE software (pages 11-13) to make sure the board is communicating with the computer.

Go to the Tools menu and verify that the correct board is selected from the Board Manager menu. If not, select the correct board from the options listed.

Then confirm that the right Port is selected. If not, select the port that lists the UNO board from the options listed.

Coding - Project Build

20 Type the "TEST SKETCH" code on the next page into the Arduino® IDE software (pages 11-13).

21 Download the Stepper library .Zip file from the book website (page 62). Install the library into the sketch.

22 Save the sketch (rename as needed).

23 Verify the sketch.

24 Upload the sketch to the UNO board.

25 Disconnect the USB connection from the board.

26 Attach the 9v barrel connector to the Metro328 board and attach a 9v battery to the snap connectors on the barrel connector.

27 Set the robot on the table top or floor to test the robot. The TEST code provided makes the robot go forward, spin, and go in reverse. If the robot does not move, check the connections and soldering.

28 Document this project, discuss the project, and complete the extensions assigned from the next page.

Project 25
Robot Race
Challenge 1

Programmed Movement

Use the code to the right on this page to direct the car through a prescribed course. The code to the right shows how to make each wheel move. It is not the course. Program the car to move within the course. The course will be set by your classroom guide and will involve making your robot move forward, turn, and go in reverse.

Change the "HIGH" and "LOW" values within the code to change the movement of the robot.

Note: The wheels may not be aligned perfectly straight. Small incremental turns can help straighten the path of the robot.

Connection - NOAA readies an Autonomous Underwater Vehicle to study water chemistry in Monterey Bay in California, USA

```
//TEST SKETCH to test the stepper motor
#include <Stepper.h>
#define STEPS 4   //Steps on your motor
Stepper stepper(STEPS, 4, 5, 6, 7); //pins attached
AIN1 and AIN2 Motor A, pins 4 and 5
//pins attached BIN1 and BIN2 Motor B, pins 6
and 7
void setup()
{
  Serial.begin(9600);
  Serial.println("Testing stepper");
  stepper.setSpeed(10);
}
void loop()
{
  Serial.println("Moves Forward");
  stepper.step(STEPS);
  Serial.println("Moves Backward");
  stepper.step(-STEPS);
}
```

```
// CHALLENGE 1 BASE CODE
#define Motor1 4 //Declare variables Motor1
#define Motor1A 5
#define Motor2 6 //Declare variables Motor2
#define Motor2B 7
int speed = 0;
void setup() {
pinMode(4, OUTPUT);
pinMode(5, OUTPUT);
pinMode(6, OUTPUT);
pinMode(7, OUTPUT);
}
void loop() {
  digitalWrite(4,HIGH); //forward
  digitalWrite(5,LOW);
  digitalWrite(6,HIGH);
  digitalWrite(7,LOW);
  delay(1000);
  digitalWrite(4,LOW); // backward
  digitalWrite(5,HIGH);
  digitalWrite(6,LOW);
  digitalWrite(7,HIGH);
  delay(1000);
  digitalWrite(4,HIGH); //forward
  digitalWrite(5,LOW);
  digitalWrite(6,HIGH);
  digitalWrite(7,LOW);
  delay(1000);
  digitalWrite(4,HIGH); //reverse pivot
  digitalWrite(5,HIGH);
  digitalWrite(6,LOW);
  digitalWrite(7,HIGH);
  delay(1000);
  digitalWrite(4,LOW); //forward pivot
  digitalWrite(5,LOW);
  digitalWrite(6,HIGH);
  digitalWrite(7,LOW);
  delay (1000);
}
```

Crash Avoidance

Use Fritzing to add the IR distance sensor to the project. Double-sided tape will affix the sensor to the breadboard. NOTES: Soldering three M/M jumper wires to the ends of the wires from the sensor helps keep the physical connections in place. Digital Pin 2 is connected to the sensor data wire with the anode(+) and cathode(-) wires connected to the respective columns on the breadboard.

Then use the code on this page to direct the robotic car to avoid obstacles. The code is only enough to show how the sensors communicate and drive the robotic car backwards from an obstacle.

Edit the code to steer the robotic car to drive around a room with obstacles.

The obstacles will be set by your classroom guide and will involve making your robotic car move forward, turn, and backward.

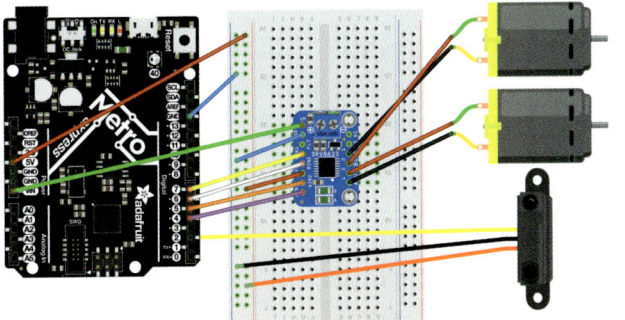

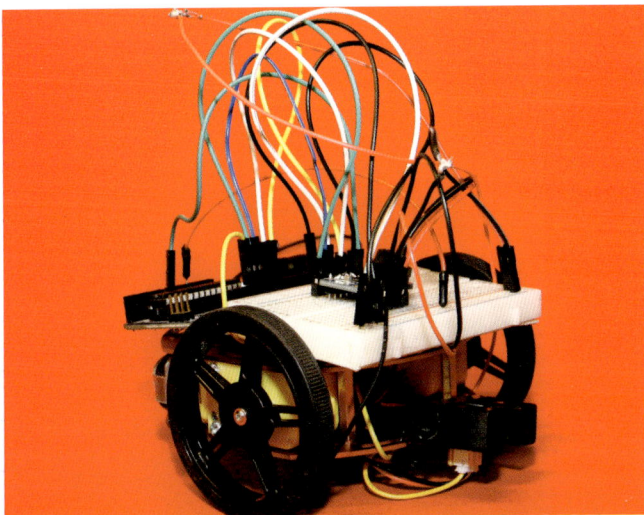

```cpp
// CHALLENGE 2 BASE CODE
int LED = 13; // Use the onboard Uno LED
int isObstaclePin = 2;  // This is our input pin
int isObstacle = HIGH;  // HIGH MEANS NO OBSTACLE
#define Motor1 4 //Declare variables Motor1
#define Motor1A 5
#define Motor2 6 //Declare variables Motor2
#define Motor2B 7
void setup() {
  pinMode(LED, OUTPUT);
  pinMode(isObstaclePin, INPUT);
  pinMode(4, OUTPUT); // Car connections
  pinMode(5, OUTPUT);
  pinMode(6, OUTPUT);
  pinMode(7, OUTPUT);
  Serial.begin(9600);
}
void loop() {
  digitalWrite(4,LOW);
  digitalWrite(5,HIGH);
  digitalWrite(6,LOW);
  digitalWrite(7,HIGH);
  isObstacle = digitalRead(isObstaclePin);
  if (isObstacle == LOW)
  {
  Serial.println("OBSTACLE!!, OBSTACLE!!");
  digitalWrite(LED, HIGH);
  }
  else
  {
  digitalWrite(4,HIGH);
  digitalWrite(5,LOW);
  digitalWrite(6,HIGH);
  digitalWrite(7,LOW);
  Serial.println("clear");
  digitalWrite(LED, LOW);
  }
  delay(1000);
}
```

A = Essential

Reflection/Summative Activity:
- Have each team member document in their project/classroom journal who was on their team, what went well, what they could improve upon, what they would do differently if they were to do the project again, and verify that every person in the group can do the project.

- Create a video of your robot racing.

Section Reflection/Summative Activity:
- Have the team create a graphic organizer summarizing the skills and knowledge they obtained in doing Projects Twenty Two through Twenty Five.

- Document in your journal key vocabulary, inputs and controls used, and skills developed in this section.

For Discussion:
- Why or in what cases would using a robot be a disadvantage?

Extensions:
- Add a light to the robot to indicate that it is performing as expected.

- Add a light to the robot to indicate that it is not performing as expected.

B = Recommended

Professional Connections:
- Research and write a short description of the work a robotics engineer does. Be sure to include the education the profession requires, any laws or regulations that guide their use, and what activities are performed.

- Interview a robotics engineer through a video conference.

Communication Connection:
- Prepare a proposal to have your robot sponsored by a local business. After approval of your proposal by your classroom guide, present the proposal to the business for sponsorship.

Ethics Connection:
- Research and write a summary about robots replacing human workers.

Extension:
- Add sound indicators to the robot to communicate its performance.

- Submit your project to the book page on the website. Register and Login to submit at: thearduinoclassroom.com

From the homepage:
Click
   Books
Click
   UNO Edition Vol. 1

C = Optional

Extensions:
- Design a robot for NASA's next planet exploration. Describe the planet, the work the robot would do, and how the robot would power itself (and how long) on the planet.

Compare and contrast your design against others submitted and defend your design to be implemented by NASA.

- Design a robot for NOAA's next mission to the deepest parts of the ocean. What work the robot would do, and how the robot would power itself (and how long) at the bottom of the ocean?

Compare and contrast your design against others submitted and defend your design to be implemented by NOAA.

Problem:
- Identify a problem that would be solved by a robot.

Include in your report the following:

Description of the problem.

Description of how the problem is solved by this project

What other resources you would need to solve the problem.

Design the solution.

A project rubric can be found on page 236.

Anticipatory Sets
(from page 226)

Anticipatory Sets:

- Identify where robots are seen. List at least three different places or situations where robots are used, describe the robots and explain the purpose of the lights.
Responses Will Vary

- Find videos of robots in use and create a playlist for the class.
Responses Will Vary

- Prepare a list of pros and cons for using robots.
Responses Will Vary

- Watch a robot competition on TV or the Internet and summarize what you saw with an emphasis on the design and capabilities of the robot.
Responses Will Vary

A

Reflection/Summative Activity:
- Have each team member document in their project/classroom journal who was on their team, what went well, what they could improve upon, what they would do differently if they were to do the project again, and verify that every person in the group can do the project.
Responses Will Vary

- Create a video of your robot

racing.
Responses Will Vary

Section Reflection/Summative Activity:
Responses Will Vary

For Discussion:
- Why or in what cases would using a robot be a disadvantage?
Responses Will Vary

Extensions:
- Add a light to the robot to indicate that it is performing as expected.
Responses Will Vary

- Add a light to the robot to indicate that it is not performing as expected.
Responses Will Vary

B

Professional Connections:
- Research and write a short description of the work a robotics engineer does. Be sure to include the education the profession requires, any laws or regulations that guide their use, and what activities are performed.
Responses Will Vary

- Interview a robotics engineer through videoconferencing.
Responses Will Vary

Communication Connections:
- Prepare a proposal to have your robot sponsored by a local business. After approval of your proposal by your classroom guide, present the proposal to the business for sponsorship.
Responses Will Vary

Ethics Connection:
- Research and write a summary about robots replacing human workers.
Responses Will Vary

Extensions:
- Add sound indicators to the robot to communicate its performance.
Responses Will Vary

- Submit your project to the book page on the website.
Responses Will Vary

C

Extensions:
- Design a robot for NASA to explore a planet. Describe the planet, the work the robot would do, and how the robot would power itself (and how long) on the planet.

Compare and contrast your design against others submitted and defend your design to be implemented by NASA.
Responses Will Vary

- Design a robot for NOAA's next mission to the deepest parts of the ocean. What work the robot would do, and how the robot would power itself (and how long) at the bottom of the ocean?

Compare and contrast your design against others submitted and defend your design to be implemented by NOAA.
Responses Will Vary

Problem:
Responses Will Vary

Glossary

A

Absorption - Light strikes a surface and some or all of its energy is transferred to the surface and stored as heat, chemical, or electrical energy.

Accuracy - How close to the correct value a measurement or average of measurements are.

Acid - a molecule that releases H^+ ions in solution.

Anode - the positive side of a circuit. The source of electrons. "+" Indicates an excess of electrons that will flow to the cathode(-) in the circuit.

Analog - Describes data that is a number, letter, sound, or other information that is not limited in value or designation.

Arduino® - An open source electronics platform based on C/C++ programming that includes hundreds of input and output devices.

B

Base - a molecule that releases OH^- or accepts H^+ ions in solution.

Binary - A choice or variable with two options. It can be a "Yes" or "No", "1" or "0", or any other condition with two possibilities.

Boolean - A type of decision making that involves a structure of combining, separating, narrowing, expanding, or deciding choices. Uses terms like "AND," "OR," "NOT," "IF," "ELSE," etc..

Breadboard - A flat board that allows for different combinations of devices to operate separately or interconnect. Depending on the connections the devices can receive electricity and inputs while interacting with each other devices and preform some tasks or create an output.

Buffer - a solution that does not change pH easily.

C

Calibration - adjustment of an input sensor's data to match expected values.

Cathode - The negative side of a circuit. The electrons flow here after leaving the anode(+).

Census- Counting the occurrence of a thing or action

Circuit - The path electricity takes through a device or devices.

Code - The computer program in the form of line by line instructions.

Color Spectrum - Visible light can be displayed as discrete colors from red to violet. Each color has its own wavelength.

Conductivity

Conductivity - The ability of a material to allow or not allow electricity to pass through it.

D

Data Array - Data defined by two ore more pieces of information. Examples include a date; July 4, 1968, positions on a chessboard Queen's Rook 3, or on the LED Matrix Row One (1), Column (3), and "On" defines the state of a particular LED.

Decibel - The unit of measure for loudness of a sound with respect to the listener.

Diastolic Phase - The time the heart is using the ventricles to pump blood to the body and lungs.

Digital - a one (1) or zero (0) logic value usually in a computer code. In a sketch or code the "HIGH" logic value is equal to a digital one (1) and "LOW" logic value is equal to a digital zero (0).

Diode - Electronic device that let a current pass through it in only one direction.

E, F

Engineering - A deliberate process based on math and science to create a technology, building, or form of transportation, or modify an environment.

Frequency - How often something happens. In the case of sound frequency indicates the note or pitch of the sound. Higher notes have higher frequencies and lower notes have lower frequencies.

G

Gas - A state of matter that has no definite volume or shape

H

Harmonic - A musical note related to a primary tone by some fractional part (1/, 1/3, 1/4, etc..).

Heart rate - the number of heart contractions over a period of time. Usually as beats per minute.

I, J, K, L

LCD - Liquid Crystal Display. A device that applies a small current to a small area of liquid crystal to make the liquid opaque.

Library - A program designed for a specific device that standardizes inputs and outputs for a sketch or code.

LED - A Light Emitting Diode. A device that gives off light when a small electrical current passes through its electronics. Usually can only emit one wavelength of light.

Glossary

M

Matrix - Something that is defined by two or more parts. The LED matrix has rows and columns to define the location of each LED.

Microcontroller - A small computer with a limited number of inputs, outputs and memory.

N, P

Parallel - When two elements of an electrical circuit share electricity by dividing the flow of electrons.

Piezoelectric - a material that changes its characteristics (sound or color) when electricity is applied to it.

pH - the measurement of OH^- or H^+ ions in a solution on a scale of 0-14 with 7 being neutral with equal amounts of OH^- and H^+ ions.

Photoresistor - A device that changes its electrical resistance based on the amount of light it senses.

Photosynthesis - The process that plants use to absorb energy from light and convert it to stored chemical energy.

Potentiometer - A device that varies the amount of resistance in a circuit.

Precision - How close a grouping of measurements are together, regardless of correct value.

Prototype - The first operating version of a machine.

Pushbutton Switch - A device that when pushed created a closed circuit. When not pushed the circuit is open.

Q, R

RGB LED - An LED with three (3) diodes inside, one red, one green, and one blue. Capable of making many different colors depending on how the colors are mixed.

Reflection - Light energy is returned to the environment as light energy

Resistor - A part in an electrical circuit that resists the flow of electrons. Usually used to avoid burning our another part in the circuit.

Robot - A device that moves based on computer controls that simulates or replaces human motions.

S

Seismology - the study of earthquakes and other movements of Earth from plate tectonics, landslides, and explosions.

Sensor - A device that reads the environment for an input (i.e. light, sound, temperature, etc..)

Series - When two elements of an electrical circuit are in sequence such that the electricity flows directly from the first element to the second.

Servo - A motor that can only spin with a limited range of motion. Usually less than 360 degrees.

Sketch - Arduino®'s name for the code.

Sonar - A process by which a sound is emitted underwater from a device that measures how long the echo from that sound takes to return to it as the sound reflects from an object. In turn a distance can be calculated based on the total trip time of the sound.

Subroutine - a portion of code that can be used many times over with inputs from different lines elsewhere in the code.

Switch - A part in an electrical circuit that through an action opens (disconnects) or closes (connects) a circuit.

Systolic Phase - The time the heart is not using the ventricles to pump blood. The heart muscles may be relaxed or the atria are pumping blood.

T

Technology - A tool, device, or machine to complete a task faster, more easily, or convert output.

Turbidity - The cloudiness or opaqueness of a liquid from material suspended in the liquid.

U

UNO - One of the many boards layouts available from Arduino®.

V

Value - A variable is assigned a number or letter(s) or other data set which may be constant or change. The number or letter(s) are the value of the variable.

W

Wavelength - Colors of light are determined by wavelength. The units for wavelengths of light are nanometer (nm) or billionths of a meter that is there are one billion nanometers in a meter.

X, Y, and Z

Project Assessment Rubric

The rubric below may be used as a guide to assess students' work for each project. Feel free to modify the rubric for your context, judgment, or needs.

✔	0 No attempt	1 Attempt unsuccessful	2 Complete and Understood	3 Complete, Understood, and Extended	4 Complete, Understood, Extended, and Connected
Engineering Design - Digital Prototype Design Code	No attempt	• Designed attempted but not complete • Code attempted but not complete or verified	• Design completed correctly • Code completed and verified	• Completed correctly, clean design • Code completed and can be run (Tinkercad® only)	• Completed correctly, very clean design • Project prototype works
Project Materials Build Code	No attempt	• Some materials assembled • Project attempted but not complete • Code attempted but not complete or verified	• All materials assembled • Project built with more than one issue • Code completed and verified	• All materials assembled and understood • Project built with no more than one issue • Code completed and uploaded	• All materials assembled, understood • Project works completely, understood, extended, and connected • Code works completely, understood.
Anticipatory Sets Discussion Project Reflection and Summary	No attempt	• Incomplete answers • Minimal participation • Minimal team interactions • Comms rules and conventions not followed*	• Answers Complete • Some participation • Some team interaction • Most communication rules and conventions followed*	• Complete with supporting evidence • All communication rules and conventions followed *	• Complete with strong supporting evidence, connections, and consensus • All communication rules and conventions followed*
Project Extension(s)	No attempt	• Extension attempted but not complete	• Extension complete and work	• Extension complete, work, and understood	• Extension complete, work, understood, and applied

* Use the grade-level standards and/or benchmarks that are approved for your curriculum to assess student language arts progress.

Also use style guides where appropriate to determine writing proficiency.

236 $ www.thearduinoclassroom.com

Bibliography and Acknowledgments

Computer Science

User. "K–12 Computer Science Framework." k12cs. Org, k12cs.org/. Computer Science Framework

Creativity

Michalko, Michael. *Thinkertoys*. Ten Speed Press, 2007.

Michalko, Michael. *Cracking Creativity: The Secrets of Creative Genius*. Ten Speed Press, 2008.

Making

Anderson, Chris. *Makers: the New Industrial Revolution*. Crown Business, 2014.

Thomas, AnnMarie P. *Making Makers*. Maker Media, 2014.

"The Maker Movement in Education: Designing, Creating, and Learning Across Contexts." Harvard Educational Review, vol. 84, no. 4, 2014, pp. 492–494., doi:10.17763/haer.84.4.b1p1352374577600.

Vossoughi, Shirin, et al. "Making Through the Lens of Culture and Power: Toward Transformative Visions for Educational Equity." Harvard Educational Review, vol. 86, no. 2, 2016, pp. 206–232., doi:10.17763/0017-8055.86.2.206.

Sheridan, Kimberly M., et al "Resourceful and Inclusive: Towards Design Principles for Makerspaces." presentation at AERA 2016.

Physics

Giancoli, Douglas C. *General Physics*. Prentice-Hall, 1985.

Teaching and Learning

Jensen, Eric. *Brain-Based Learning: The New Paradigm of Teaching*. Corwin Press, 2008.

Postman, Neil. *Technopoly: the Surrender of Culture to Technology*. Vintage Books, 1993.

Technology and Learning

Bocconi, S., Chioccariello, A. and Earp, J. (2018). The Nordic approach to introducing Computational Thinking and programming in compulsory education. Report prepared for the Nordic@BETT2018 Steering Group. https://doi.org/10.17471/54007

Csikszentmihalyi, Mihaly. Flow: The Psychology of Optimal Experience. Harper Row, 2009
Naisbitt, John, et al. High Tech, High Touch: Technology and Our Accelerated Search for Meaning. Nicholas Brealey Publ., 2001.

Mcdaniel, Rhett. "Bloom's Taxonomy." Vanderbilt University, Vanderbilt University, 13 Aug. 2018, wp0. vanderbilt.edu/cft/guides-sub-pages/blooms-taxonomy/.

Macaulay, D. (2016). *Way Things Work*. Dorling Kindersley.

Postman, N., & Weingartner, C. (1973). *Teaching as a subversive activity*. New York: Dalacorte Press.

Miscellaneous

Dowd, H., & Green, P. (2016). *Classroom management in the digital age*. Irvine, CA: EdTechTeam Press.

Waldrop, M. Mitchell. *Complexity: The Emerging Science at the Edge of Order and Chaos*. Simon & Schuster Paperbacks, 2008.

For our complete bibliography see https://www.thearduinoclassroom.com/about/bibliography.html

We wanted to thank Susan Haydock Ph.D. for her support, guidance, and consultation. We are especially grateful for her reading our early drafts, commenting on the pedagogy, and consultation on the classroom assessment parts of this book. We also wanted to thank her for watching Cookie and Tesla as we traveled to present our book.

Materials List and Project Guidance

Number	Part (Ordering links are available on the book web page)
1	Arduino® UNO R3 or UNO Compatible board
1	USB Cable (Connects board and computer)
1	short breadboard
1	long breadboard
25	6 in. jumper wire (Male to Male)
6	6 in. jumper wire (Male to Female)
1	9v Battery Barrel Plug
2	10k Potentiometer
6	220 Ohm Resistor
1	100 Ohm Resistor
2	Micro Servo
1	20 mm 8x8 LED Matrix
7	LED (various colors, at least one red)
1	RGB LED (Anode+)
1	16x2 LCD (16 pin)
1	Piezoelectric Buzzer
6	Small Pushbutton (2 pin)
1	Four Digit Seven Segment Display
1	Photoresistor LDR
1	Tilt Ball Sensor
1	TCS3200 Color Sensor
1	HC-SR04 Sensor
1	Heartbeat Sensor
1	Soil Moisture Sensor Kit
1	MQ4 Gas Sensor
1	TMP36 Temperature Sensor
1	pH Sensor Kit
1	Turbidity Sensor Kit
1	Analog Sound Sensor
1	4x4 Keypad
1	5v Computer Fan
1	Smart Robot Kit
1	Stepper Motor /Breakout Board (DRV8833)
1	IR Distance Sensor

Materials, tools, and supplies not included.

Universal Tools and Supplies:
- PC or Mac with Internet access and the IDE software installed.
- Soldering Iron (with 80/20 resin core solder)
- Phillips Head Screwdriver (#1)
- 9v battery

Check each project for materials unique to that project.

There are some lessons learned along the way that we wish to share with you from our start in 2016 with the UNO microcontroller. This book will develop your skills in tested lessons that build success and your appreciation for the UNO platform. We also want to prepare you for going online to try new projects. We have found both inspiration and frustration online. Our guidance and experience will give you more of the former and hopefully none of the latter.

The online sites seemingly include everything needed to make the projects work. However, we have found most projects have at least one fatal flaw preventing your success including:
- mistakes in the code
- lack of documentation
- discrepancies between the diagrams and photos
- lack of cautions about materials.

As the authors our goal is:
- to walk you through each step
- help you plan your student's success
- recommend the best materials to use
- document the nuances detailing how to change your code, the circuits, or design for successful completion of the project.

The list of resources on this page should guide your purchase or materials. Go online to our website for links to suppliers that will allow you to have success with the projects with your students.

Please also visit our website and register:
www.thearduinoclassroom.com
You will be joining a dedicated community of dedicated and focused users.

Isabel and Peter